Biology
for *CSEC*®

NEW EDITION

Phil Bradfield, Alexcia Morris, Steve Potter

HODDER
EDUCATION
AN HACHETTE UK COMPANY

Orders: please contact Hachette UK Distribution, Hely Hutchinson Centre, Milton Road, Didcot, Oxfordshire, OX11 7HH. Telephone: +44 (0)1235 827827. Email: education@hachette.co.uk. Lines are open from 9 a.m. to 5 p.m., Monday to Friday. You can also order through our website: www.hoddereducation.com

ISBN 978 14479 5219 0

Produced by DTP Impressions
Cover design by Firelight Studio

Picture Credits

The publisher would like to thank the following for their kind permission to reproduce their photographs:

(Key: b-bottom; c-centre; l-left; r-right; t-top)

Alamy Images: Art Directors & TRIP 371c, Blend Images 228t, Charles O. Cecil 342tr, David Hosking 373tl, Deborah Davis 286cl, FLPA 381bl, GoGo Images Corporation 349r, M I (Spike) Walker 89tl, mediacolor's 61t, 321cr, Steve Murray 24tr; **Bigstock:** Ale-ks 111t, alexsol 341cr, Anandkrish16 342tc, AndamanSE 8 (Flatworm), anphotos 24br, aorlemann 77br, Artush 12cr, bejim 10 (Parrot), c photo 9 (centipede), danaseilhan 13 (Fern spores), davespates 10, EcoShot 8 (crab), epantha 11 (Newt), Epixx 77tl, Eugene Sergeev 13 (Fern), Ian Redmond 10 (Dolphin), inaquim 14 (Oranges), interactimages 9 (millipede), jaimepharr 70b, JMcLoughlin 8 (Earthworm), kadmy 291t, Kesu01 187bl, kikkerdirk 11 (frog), lawcain 7 (sand dollars), Life on White 8 (Snail), lonelyblueart 11 (Crocodile), Mariusz Prusaczyk 1t, Marmion 324t, Mbongo, 221t, merulaki 7 (jellyfish), mhathorn 7 (sea anemone), microdac 14 (Breadfruit), nico64 9 (spider), nikitos77 367t, nonillion 14 (banana), Nosnibor137 222c, Olgany 148t, prill 373tc, Rich Carey 76c, smithore 68cr, snaphappysara 308t, surpasspro 167t, Swims with Fish 11 (Turtle), vtupinamba 14 (Sugarcane), witold krasowski 224bc; **Bridgeman Art Library Ltd:** English Heritage Photo Library / The Bridgeman Art Library 373cr; **Corbis:** Clifford White 287tl, David Cavagnaro / Visuals Unlimited 12br, John Harper 23tl, Lester V. Bergman 157cr, Pablo Corral 75t, sampics 241c, Tom Brakefield 45cr, Wally Eberhart / Visuals Unlimited 269cr; **DK Images:** 266c, 382 (poodle), 382 (wolf), Alistair Duncan 26tl, Dave King 208t, 382 (Dachshund), David Munns 222bc, Frank Greenaway 7 (Sea Urchin), 8 (Leech), 9 (scorpion), Geoff Brightling 9, Jerry Young 11 (damsel fish), 314l, Jonathan Buckley 25l, Steve Gorton 342tl; **Big Funky Pictures Ltd:** Rich Carey 11 (parrot); **Getty Images:** Nancy R. Cohen 243bl, Oxford Scientific 268tl; **Imagemore Co., Ltd:** 8 (Oyster); **Imagestate Media:** Ian Cartwright 7 (Starfish); **Pearson Education Ltd:** Sozaijiten 39t; **Photoshot Holdings Limited:** NHPA 311bl; © **Rough Guides:** Alex Robinson 19c, Ian Cummings 51t; **Science Photo Library Ltd:** 185cr, Adam hart-davis 65tr, 229bl, aj photo / hop americain 238b, andy crump, tdr, who 149br, art wolfe 26tr, astrid & hanns-frieder michler 82cl, biology media 202cl, biophoto associates 97t, 150bl, blair seitz 123cr, bsip 328 (b), claude nuridsany & marie perennou 374bl, clouds hill imaging ltd 341bl, cordelia molloy 326tr, d. Phillips 291br, dept. Of clinical cytogenetics, addenbrookes hospital 361tl, 361tc, dr gopal murti 2br, 86tr, 282tl, 341tr, dr jeremy burgess 54tl, 214tl, edward kinsman 281t, eric grave 83br, eye of science 161cr, 176, 328 (c), f003 / 3864 82t, francis leroy, biocosmos 383c, frans lanting, mint images 10 (monkey), fred mcconnaughey 341bc, gene cox 283cr, j.C. Revy, ism 86cr, 103r, 208cl, james kingholmes 344bl, jim varney 298bl, john durham 378tl, juergen berger 86bl, kevin & betty collins, visuals unlimited 337t, kwangshin kim 328bl, laguna design 138tl, leonard lessin 242cl, 242bl, m.I. Walker 2bl, mark williamson 56t, martin dohrn 331, maximilian stock ltd 149tr, michael abbey 328 (d), michael p. Gadomski 65br, michael w. Tweedie 374cl, moredun animal health ltd 86cl, nigel cattlin 334c, norm thomas 5bl, omikron 359br, peter muller 128t, power and syred 3cl, 171br, ron bass 24bl, saturn stills 326bl, science pictures limited 292tl, sidney moulds 217bl, sinclair stammers 2bc, st. Mary's hospital medical school 301br, stephen & donna o'meara 77bl, steve gschmeissner 104r, 345cr, Susumu Nishinaga 104l, university of medicine & dentistry of new jersey 301tr, william ervin 45tr

Cover images: *Front:* **Alamy Images:** Rolf Nussbaumer Photography

All other images © Hodder & Stoughton Limited

Every effort has been made to trace the copyright holders and we apologise in advance for any unintentional omissions. We would be pleased to insert the appropriate acknowledgement in any subsequent edition of this publication.

Printed in the UK

Introduction

How this book is organised

This brand-new text from Pearson has been written for the latest CSEC syllabus which will be first examined in June 2015, and contains many features that will help you achieve your best in the CSEC Biology exam and SBA.

Each chapter begins with a set of **Objectives** which set the scene for what you will be studying. At the end of each chapter, there are some longer **Summary** notes which will be useful to you for exam revision.

The chapters contain all the teaching and **Activities** you will need to cover the requirements of the syllabus and master the scientific principles you are being taught.

Next to each Activity are icons which indicate which of the **SBA skills** can be assessed with that Activity: Observation, Recording and Reporting *(ORR)*; Manipulation and Measurement *(MM)*; Analysis and Interpretation *(AI)*; Planning and Designing *(PD)*, and Drawing *(Dr)*. When the boxes containing the abbreviations of these skills are highlighted in *white*, this indicates that the skill can be tested with this particular Activity. Your teacher will sometimes use this information to decide which Activities should be used to assess the different SBA skills.

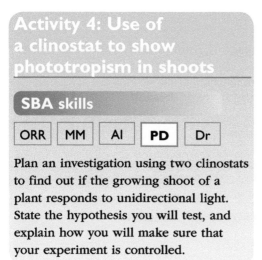

Activity 4: Use of a clinostat to show phototropism in shoots

SBA skills

| ORR | MM | AI | **PD** | Dr |

Plan an investigation using two clinostats to find out if the growing shoot of a plant responds to unidirectional light. State the hypothesis you will test, and explain how you will make sure that your experiment is controlled.

Within each chapter are short **Exercises** designed to help your teacher check as you go along that you understand the key concepts being taught.

Each chapter finishes with full **CSEC-style questions**. First, there are some multiple choice questions, like those in Paper 1 of the exam. These are followed by short-answer questions, like those in Paper 2 of the exam, which tend to test your use of knowledge.

The book ends with some important Appendices. The first two are an **SBA Guide** and an **SBA Skills-matching Chart**, which your teacher may use to decide how the SBA should be approached and marked. The third provides some useful examinations tips.

Answers to the multiple choice questions and to the short answer questions are provided in the supporting CSEC Science website.

Contents

Contents

Section C: Continuity and Variation

Section A: Living Organisms in the Environment

Chapter 1: The Diversity of Life

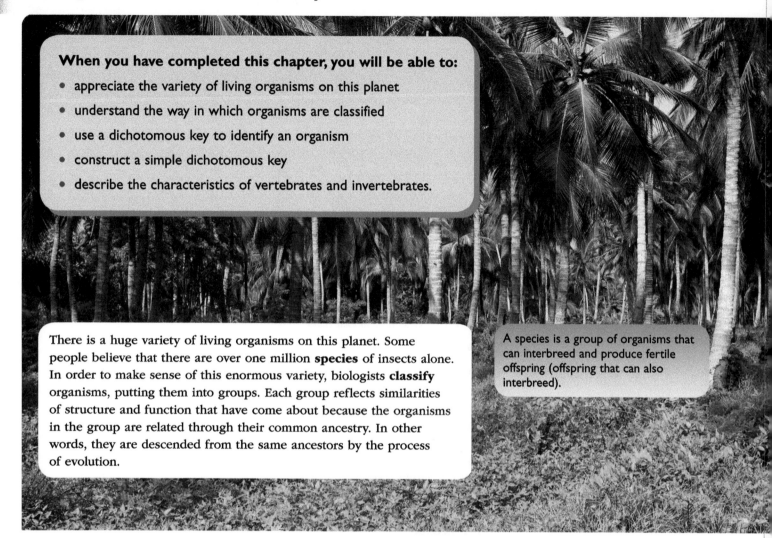

When you have completed this chapter, you will be able to:

- appreciate the variety of living organisms on this planet
- understand the way in which organisms are classified
- use a dichotomous key to identify an organism
- construct a simple dichotomous key
- describe the characteristics of vertebrates and invertebrates.

There is a huge variety of living organisms on this planet. Some people believe that there are over one million **species** of insects alone. In order to make sense of this enormous variety, biologists **classify** organisms, putting them into groups. Each group reflects similarities of structure and function that have come about because the organisms in the group are related through their common ancestry. In other words, they are descended from the same ancestors by the process of evolution.

A species is a group of organisms that can interbreed and produce fertile offspring (offspring that can also interbreed).

Classifying organisms

Species that are obviously very similar are placed together in larger groups, each called a **genus**. The name of the genus and the name of the species an organism belongs to give the organism its scientific name. For example, humans belong to the genus *Homo* and the species *sapiens*. The scientific name is *Homo sapiens*.

Similar genera can then be placed into **families** and so on into ever-larger groups. The science of placing organisms into the correct groups is called **taxonomy**. Each group has its own set of features that, together, define the group. If we find an organism we have not seen before, we can use these sets of features to classify the organism, right down to the species level. Figure 1.1 on page 2 shows the main groups of organisms.

The major groups of organisms are **bacteria**, **protoctists**, **fungi**, **viruses**, **animals** and **plants**.

1

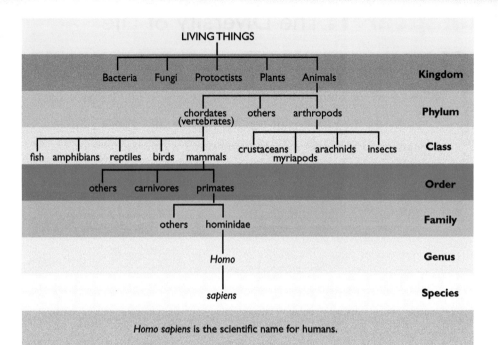

Figure 1.1 *Classification of living things.*

Bacteria

Bacteria are small single-celled organisms. However, bacterial cells are much smaller than those of animals, plants, or protoctists, and have a much simpler structure. To give you some idea of their size, a typical animal cell might be 10 μm to 50 μm in diameter (1 μm, or one micrometre, is a millionth of a metre, or a thousandth of a millimeter). Compared with this, a typical bacterium is only 1 μm to 5 μm in length and its volume can be thousands of times less than the larger cell.

Bacteria are present everywhere in the soil, in water and in the air.

Protoctists

Protoctists are sometimes called the 'dustbin kingdom', because they are a mixed group of organisms that don't fit into the plant, animal or fungus groups. Most protoctists are microscopic single-celled organisms (Figure 1.2). Some look like animal cells, such as *Amoeba*, which lives in pond water. These are known as **protozoa**. Other protoctists have chloroplasts and carry out photosynthesis, so are more like plants. These are called **algae**. Most algae are unicellular, but some species, such as seaweeds, are multicellular and can grow to a great size. Some protoctists are the agents of disease, such as *Plasmodium*, the organism that causes malaria.

> Protoctists form a very diverse group. It is often regarded as a taxonomic dumping ground – 'if it doesn't fit anywhere else, it must be a protoctistan'. The group includes unicells like amoeba and also giant kelp seaweeds that can be several metres long.

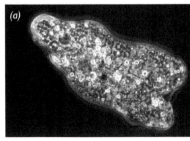 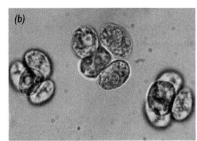 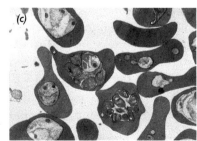

Figure 1.2 (a) Amoeba, *a protozoan that lives in ponds.* (b) Chlorella, *a unicellular freshwater alga.* (c) *Blood cells containing the protoctist parasite* Plasmodium, *the organism responsible for causing malaria.*

Fungi

Fungi include mushrooms and toadstools, as well as moulds. These groups of fungi are multicellular. Another group of fungi are the yeasts, which are **unicellular** (made of single cells). Different species of yeasts live everywhere – on the surface of fruits, in soil, water, and even on dust in the air. The yeast powder used for baking contains millions of yeast cells (Figure 1.3). The cells of fungi never contain chloroplasts, so they cannot photosynthesis. Their cells have cell walls, but they are not composed of cellulose (Figure 1.4).

A mushroom or toadstool is the reproductive structure of the organism, and is called a fruiting body (Figure 1.5). Under the soil, the mushroom has many fine thread-like filaments called **hyphae** (pronounded high-fee). A mould is rather like a mushroom without the fruiting body. It just consists of the network of hyphae (Figure 1.6). The whole network is called a **mycelium** (pronounced my-sea-lee-um). Moulds feed by absorbing nutrients from dead (or sometimes living) material, so they are found wherever this is present, for example, in soil, rotting leaves or decaying fruit.

Because fungi have cell walls, they were once thought to be plants that had lost their chlorophyll. We now know that their cell wall is not made of cellulose as in plants, but of a different chemical called **chitin** (the same material that makes up the outside skeleton of insects). Fungi are quite different from plants in many ways (the most obvious is that they do not photosynthesis) and they are not closely related to plants at all.

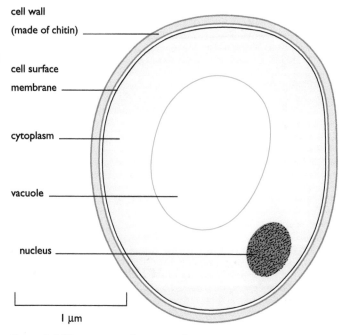

cell wall (made of chitin)

cell surface membrane

cytoplasm

vacuole

nucleus

1 μm

Figure 1.4 *The structure of a yeast cell.*

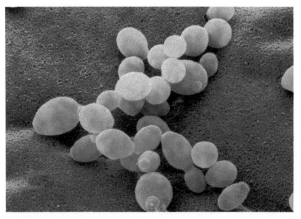

Figure 1.3 *Yeast cells, highly magnified.*

Figure 1.5 *Toadstools growing on a rotting log.*

Figure 1.6 *The 'pin mould' Mucor, growing on a piece of bread. The dark spots are structures that produce spores for reproduction.*

If you leave a piece of bread or fruit exposed to the air for a few days, it will soon become mouldy. Mould spores carried in the air have landed on the food and grown into a mycelium of hyphae (Figure 1.7).

The thread-like hyphae of *Mucor* have cell walls surrounding their cytoplasm. The cytoplasm contains many nuclei, in other words the hyphae are not divided up into separate cells.

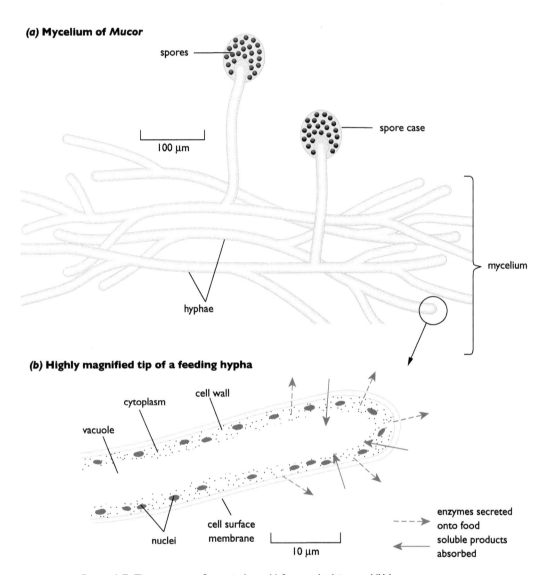

(a) Mycelium of Mucor

spores

spore case

100 μm

mycelium

hyphae

(b) Highly magnified tip of a feeding hypha

cell wall

cytoplasm

vacuole

nuclei

cell surface membrane

10 μm

enzymes secreted onto food

soluble products absorbed

Figure 1.7 *The structure of a typical mould fungus, the 'pin mould'* Mucor.

When a spore from *Mucor* lands on the food, a hypha grows out from it. The hypha grows and branches again and again, until the mycelium covers the surface of the food. The hyphae secrete digestive enzymes onto the food, breaking it down into soluble substances such as sugars, which are then absorbed by the mould. Eventually, the food is used up and the mould must infect another source of food, by producing more spores.

When an organisms feeds on dead organic material in this way, and digestion takes place outside of the organism, this is called **saprotrophic** nutrition. Enzymes that are secreted out of cells for this purpose are called **extracellular** enzymes (see Chapter 11, page 138).

Viruses

All viruses are parasites, and can only reproduce inside living cells. The cell in which a virus lives is called the host. There are many different types of viruses. Some live in the cells of animals or plants, and there are even viruses which infect bacteria. Viruses are much smaller than bacterial cells: most are between 0.01 μm and 0.1 μm in diameter.

Notice that we say 'types' of virus, and not 'species'. This is because viruses are not made of cells. A virus particle is very simple. It has no nucleus or cytoplasm, and is composed of a core of genetic material surrounded by a protein coat (Figure 1.8). The genetic material can be either DNA, or a similar chemical called RNA (see Chapter 25). In either case, the genetic material makes up just a few genes – all that is needed for the virus to reproduce inside its host cell.

Sometimes a membrane, called an **envelope**, may surround a virus particle, but the virus does not make this. Instead it is 'stolen' from the surface membrane of the host cell.

Viruses do not feed, respire, excrete, move, grow or respond to their surroundings. They do not carry out any of the normal 'characteristics' of living things except reproduction, and they can only do this parasitically. This is why some scientists think of viruses as being on the border between a living organism and a non-living chemical.

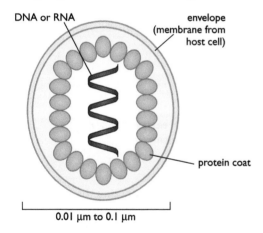

DNA or RNA

envelope
(membrane from host cell)

protein coat

0.01 μm to 0.1 μm

Figure 1.8 *The structure of a typical virus, such as the type causing influenza (flu).*

Figure 1.9 *Discoloration of the leaves of a tobacco plant, caused by infection with tobacco mosaic virus*

Viruses don't just parasitise animal cells. Some infect plant cells, such as the tobacco mosaic virus (Figure 1.10), which interferes with the ability of the tobacco plant to make chloroplasts, causing mottled patches to develop on the leaves (Figure 1.9).

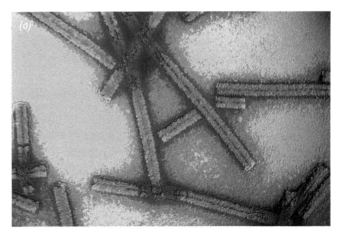

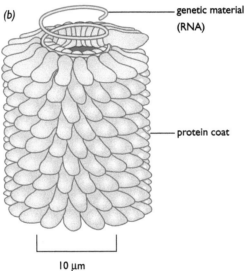

genetic material (RNA)

protein coat

10 μm

Figure 1.10 *(a) The tobacco mosaic virus (TMV), seen through an electron microscope. (b) The structure of part of a TMV particle, magnified 1.25 million times.*

Animals

You will be even more familiar with this kingdom, as it contains the species *Homo sapiens*, i.e. humans! The variety in the animal kingdom is enormous. It contains the phyla Cnidaria, Echinodermata, Mollusca, Platyhelminthes, Annelida, Arthropoda and Chordata, as well as many others.

The chordates in the animal kingdom include fish, amphibians, reptiles, birds and mammals. Chordates are all **vertebrates**, which means that they have a vertebral column, or a backbone. All other animals lack this feature, and so are called **invertebrates**.

Animals are also multicellular organisms. Their cells never contain chloroplasts, so they are unable to carry out photosynthesis. Instead, they gain their nutrition by feeding on other animals or plants. Animal cells also lack cell walls. This allows the cells to change shape – an important feature for organisms that need to move from place to place. Movement in animals is achieved in various ways, but often involves coordination by a nervous system (see Chapter 18). Another feature common to most animals is that they store carbohydrates in their cells as the compound glycogen.

Table 1.1 below and on pages 8 and 9 lists characteristics of different invertebrates, and shows examples.

Phylum	Characteristics	Examples
Cnidaria	• sac-like body with a single opening • stinging tentacles	 Figure 1.11 *A jellyfish.*  Figure 1.12 *A sea anemone.*
Echinodermata	• body is evenly divided into five parts • spiny skin • all are marine organisms	 Figure 1.13 *A starfish.* 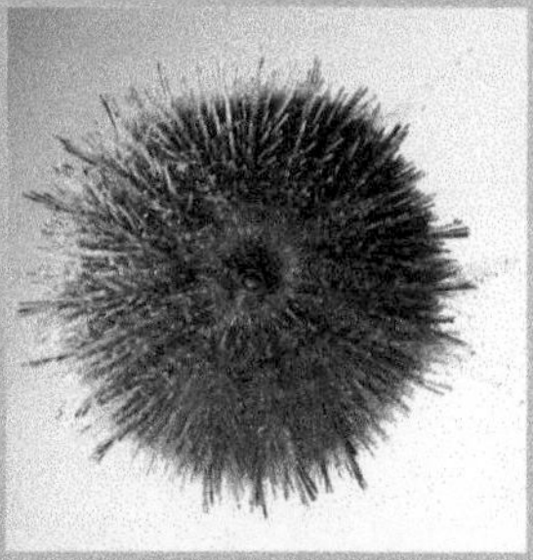 Figure 1.14 *A sea urchin.*  Figure 1.16 *A sand dollar.*

Phylum	Characteristics	Examples
Mollusca	• soft, unsegmented body • some have one or two shells for protection • a muscular 'foot' to enable movement	 Figure 1.16 A *nail*. 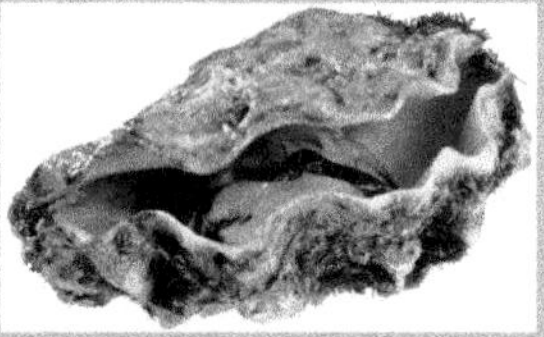 Figure 1.17 *An oyster*.
Platyhelminthes	• flat body • some inhabit fresh water • some are parasitic to other animals	 Figure 1.18 A *planarian flatworm*.
Annelida	• long, tubular body that is segmented • body has bristles, called setae	 Figure 1.19 *An earthworm*.  Figure 1.20 A *leech*.

Phylum	Characteristics	Examples
Arthropoda	• they have an exoskeleton • the body is segmented • the legs are jointed There are four classes in the phylum: • **Crustacea:** • three body segments • two pairs of antennae • four pairs of legs or more, but fewer than twenty pairs	Figure 1.21 *A crab.* Figure 1.22 *A lobster.*
	• **Arachnida** • two body segments • no antennae; some have poisonous fangs • four pairs of legs	Figure 1.23 *A spider.* Figure 1.24 *A scorpion.*
	• **Insecta** • three body segments • one pair of antennae • three pairs of legs • two pairs of wings, usually	Figure 1.25 *A cockroach.*
	• **Myriapoda** • long body that has several segments • one pair of antennae • the number of legs depends on the number of body segments	Figure 1.26 *A millipede has two pairs of legs on each body segment.* Figure 1.27 *A centipede has one pair of legs on each body segment.*

Table 1.1: *Characteristics of different invertebrates.*

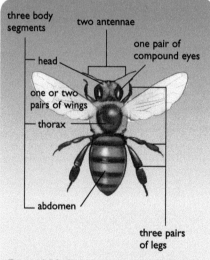

three body segments — two antennae — one pair of compound eyes — head — one or two pairs of wings — thorax — abdomen — three pairs of legs

Figure 1.28 *The body parts of an insect.*

Table 1.2 below and on page 11 lists characteristics of different vertebrates, and shows examples.

Class	Characteristics	Examples
Mammalia	• fur or hair • mammary glands • use lungs to breathe • internal fertilisation, with young born developed	 Figure 1.29 *A green monkey.*  Figure 1.30 *A dolphin.*
Aves (birds)	• feathers and wings, and most are able to fly • a beak • lay eggs	 Figure 1.31 *A parrot.*  Figure 1.32 *A pelican.*

Class	Characteristics	Examples
Reptilia	• dry, scaly skin to prevent desiccation • internal fertilisation, and lay eggs with leathery shells	 Figure 1.33 A crocodile.  Figure 1.34 A green sea turtle.
Amphibia	• moist skin • in water, they breathe through their skin; on land, they use their lungs • external fertilisation of eggs, which hatch as tadpoles	 Figure 1.35 A frog.  Figure 1.36 A newt.
Fish	• skin covered with scales • use gills to breathe • fins for swimming	 Figure 1.37 A damsel fish.  Figure 1.38 A parrot fish.

Table 1.2: *Characteristics of different vertebrates.*

Plants

You will no doubt be familiar with many different types of plants. The plant kingdom contains **non-flowering plants** (plants that do not produce flowers) and **flowering plants** (plants that produce flowers, which bear the reproductive organs of the plant). Non-flowering plants include **mosses**, **liverworts**, **conifers** and **ferns**. Flowering plants are divided into **monocotyledons** (the seeds have one cotyledon) and **dicotyledons** (the seeds have two cotyledons).

All plants are **multicellular**, which means that their 'bodies' are made up of many cells. Their main distinguishing feature is that their cells contain chloroplasts, which carry out photosynthesis: the process that uses light energy to convert simple inorganic molecules such as water and carbon dioxide into complex organic compounds (see Chapter 9). One of these organic compounds is the carbohydrate **cellulose**, and all plants have cells walls made of this material. Plants can make a range of organic compounds using photosynthesis. One of the first compounds that they make is the storage carbohydrate **starch**, which is often found inside plant cells. Another is the sugar **sucrose**, which is transported around the plant and is sometimes stored in fruits and other plant organs.

Table 1.3 below and on page 13 lists characteristics of different non-flowering plants, and shows examples.

Phylum	Characteristics	Examples
mosses and liverworts (Bryophyta)	• grow in moist areas e.g. in cracks in rocks • have neither xylem vessels nor phloem tubes; water moves by osmosis in the plant • root-like structures, called rhizoids • simple leaves and stems	Figure 1.39 *Moss growing on rocks.* Figure 1.40 *A liverwort.*

Phylum	Characteristics	Examples
conifers (Pinophyta)	• many are evergreen • leaves are modified into needle-like structures to reduce water loss • have xylem vessels and phloem tubes • produce male and female cones	 Figure 1.41 *Pine trees.*  pine tree leaves, called needles pine cone Figure 1.42 *A Pine tree cone and leaves.*
ferns (Pteridophyta)	• have xylem vessels and phloem tubes • spores for reproduction are borne on the underside of the leaves • stems and roots	 Figure 1.43 *A fern.*  Figure 1.44 *Fern spores germinate into new fern plants.*

Table 1.3: *Characteristics of different non-flowering plants.*

Table 1.4 lists characteristics of different flowering plants, and shows examples.

Phylum	Characteristics	Examples
Monocotyledons	<ul><li>narrow leaves</li><li>leaves have parallel veins</li><li>fibrous root system</li></ul>	Figure 1.45 *Sugarcane.* Figure 1.46 *A banana plant.*
Dicotyledons	<ul><li>broad leaves</li><li>leaves have net-like veins</li><li>taproot system</li></ul>	Figure 1.47 *A breadfruit tree.* Figure 1.48 *An orange tree.*

Table 1.4: *Characteristics of different flowering plants.*

Dichotomous keys

We can use a **dichotomous key** to help us to decide which group to put an organism in. For example, a dichotomous key to place an organism in the correct kingdom might be:

1. Organism has leaves	Y	–	**Plant**
	N	–	Go to 2
2. Organism is single-celled with no true nucleus	Y	–	**Bacterium**
	N	–	Go to 3
3. Main body of organism consists of threads with some spore cases	Y	–	**Fungus**
	N	–	Go to 4
4. Organism has legs/fins/wings	Y	–	**Animal**
	N	–	Go to 5
5. Organism does not belong to any of the kingdoms above	Y	–	**Protoctist**

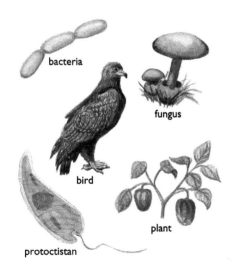

Figure 1.49 *Examples of organisms from the different kingdoms (not to scale).*

A dichotomous key to place an animal in the correct class might be:

1. Animal has a backbone	Y	–	Go to 2
	N	–	Go to 7
2. Animal has fins and gills	Y	–	**Fish**
	N	–	Go to 3
3. Animal has slimy skin and webbed feet	Y	–	**Amphibian**
	N	–	Go to 4
4. Animal has dry, scaly skin	Y	–	**Reptile**
	N	–	Go to 5
5. Animal has feathers	Y	–	**Bird**
	N	–	Go to 6
6. Animal has hair or fur	Y	–	**Mammal**
7. Body has 3 sections and 3 pairs of jointed legs on middle section	Y	–	**Insect**
	N	–	Go to 8
8. Body has 2 sections with 4 pairs of jointed legs on second section	Y	–	**Arachnid**
	N	–	Go to 9
9. Body made of many segments with 1 or 2 pairs of jointed legs on each	Y	–	**Myriapod**
	N	–	Go to 10
10. Body made of segments covered by a shell	Y	–	**Crustacean**
	N	–	Animal belongs to a class not covered by this key

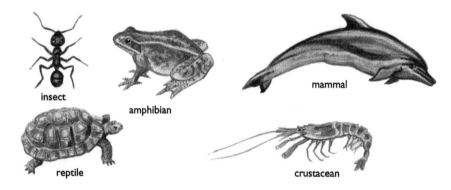

insect

amphibian

mammal

reptile

crustacean

Figure 1.50 *Examples of animals from some of the different classes (not to scale).*

Activity 1: Using a dichotomous key to identify small animals

SBA skills

ORR	MM	AI	PD	Dr

You will need:

- a hand lens or viewer
- a collection of small animals

Carry out the following:

For each animal in turn, use the appropriate dichotomous key above to find which class of animals it belongs to.

SBA skills

ORR	MM	AI	PD	Dr

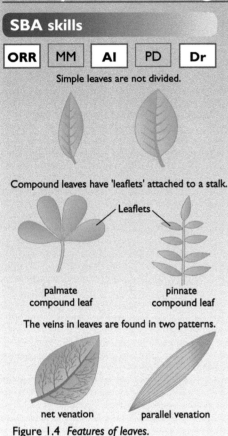

Simple leaves are not divided.

Compound leaves have 'leaflets' attached to a stalk.

Leaflets

palmate compound leaf

pinnate compound leaf

The veins in leaves are found in two patterns.

net venation

parallel venation

Figure 1.4 *Features of leaves.*

You will need:

- a hand lens or viewer
- a collection of different leaves

Carry out the following:

1. Choose several different leaves from the collection. Identify each one or give each a label, e.g. leaf A, leaf B, etc.

2. Make a careful, labelled drawing of each leaf.

3. Construct a table like the one below and fill in the details for each leaf.

Leaf	Shape	Simple/compound	Venation	Other features
A				
B				
C				
D				

5. Use the features you have described to construct a dichotomous key to identify each leaf.

You may also be asked to construct a dichotomous key to identify some animals. You should carry out this activity in the same sequence. First, construct and complete a table using observable features that distinguish each animal. Then construct a key using those observable features.

Chapter summary

In this chapter you have learnt that:

- there is a wide variety of organisms on the Earth
- all these organisms can be classified into appropriate groups
- the science of classifying these organisms is called taxonomy
- the main taxonomic groups (starting with the largest) are: Kingdom, Phylum, Class, Order, Family, Genus, Species
- the genus and species names of an organism together give the organism its scientific name
- living things can be classified into five kingdoms: bacteria, protoctists, fungi, animals and plants
- plants are classified into non-flowering and flowering plants
- animals are classified into invertebrates and vertebrates
- a dichotomous key consists of pairs of contrasting statements and is used to identify organisms or groups of organisms.

Questions

1. The science of placing organisms in their correct groups is known as:

   A classification
   B keying
   C taxonomy
   D observation

2. Each organism has a scientific name in two parts. These are:

   A the genus and class names of the organism
   B the genus and species names of the organism
   C the genus and family names of the organism
   D the family and class names of the organism

3. The correct sequence for the taxonomic groups, starting with the largest group is:

   A genus, species, family
   B species, genus, family
   C family, species, genus
   D family, genus, species

4. Which is the best description of a dichotomous key?

   A a key with pairs of contrasting statements
   B a key to identify animals
   C a key to identify plants
   D a key to place animals and plants in the correct kingdom

5. Which one of the following classes of animals has an exoskeleton and four pairs of jointed limbs?

   A reptiles
   B insects
   C arachnids
   D crustaceans

6. The diagram shows some fruits that are dispersed by wind.

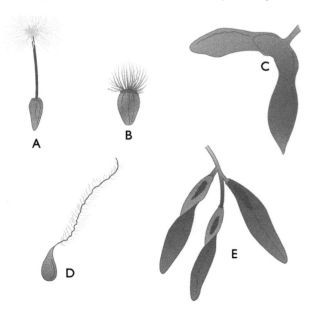

a) Describe the structure of each of the fruits. (5)

b) Use the dichotomous key below to identify the fruits. (5)

1	Fruit has 'hairs'	Go to 2
	Fruit has 'wings'	Go to 4
2	Hairs are attached directly to the fruit	Thistle
	Hairs are on a stalk	Go to 3
3	Hairs are only found at the end of the stalk	Dandelion
	Hairs are found all along the stalk	Clematis
4	Fruits are in pairs	Sycamore
	Fruits are in threes	Ash

7. The diagram shows four different insects.

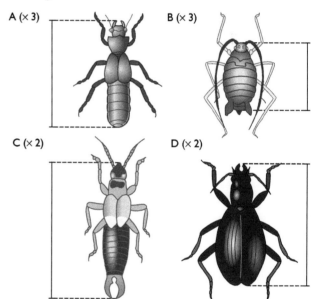

A (× 3) B (× 3)

C (× 2) D (× 2)

a) i) State *two* differences between insect A and insect C. (2)

   ii) State *two* features that are common to all the insects. (2)

b) i) Copy and complete the following table (the values for insect A are entered for you). (3)

Insect in diagram	Length in diagram (mm)	Magnification	Actual length (mm)
A	28	×3	9.3
B			
C			
D			

ii) Place the insects in size order, starting with the largest. *(1)*

iii) Construct a dichotomous key using observable characteristics other than size to identify the four insects in the diagram. *(4)*

8. The diagram shows six different organisms found in or on soil. Identify two features of each organism that could be used to classify it. *(2)*

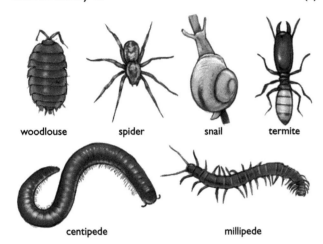

woodlouse　　　spider　　　snail　　　termite

centipede　　　　　millipede

Chapter 2: Living Organisms and the Environment

When you have completed this chapter, you will be able to:

- describe the main components of an ecosystem
- understand the difference between a population and a community
- understand what is meant by the term environment
- appreciate that the environment of an organism can supply many of its needs
- list the components of soil and explain how the balance of the different components can affect the properties of the soil
- describe how to estimate the numbers of a specified organism in an area.

What is ecology?

Ecology is the study of how organisms interact with their environment. It includes:

- the interaction of all organisms that live in that environment
- how the organisms impact one another, as well as their environment
- how that environment impacts organisms' behaviour.

As we learn more about these complex interactions, ecology is seen to be an increasingly important area of science. Plants and animals in particular habitats interact with each other and if changes occur to one set of organisms, the effects on the others can be far-reaching and not easy to predict. For example farming practices where only one type of crop is grown, monocultures, have a great effect on the kinds of birds and insects that can live there. Similarly, the ways in which humans change the environment has a profound effect on other plants and animals. In order not to damage our environment and the plants and animals in it, we need a good understanding of ecology.

The components of ecosystems

An **ecosystem** is a distinct, self-supporting system of organisms interacting with each other and with a physical environment. An ecosystem can be small, such as a garden pond, or large, such as a large mangrove swamp.

Whatever their size, ecosystems usually have the same components:

- **producers** – plants which photosynthesise to produce food

- **consumers** – animals that eat plants or other animals

- **decomposers** – organisms that decay dead material and help to recycle nutrients

- a **physical environment** – the sum total of the non-biological components of the ecosystem; for example, the water and soil in a pond or the soil and air in a forest.

The environment is everything that surrounds the organisms in an ecosystem. Both the physical (i.e. non-living, or abiotic) components and the living (biotic) components contribute to the environment of an ecosystem. How the abiotic and biotic components of an ecosystem interact with one another is very important.

Some examples of biological and non-biological components of ecosystems are given in Table 2.1.

Ecosystem	Biological components	Physical environment
rainforest	<ul><li>trees</li><li>other plants</li><li>animals living in the forest</li><li>decomposers in the soil/water</li></ul>	<ul><li>soil</li><li>air in/around the forest</li><li>any water in rivers/streams/ponds</li></ul>
pond	<ul><li>plants and algae living in and around the pond</li><li>animals living in the pond</li><li>decomposers living in the water and soil at the bottom the pond</li></ul>	<ul><li>water</li><li>air above the water</li><li>soil at the bottom of the pond</li></ul>
mangrove swamp	<ul><li>mangroves and other plants living in the swamp</li><li>animals living in the swamp</li><li>decomposers</li></ul>	<ul><li>water</li><li>air above the swamp</li><li>soil at the bottom of the swamp</li></ul>

Table 2.1: *Biological and physical components of ecosystems.*

The physical environment is important to the living organisms in a number of ways. It can supply:

- habitats for living organisms

- a supply of nutrients, such as mineral ions, for plants

- gases: oxygen for animals, carbon dioxide and oxygen for plants, and nitrogen for nitrogen-fixing bacteria in soil.

Within each ecosystem there is a range of **habitats** – these are the places where specific organisms live. Some are provided by the physical environment. For example, in a pond ecosystem, the habitat of many of the plants is provided partly by the soil at the bottom of the pond (where the roots penetrate) and partly by the water itself (where the stem, leaves and flowers grow). Tadpoles spend most of their time swimming in the surface waters of a pond, and that is their habitat.

Other habitats are provided by organisms themselves. For example, dead vegetation provides a habitat for many decomposers.

Groupings of organisms

Living things are classified based on similar characteristics they possess. The smallest grouping is called the species. A **species** is a group of organisms that share similar characteristics and can produce offspring that are capable of producing their own offspring.

All the organisms of a particular species found in an ecosystem at any one time form the **population** of that species in that ecosystem. For example, in a pond, all the tadpoles swimming in it form a population of tadpoles; all the *Elodea* plants growing in it make up a population of *Elodea*. A population is limited to the boundary of the habitat. Therefore, different populations of the same species can live separately in different habitats.

The populations of all the species (all the animals, plants and decomposers) found in a particular ecosystem at any one time form the **community** in that ecosystem. A garden pond is a good example of these interrelationships. Figure 2.1 illustrates the main components of a pond ecosystem.

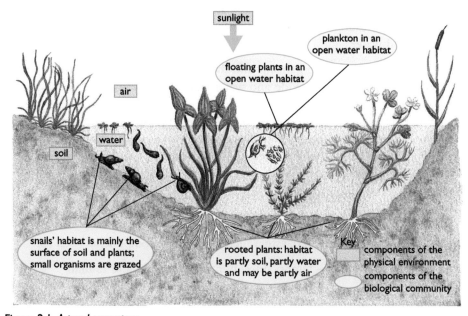

Figure 2.1 *A pond ecosystem.*

Niches in ecosystems

Different populations interact with one another in a habitat. For example, in a pond ecosystem, *Elodea* (water weed) and the water lily trap energy from sunlight to make their own food. *Elodea* produces oxygen, which is needed by the aquatic animals, and the water lily provides food for the snails. Microscopic plant organisms may live on the surface of the plants' leaves. (Plants that grow on other plants without harming them are called **epiphytes**.) Egested and excreted materials from the snails, tadpoles and fish return nutrients to the soil. The water lily is a producer as well as a habitat for other organisms, and the snail is a primary consumer and helps to recycle nutrients.

Each organism, therefore, has a role to play in its habitat. This role within the habitat is referred to as the organism's **niche**. The above examples show that an organism's niche is not limited to its place in the food web. For example, plants, which are producers, may also provide nesting and spawning sites for animals. No two species will occupy the same niche within a habitat.

1. What are the three living components of an ecosystem?
2. Explain what is meant by the term habitat.
3. Explain the difference between a population and a community.
4. What is the difference between a species and a population?
5. Explain what is meant by the term niche.

Interactions in ecosystems

The organisms (biotic factors) in an ecosystem interact continually with one another and with their physical environment (abiotic factors). Some interactions include:

- Feeding among the organisms – the plants, animals and decomposers are continually recycling the same nutrients through the ecosystem.

- Competition among the organisms – animals compete for food, shelter, mates and nesting sites; plants compete for carbon dioxide, mineral ions, light and water.

- Interactions between organisms and the environment – plants absorb mineral ions, carbon dioxide and water from the environment; plants also give off water vapour and oxygen into the environment; animals use materials from the environment to build shelters; the temperature of the environment can affect processes occurring in the organisms; processes occurring in organisms can affect the temperature of the environment (all organisms give off some heat).

Don't forget that plants take in carbon dioxide and give out oxygen only when there is sufficient light for photosynthesis to occur efficiently. When there is little light, plants take in oxygen and give out carbon dioxide. You should be able to explain why – if not, see Chapter 9.

Abiotic factors in ecosystems

The abiotic factors in an ecosystem include physical components such as:

- **climatic** conditions – temperature, humidity, sunlight and rainfall
- **topography** – the physical shape of the land
- **aquatic** factors – water salinity, oxygen concentration and pH
- **edaphic** factors – soil and its components.

The physical characteristics of the environment determine the type of organisms present in an ecosystem.

Temperature

One of the main climatic factors is temperature. The temperature range is much wider in terrestrial habitats than in aquatic ones, so temperature has a greater effect on a terrestrial habitat. This affects the types of flora that will be available for food and shelter, and this directly determines the types of fauna that will be present. For example, the Northern Range tropical forest of Trinidad and Tobago (Figure 2.2) has warm temperatures. Monkeys, butterflies and plants are all able to tolerate this climate.

Figure 2.2 *The Northern Range, a tropical rainforest in Trinidad and Tobago, provides habitat for a wide variety of flora and fauna.*

Humidity

Humidity is the measure of the amount of moisture in the atmosphere. The humidity of a habitat may vary, depending on temperature and wind. For this reason, humidity influences the distribution of organisms in a habitat.

When the humidity is low, the rate of water loss from an organism's body increases; when humidity is high, water loss decreases. In plants, low humidity causes wilting as the cells become plasmolysed. High humidity results in cells becoming turgid. Some mammals sweat to keep their bodies cool when there is an increase in humidity. So, water loss in plants and animals is related to humidity.

Horses and hippopotamuses sweat to keep their bodies cool.

Light

Sunlight provides the Earth with warmth and with energy that supports all life on the Earth. Sunlight provides the energy that is trapped by plants and used in a process called photosynthesis to produce food. The food manufactured during photosynthesis feeds almost all the organisms found within an ecosystem.

Light intensity and the amount of light present in a terrestrial ecosystem will determine the types of organisms found there. For example, Fern Gully in Jamaica (Figure 2.3, page 24) is home to hundreds of fern species. Because of the valley, most of the sunlight is blocked from reaching the gully, creating a shaded area in which ferns thrive. Orchids are plants that also grow best in areas with low light. By contrast, many agricultural crops, such as sweet potatoes and yams (Figure 2.4, page 24), prefer direct sunlight for growth.

Figure 2.3 *The shaded valley of Fern Gully in Jamaica.*

(a) Yam

(b) Sweet potato

Figure 2.4 *Some plants need bright light to grow.*

In the tropics, the flowering of some plants depends on the day length. For example, poinsettia, gungo peas and sorrel will flower when the day length decreases. This is generally around the time when many people celebrate Christmas and, traditionally, these plants are associated with Christmas (Figure 2.5, page 25).

In aquatic habitats, the amount of light available is limited by the depth of the water. The deeper the water, the less light there is available. This affects the number of plants that will grow at the bottom of the habitat: fewer plants grow in deeper water. The number of animals present at greater depths will also decrease, as animals depend on plants for food.

Light also affects animals in different ways. Invertebrates, such as the earthworm and the woodlouse, move away from light, while others enjoy light and are active in the daytime.

Sorrel has been genetically modified to grow all year round.

(a) Poinsettias are used as Christmas decorations.   *(b)* Gungo beans are used in traditional Christmas dishes.

Figure 2.5 *Poinsettiass (a) and gungo peas (b) are plants traditionally used in Christmas celebrations.*

Water

Water is an important abiotic factor in ecosystems. All organisms need water to carry out their life processes. For example, water is a key reactant in the process of photosynthesis in plants, and it serves as a medium in which metabolic reactions occur. Because of water's chemical properties, it can serve as a habitat for both plants and animals.

The amount of water available in an ecosystem is based on the amount of rainfall it receives and its underground water sources. Various water bodies, such as lakes and rivers, are generated as a result of this water in the ecosystem. The water available in an ecosystem determines the diversity and number of the organisms that can live in that ecosystem (Figure 2.6, page 26). Plants become specially adapted to survive on the amount of water that the habitat provides.

Water bodies are differentiated based on the concentration of dissolved salts, or salinity, of the water. Water bodies may be marine, which has a high salinity, or fresh water, which has a low salinity. Aquatic habitats are affected by water's salinity, pH, flow rate and depth, and how much light penetrates the water.

(a) The limited amount of water in a desert means that there is a low diversity of organisms.

(b) A rainforest has a greater diversity and number of organisms as more water is available.

Figure 2.6 *The amount of water available in an ecosystem determines the number and diversity of the organisms living there.*

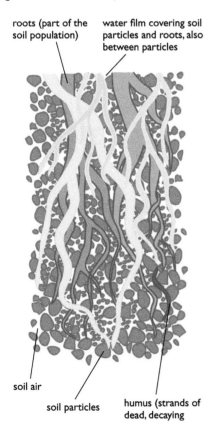

roots (part of the soil population)

water film covering soil particles and roots, also between particles

soil air

soil particles

humus (strands of dead, decaying

Figure 2.7 *The main components of a typical soil.*

Oxygen

The amount of oxygen present in an aquatic habitat is determined by several factors. These include the number of plants present, temperature, and the water's flow rate and depth. The level of oxygen in an aquatic environment has a great effect on the types of flora and fauna that can survive in that environment.

Soil

Soil is an important component of the physical environment of many ecosystems. It provides a habitat for many microorganisms and animals as well as for the roots of many plants. The main components of a typical soil are listed below and illustrated in Figure 2.7.

- **Mineral particles**, formed from rock by **weathering** – the rock is broken into ever smaller pieces by the action of rain, wind, frost in some countries and physical contact between pieces of rock. There are different sized mineral particles:

  - gravel – particles bigger than 2 mm diameter

  - sand – particles between 2 mm and 0.02 mm diameter

  - silt – particles between 0.02 mm and 0.002 mm diameter

  - clay – particles less than 0.002 mm diameter

- **Humus** (organic material) formed from the bodies of dead organisms

- **Water**, held in the spaces between the particles that make up the soil

- **Air**, also in the spaces between the particles
- A **soil population** comprising all the animals, plants and decomposers for which soil forms a habitat.

Although all soils have the same five components, the proportions differ and form a range of different soils (Figure 2.8). For example:

- Soils with a large proportion of sand particles will drain freely but hold water poorly.
- Soils with a large proportion of silt and clay will particles drain poorly but hold a lot of water.
- Soils with a mixture of all particles drain well and hold water well.
- Soils with a lot of humus will absorb heat easily and warm up quickly.

(a) Sand

(b) Clay

(c) Loam

Figure 2.8 *Different soil particles*

Activity 1: Investigating abiotic and biotic components of a habitat

SBA skills

| ORR | MM | AI | PD | Dr |

1 – Relationships between abiotic and biotoic components

You will need:

- thermometer
- hand lens or magnifying glass

Carry out the following:

1. Select a habitat, and then describe it by stating the site name or location, and the current weather conditions.
2. Measure the temperature of the habitat.
3. List as many organisms that are present in the habitat as you can. (Use a magnifying glass to view smaller organisms.)
4. Describe how five of these organisms relate to important factors in their natural environment.
5. Suggest relationships that may exist between different organisms in the habitat.

2 – The pH of soil

You will need:

- beaker
- measuring cylinder
- water
- pH paper
- filter paper
- filter funnel
- conical flask

Carry out the following:

1. Collect 100 ml of soil, and pour it into a beaker.

2. Pour 50 ml of water into a measuring cylinder. Test the water pH using pH paper. Record the pH of the water.

3. Fold a piece of filter paper and insert it into the filter funnel. Place the end of the filter funnel into the conical flask.

4. Pour the soil from the beaker into the filter funnel, and then pour the water over it.

5. Test the pH of the filtrate. Record the measurement as the pH of the soil.

Activity 2: Investigating the main components of soils

SBA skills

ORR	MM	AI	PD	Dr

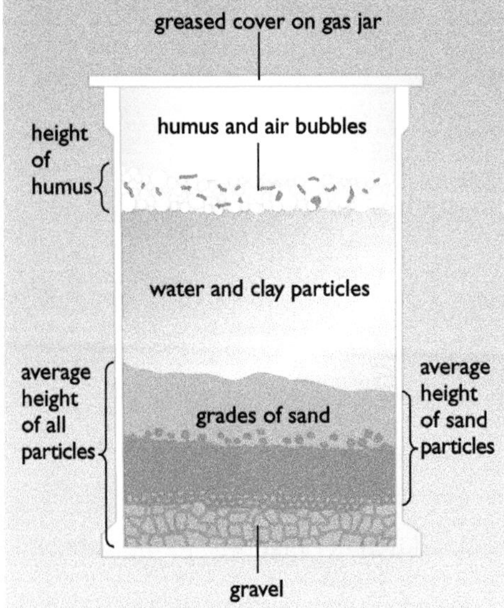

greased cover on gas jar

height of humus

humus and air bubbles

water and clay particles

average height of all particles

grades of sand

average height of sand particles

gravel

Figure 2.9 *Soil layers after settling.*

1 – The main solid components of soil

You will need:

- two gas jars and covers
- two soil samples
- some graph paper
- sticky tape

Carry out the following:

1. Place a sample of soil into each gas jar. The gas jars should contain the same amount of soil and be about one-third full.

2. Pour water into each gas jar until it is nearly full. Each gas jar should contain the same amount of water.

3. Cover one of the gas jars and shake the jar thoroughly for 1 minute. Repeat with the other gas jar.

4. Allow the jars to stand for 10 minutes. The mineral particles in the soil will settle to the bottom of the gas jar. The biggest ones (gravel) will settle first. The smallest particles to settle (silt) will form a pale layer on top of the other particles. You will not be able to see the individual silt particles. Clay particles will not normally settle out of water. They will remain suspended in the water and make it look cloudy. The humus will float.

5. Stick some graph paper to the side of each gas jar.

6. Estimate the percentage of each type of particle in each soil, using the following formula (using sand as an example):

$$\text{percentage of sand} = \frac{\text{height of sand}}{\text{total height of mineral particles} + \text{height of humus}}$$

Activity 2: Investigating the main components of soils (continued)

2 – The water content of soil

You will need:

- a balance
- an incubator
- two metal containers
- two soil samples

Carry out the following:

1. Construct a table like the one on the left to record your results.

2. Weigh out 10 g of a soil sample into a metal container. Weigh 10 g of the second soil sample into the other metal container.

3. Label the containers, place each into an incubator set at 105°C and leave for 24 hours.

4. Reweigh both samples and record the masses in the table. Replace the samples in the incubator for a further 24 hours.

5. Reweigh both samples and record the masses in the table. If the masses are the same as those after 24 hours, you need make no further weighings. If the masses are not the same as those after 24 hours, replace the samples in the incubator for a further 24 hours and reweigh.

6. Keep reweighing until you have two successive masses that are the same. Calculate the percentage of water in the sample using the formula:

$$\text{percentage of water} = \frac{(\text{initial mass} - \text{final mass})}{10} \times 100$$

Soil A or B	Mass (g)
mass of empty tin	
mass of tin + soil	
mass of tin + soil	
after 24 hours	
after 48 hours	
after 72 hours	

Activity 3: Investigating the water-holding capacity of different soils

SBA skills

ORR	MM	AI	PD	Dr

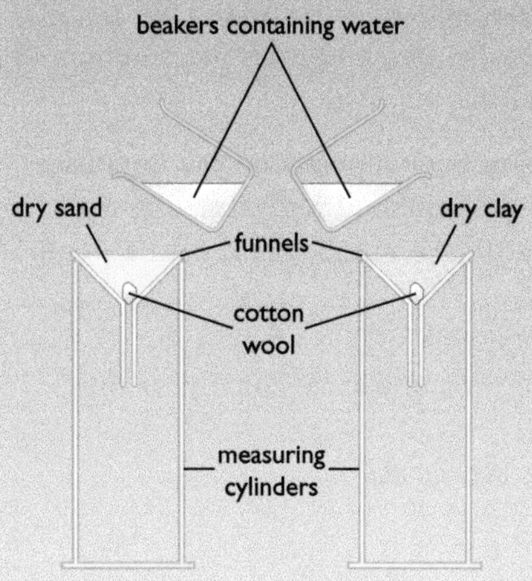

Figure 2.10 *Apparatus for Activity 3.*

You will need:

- samples of a clay soil and a sandy soil
- two 100-cm³ measuring cylinders
- two funnels
- cotton wool

Carry out the following:

1. Set up the apparatus as shown in Figure 2.10.

2. Place 20 g of dry clay soil into one funnel and 20 g of dry sandy soil into the other funnel.

3. Pour 100 cm³ of water over each soil sample.

4. Allow the water to drain through and measure how much collects in each measuring cylinder.

5. Calculate how much water has been retained by the soil.

6. Work out how much water would be retained by 100 g of the soil; this is the water-holding capacity of the soil.

7. Write up your experiment and try to explain any difference in the water-holding capacity of the two soils.

Activity 4: Comparing the humus (organic matter) content of two soils

SBA skills

ORR	MM	AI	PD	Dr

Stage	Mass (g)	
	Clay soil	Sandy soil
container + 10 g soil		
after 2 days in incubator		
after 3rd day in incubator (x)		
after 10 min heating		
after 20 min heating		
after 30 min heating (y)		
mass of humus in 10 g soil ($x - y$)		

You will need:

- samples of a clay soil and a sandy soil
- an incubator/oven
- a balance
- a Bunsen burner, tripod and gauze
- two small metal containers
- tongs

Carry out the following:

1. Construct a table like the one on the left.

2. Weigh out 10 g of each soil into a metal container.

3. Place each container into an incubator/oven set at 50°C for 2 days.

4. Reweigh each container and note the weight.

5. Place the containers back in the oven for a further day.

6. Reweigh the containers; this weight should be the same as at stage 3; if not, repeat stages 4 and 5 until there is no change in weight.

7. Stand each container of soil on a tripod and gauze and heat for 10 minutes *from above* with the Bunsen burner so that the soil burns. (**WARNING:** Wear safety goggles as the soil may spit.)

8. Allow to cool, then, using the tongs transfer the container to the balance and reweigh.

9. Repeat stages 7 and 8 until two reweighings give the same result.

10. Calculate the mass of humus in 10 g of soil as shown in the table.

11. Now calculate the percentage of soil that is humus.

12. Write up your experiment and try to explain any differences between the humus content of the two soils.

Investigating numbers of organisms

To find out whether a population is increasing or decreasing, we must be able to count the numbers of organisms in the population. It is usually not possible to count all the organisms. We must sample the area and from the numbers in our sample, make an estimate of the size of the population.

The method of estimating the population numbers depends on the organism and the environment – we cannot use the same method for estimating the number of grass plants in a field as we can for the number of fish in a pond.

> 6. What is meant by the term environment?
> 7. List three things that the physical environment can supply to the organisms living in an area.
> 8. List the five main components of a typical soil.
> 9. What is humus?
> 10. Why do sandy soils drain freely?

Estimating populations on land

Different techniques are needed for organisms that are static and organisms that are mobile. For static organisms we would normally use **quadrat** sampling, while for mobile organisms we would normally use the **mark-release-recapture** technique.

The quadrat

A quadrat is a square frame that is usually made of wood, metal or plastic. The size can vary, but the most commonly used frames are either 0.5 m² or 1.0 m². They are placed randomly several times within the area to be investigated, and the numbers of the particular organism under investigation are counted and recorded. The average number of times each organism appears is calculated. From the result, the frequency and density of a species can be estimated.

Determining the frequency and density of organisms within a habitat

The population characteristics of a habitat must be known when an area is studied. These characteristics include the species density and the species frequency.

Species frequency

The percentage frequency of a species is the percentage of the sample that the species is present in. The percentage frequency is determined using this formula:

number of times an organism occurs ÷ number of quadrat throws × 100 = species frequency

For example, an organism present in four out of the five quadrats thrown gives a percentage frequency of 80%:

4 ÷ 5 × 100 = 80%

Species density

The density of a species is the mean (average) number of individuals of the species in a given area. Species density is determined using this formula:

mean number of individuals of the species ÷ area of quadrat

For example, a species has a total number of 133 individual organisms from 10 quadrat throws:

average number of species: 133 ÷ 10 = 13.3

area of the quadrat = 1 m²

∴ species density = 13.3 ÷ 1 m² = 13.3 m²

Figure 2.11 *Students using a quadrat.*

Mark-release-recapture

This method requires you to capture a sample of the organism being investigated. The individual organisms are marked with paint in an inconspicuous area, and then released. The organisms are given time to reassemble with unmarked organisms. A second sample is then captured. The number of marked organisms and the total number captured are recorded.

Bottles

Bottles can be used to collect samples of aquatic and terrestrial organisms. The volume of the bottles used must be known. In aquatic habitats, samples of the water can be removed and the number of organisms in the sample counted. In terrestrial habitats, pitfall traps can be used to collect animals.

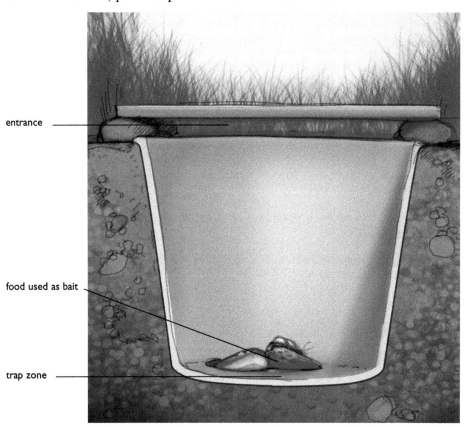

entrance

food used as bait

trap zone

Figure 2.12 *A pitfall trap, used to collect small animals.*

Nets

Nets used to collect samples are usually a mesh or a fine piece of cloth attached to a handle. The size of the mesh may vary and determines the size of the organisms collected. The sweep net can be used to skim the surface of water or to sweep through the water to collect organisms. The sweep net is also used to dislodge animals in the field in order to collect them.

Investigating changes in abundance across an area

As the topography of the environment changes across an area, so may the distribution of the organisms found in the area. For example, as you move from land to sea, the conditions change from dry to wet, and the soil type may also change, for example from loam to sandy soil. Such factors will determine where particular organisms are found in a given area. In this case, you may try to show how the numbers of, say, crabs change from the top of a beach to the bottom of a beach. To do this we must lay out a **transect**.

Transects

A transect is a long tape measure or a rope marked at regular intervals that is stretched across the area under investigation. The organisms that touch the transect at specific intervals are recorded. The transect is drawn to scale to represent the organisms that touched it. Symbols are used to represent organisms. This type of transect is known as a line transect.

A second type of transect is the belt transect. It comprises two line transects laid parallel to each other. The area that lies between the two transects is then sampled.

(a) A line transect.

(b) A belt transect.

Figure 2.13 *Transects are used to investigate changes in abundance across an area.*

11. What technique would you use to estimate the numbers of:
    (a) a mobile terrestrial organism
    (b) a static terrestrial organism
    (c) an aquatic organism?

Activity 5: Estimating the numbers of static organisms in an area

You will need:

- a quadrat
- a calculator

Carry out the following:

1. Divide the area into a grid (Figure 2.14).

2. Use the random number generator on a calculator to produce a pair of coordinates or pull two numbers out of different containers to produce the coordinates. For example, A5 and B6 would produce the point indicated in Figure 2.14.

3. Place the quadrat with its top left-hand corner on the point.

4. Count the number of organisms of the species under investigation in the quadrat.

5. Repeat steps 2 to 4 for the number of quadrats you have decided to use.

6. Now find the mean (average) number of organisms per quadrat.

7. Calculate the area of the quadrat.

8. Estimate the area of the environment under investigation in the same units as the quadrat area, (e.g both in cm² or both in m²).

9. You can now make an estimate of the population size using the formula:

$$\text{population size} = \frac{\text{mean number of organisms per quadrat} \times \text{area of environment}}{\text{area of quadrat}}$$

If you do not have access to a random number generator, then instead of the procedure given in steps 2 and 3, stand with your eyes closed and throw the quadrat over your left shoulder. (Make sure no one is standing behind you!) Then carry on from step 4.

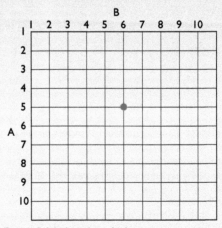

Figure 2.14 A grid marked out.

Activity 6: Estimating the numbers of a mobile species

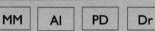

SBA skills

ORR	MM	AI	PD	Dr

You will need:

- container
- paint

Carry out the following:

1. Collect a small sample of the animals from the area under investigation and count them (N_1).

2. Put a small paint mark in an unobtrusive place on each animal.

3. Release them and allow sufficient time for them to disperse among the population.

4. Collect a second sample and note:

   - the total size of the sample (N_2), and

   - the number that are marked (n).

5. Estimate the population size using the formula:

$$\text{population size} = \frac{(N_1 \times N_2)}{n}$$

Activity 7: Estimating numbers of a population of aquatic organisms

SBA skills

ORR	MM	AI	PD	Dr

You will need:

- a bottle or container

Carry out the following:

The following procedure can be used to estimate the number of tadpoles in a pond.

1. Immerse a container of known volume in the pond.

2. Count the number of tadpoles in the container.

3. Repeat the procedure at several other randomly determined points in the pond.

4. Find the mean (average) number of tadpoles per container.

5. Estimate the volume of the pond (in the same units as the container).

6. Estimate the population size using the formula:

$$\text{population size} = \frac{\text{mean number of tadpoles per container} \times \text{volume of pond}}{\text{volume of container}}$$

Activity 8: Investigating changes in abundance across an area

You will need:

- a tape measure or a rope marked at regular intervals

Carry out the following:

1. Lay out a tape measure across the area under investigation.

2. Mark the tape measure at regular intervals (this will depend on the size of the area and the time available – every 4 metres in Figure 2.15).

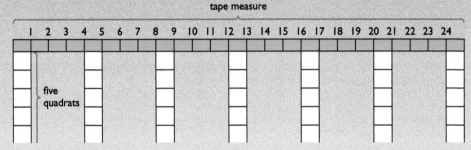

Figure 2.15 *A belt transect.*

3. Estimate the abundance of the organism in five quadrats at each sampling point by either:

- recording the presence or absence of the organism in each of the five quadrats at a sampling point and converting this to a percentage frequency of occurrence, e.g. organism present in four of the five quadrats gives a percentage frequency of 80%

- estimating the percentage of the quadrat that is covered by the organism

- counting the actual numbers of the organism in each quadrat.

4. This information can then be used to see how the frequency, area covered or actual numbers change across the area.

Chapter summary

In this chapter you have learnt that:

- the main components of an ecosystem are producers, consumers, decomposers and the physical environment

- all the organisms of one species form a population in an ecosystem

- all the organisms of all species form the community of that ecosystem

- each organism has its own specific habitat within the ecosystem

- the physical environment of an organism is the sum of all the non-biological components of the ecosystem where the organism is found

- the environment can supply many resources including mineral ions, gases, water, as well as habitats for individual organisms

- all soils comprise mineral particles (sand, silt and clay), humus (organic matter), water, air and a soil community

- sand particles improve drainage but make water holding worse

- clay particles improve water holding but make drainage worse

- a population is the number of organisms of one species in an ecosystem at one time

- the size of a population is influenced by the availability of resources, disease and predation

- the numbers of organisms in an area can be estimated by quadrat sampling for static terrestrial organisms, mark-release-recapture for mobile terrestrial organisms and volume sampling for aquatic organisms.

Questions

1. Which of the following is **not** a component of an ecosystem?

   A producer
   B consumer
   C population
   D decomposer

2. The physical environment of an organism is best defined as:

   A the air and water that surround the organism
   B all the biological factors that surround the organism
   C all the non-biological factors that surround the organism
   D the air, water and soil that surround the organism

3. What is the main source of energy for ecosystems?

   A food made by plants
   B light from the Sun
   C nutrients from the soil
   D oxygen from the air

4. Which of the following methods is most appropriate for estimating the number of organisms of a static species in an area?

   A mark-release-recapture
   B volume sampling
   C a line transect
   D random quadrat sampling

5. A soil which is free-draining will probably contain:

   A a large proportion of silt and clay particles and few sand particles
   B a large proportion of sand and clay particles and few silt particles
   C a small proportion of silt and clay particles and many sand particles
   D a small proportion of sand and clay particles and many silt particles

6. For each of the following, distinguish between the pair of terms and give an example of each item:

   a) abiotic and biotic factors

   b) niche and habitat

   c) population and species.

7.

Plant organisms	Quadrat number									
	1	2	3	4	5	6	7	8	9	10
Clusters of grass	5	8	2	8	4	1	0	5	2	3
Small herbaceous plant	0	0	3	7	9	10	15	9	6	0
Small shrub	5	7	3	9	1	7	4	9	3	2

A group of students used a 1-m² quadrat to collect data to determine the distribution of three plant species. The results are shown in the table above.

a) Calculate the species density of each plant species. *(3)*

b) Calculate the species frequency of each plant species. *(3)*

8. The table gives information about the mineral particles found in four soils.

Type of mineral particle	Percentage of each mineral particle in soil samples			
	Soil 1	Soil 2	Soil 3	Soil 4
sand	86	55	30	10
silt	8	27	20	45
clay	6	18	50	45

a) Plot a bar chart to compare the percentage of sand in the four soils. *(3)*

b) Which soil would you expect to have the worst drainage? Explain your answer. *(3)*

c) Which soil would you expect to be the most fertile? Explain your answer. *(4)*

9. The diagram shows two measuring cylinders which collected water that had drained through two different soil samples – sandy soil and clay soil. Fifty cubic centimetres of water was added to both soil samples.

a) Which measuring cylinder (1 or 2) collected water from:

   i) the sandy soil

   ii) the clay soil? *(2)*

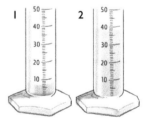

b) Give reasons for your choice using information about the characteristics of each soil type. *(2)*

10. A student carried out an investigation to find the water content of a sample of soil. He placed the sample in a metal dish and weighed it before and after drying. His results were as follows:

Mass of metal dish = 12.4 g

Mass of dish plus moist soil = 19.6 g

Mass of dish plus dry soil = 18.3 g

a) How could the student have dried the soil sample? *(2)*

b) How could he make sure that the sample was completely dry? *(2)*

c) Calculate the percentage of water in the sample of soil. *(2)*

Chapter 3: Food Chains and Food Webs

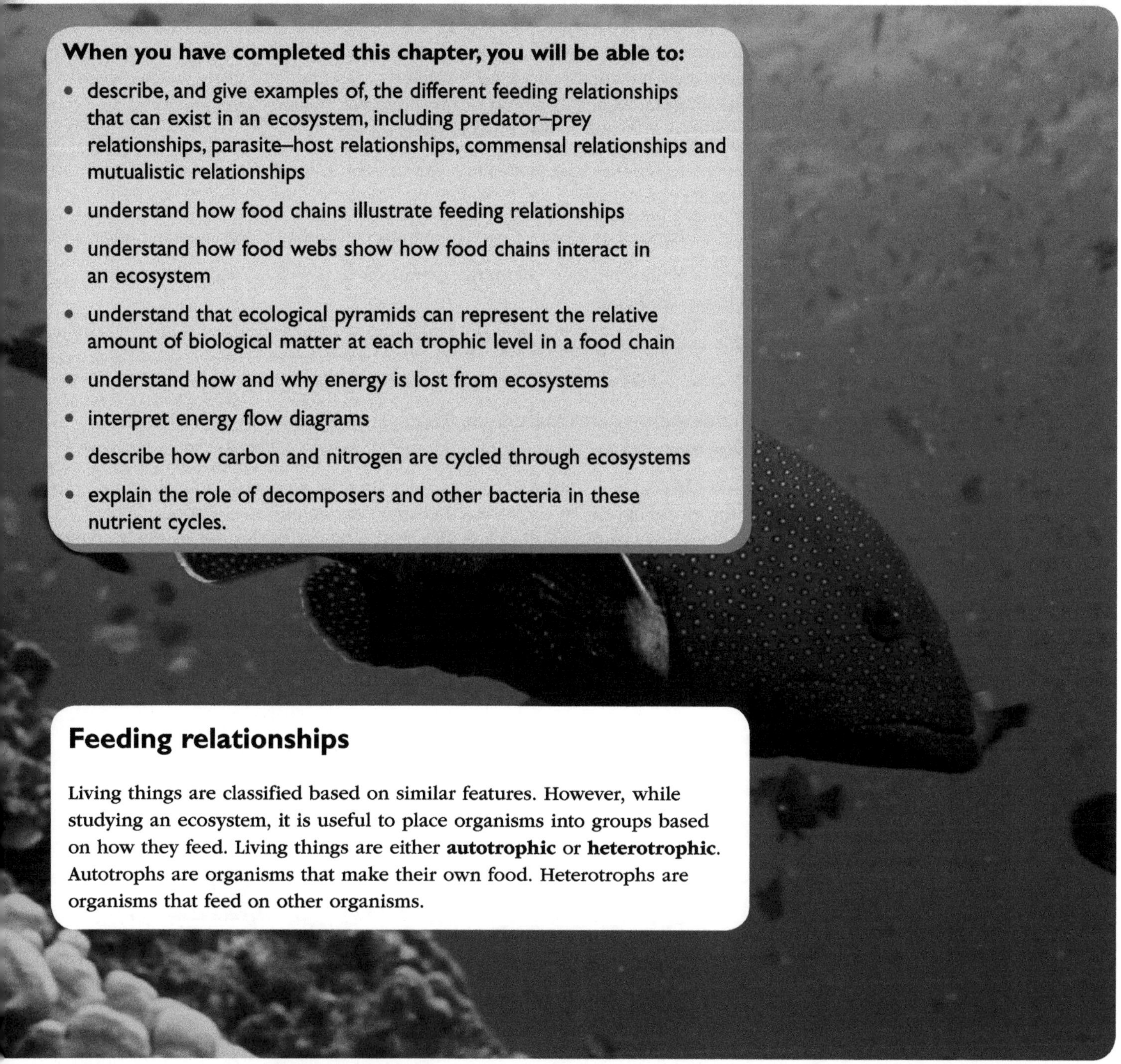

When you have completed this chapter, you will be able to:

- describe, and give examples of, the different feeding relationships that can exist in an ecosystem, including predator–prey relationships, parasite–host relationships, commensal relationships and mutualistic relationships

- understand how food chains illustrate feeding relationships

- understand how food webs show how food chains interact in an ecosystem

- understand that ecological pyramids can represent the relative amount of biological matter at each trophic level in a food chain

- understand how and why energy is lost from ecosystems

- interpret energy flow diagrams

- describe how carbon and nitrogen are cycled through ecosystems

- explain the role of decomposers and other bacteria in these nutrient cycles.

Feeding relationships

Living things are classified based on similar features. However, while studying an ecosystem, it is useful to place organisms into groups based on how they feed. Living things are either **autotrophic** or **heterotrophic**. Autotrophs are organisms that make their own food. Heterotrophs are organisms that feed on other organisms.

The simplest way of showing feeding relationships within an ecosystem is by using a **food chain** (Figure 3.1).

grass   grasshopper   lizard

Figure 3.1 *A simple food chain.*

You can find out more about herbivores and carnivores in Chapter 12.

In any food chain, the arrow (→) means 'is eaten by'. In the food chain in Figure 3.1, the grass is the **producer**. It is a plant so it can photosynthesise and produce food materials. The grasshopper is the **primary consumer**. It is an animal which eats the producer and is also a **herbivore**. The lizard is the **secondary consumer**. It eats the primary consumer and is also a **carnivore**. The different stages in a food chain (producer, primary consumer and secondary consumer) are called **trophic levels**.

Many food chains have more than three links in them. This is one example of a longer food chain:

> filamentous algae → mayfly nymph → caddis fly larvae → salmon

In this freshwater food chain, the extra link in the chain makes the salmon a **tertiary consumer**.

This is anther example of a longer food chain:

> hibiscus → caterpillar → carnivorous insect → rat → owl

In this terrestrial food chain, the fifth link makes the owl a **quaternary consumer**. Because nothing eats the owl, it is also called the **top carnivore**.

A food web shows the feeding relationships among several different producers and consumers within an ecosystem.

Food chains are a convenient way of showing the feeding relationships between a few organisms in an ecosystem, but they oversimplify the situation. The marine food chain above implies that only crustaceans feed on plankton, which is not true. Some whales and other mammals also feed on plankton. For a fuller understanding, you need to consider how the different food chains in an ecosystem relate to each other. Figure 3.2 gives a clearer picture of the feeding relationships involved in a wetland ecosystem in which salmon are the top carnivores. The diagram shows the **food web** of the egret.

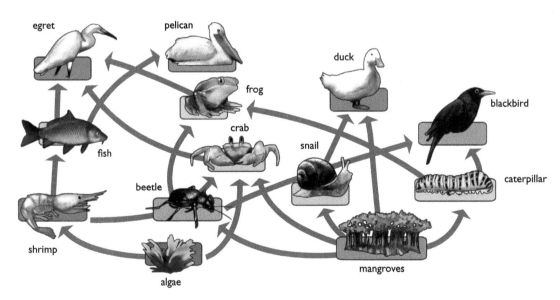

Figure 3.2 *The food web of the egret.*

In the food web, another type of consumer is identified: the duck is an **omnivore**. It is an animal which eats both producers and consumers. Humans are also omnivores.

This is still a simplification of the true situation, as some feeding relationships are not shown. It does, however, give some indication of the interrelationships that exist between food chains in an ecosystem. With a little thought, you can predict how changes in the numbers of an organism in one food chain in the food web might affect those in another food chain. For example, if the frog population were to decline through disease, there could be several possible consequences:

- The blackbird population could increase as there would be more beetles and caterpillars to feed on.

- The beetle and the caterpillar population could decrease as the blackbird would eat more of them.

- The numbers could stay static due to a combination of the above.

Although food webs give us more information than food chains, they don't give any information about how many, or what mass of organisms is involved. Neither do they show the role of the decomposers. To see this, we must look at other ways of presenting information about feeding relationships in an ecosystem.

Decomposers

Decomposers are bacteria and fungi which play a key role in recycling. They break down complex organic molecules in the bodies of dead animals and plants into simpler substances, which they release into the environment. This makes these nutrients available to producers.

Activity 1: Constructing a food web

SBA skills

| ORR | MM | AI | PD | Dr |

You will need:

- containers in which to store the organisms temporarily
- labels for the containers
- a hand lens or viewer
- forceps

Carry out the following:

1. Construct a table like the one below.

Name of organism	Where found	Plant or animal	If animal, is it a herbivore or carnivore?	Other comments

Be clear in your mind about the area that you have to study; your teacher will explain precisely the limits of your area. All the organisms that you will handle are living organisms and should still be living organisms at the end of your investigation. You do not have the right to harm them in any way.

2. Collect the organisms carefully, using forceps where necessary to handle them.

3. Observe each organism carefully, using the hand lens where needed, and complete the table. The following may help you to decide if an animal is a herbivore or carnivore:

- Carnivores are often fast moving; herbivores tend to move more slowly.

- The mouthparts of a carnivore will be capable of piercing or cutting the bodies of other animals; those of herbivores are usually adapted for grinding plant material (however, caterpillars that feed on leaves often have cutting mouthparts).

Activity 2: Examining the role of decompsers

You will need:

- four cups or transparent plastic bags
- a slice of fresh bread
- a slice of mouldy bread
- cotton swabs
- plastic wrap
- rubber bands
- a dropper
- distilled water
- a hand lens

Carry out the following:

1. Label each cup with your name, and then label the cups A, B, C and D.

2. Cut the slice of fresh bread into quarters.

3. Place one quarter into cup A and one into cup C.

4. Brush a swab gently over the surface of the mouldy bread, and then gently rub the swab across the bread quarter in cup A and cup C.

5. Cover cups A and C with plastic wrap, and use rubber bands to secure the plastic.

6. Place one quarter of the bread into cup B and one into cup D.

7. Gently rub the surface of the bread in cup B and cup D with the mould swab. Moisten the bread in both cups with the distilled water, using the dropper.

8. Cover cups B and D with plastic wrap, and use rubber bands to secure the plastic.

9. Place cups A and B in a well-lit area.

10. Place cups C and D in a dark area.

11. Examine the bread quarters each day for seven days, without removing the plastic wrap.

12. Measure the total area of the mould growth on each piece of bread each day. Record the results in a table.

Special feeding relationships in ecosystems

Some relationships are a little more complex than the situation in which one animal simply eats a plant or another animal. Sometimes two organisms form some kind of **feeding association**. The main feeding associations are:

- **parasitism** – one organism (the parasite) feeds from another (the host), which is harmed by being deprived of food

- **commensalism** – one organism derives benefit from the association, but the other is not harmed

- **mutualism** – both organisms derive benefit from the association.

Parasitism

There are many examples of parasites. Humans have many parasites; some internal (endoparasites), like the tapeworm, and some external (ectoparasites), like the flea. Figure 3.3 illustrates the association between a human and a tapeworm, and Figure 3.4 shows the life cycle of a tapeworm.

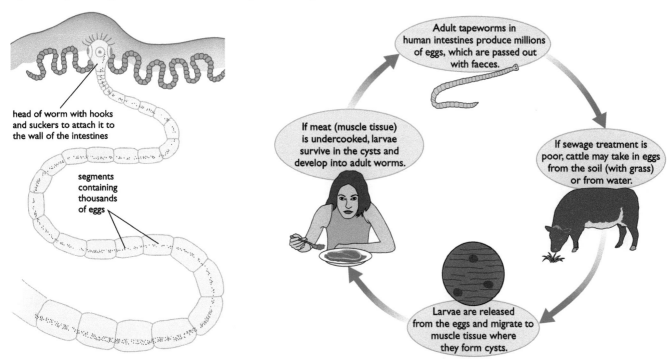

head of worm with hooks and suckers to attach it to the wall of the intestines

segments containing thousands of eggs

Adult tapeworms in human intestines produce millions of eggs, which are passed out with faeces.

If meat (muscle tissue) is undercooked, larvae survive in the cysts and develop into adult worms.

If sewage treatment is poor, cattle may take in eggs from the soil (with grass) or from water.

Larvae are released from the eggs and migrate to muscle tissue where they form cysts.

Figure 3.3 *A tapeworm in the human gut.*

Figure 3.4 *The life cycle of the beef tapeworm.*

Some plants around the Caribbean are parasitised by the *Cuscuta*, also known as the love bush plant (see Figure 3.5). *Cuscuta* uses its haustoria to pierce the phloem of another plant and absorb its nutrients.

Figure 3.5 *A Jamaican love bush parasitising another plant.*

Commensalism

There are many examples of commensalism in the Caribbean coral-reef ecosystem. One type of shrimp is immune to the stings from the tentacles of the sea anemone and lives, protected, among these tentacles (Figure 3.6). It emerges from time to time to feed by picking ectoparasites from passing fish. The shrimp benefits by being protected, but does not affect the sea anemone in any way.

Some plants may be found living on other plants. Plants that live on other plants without harming them are epiphytes. For example, bromeliads are found growing on tree branches (Figure 3.7); the tree gives them support and remains unharmed.

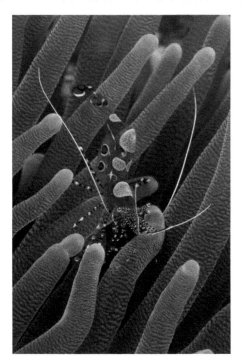

Figure 3.6 *An example of commensalism.*

Figure 3.7 *Old man's beard, a bromeliad, growing on a tree.*

Mutualism

In this type of association, both organisms benefit. In the shrimp/sea anemone example of commensalism just described, the shrimp forms a temporary mutualistic association with the fish it visits to remove ectoparasites. The shrimp gets a supply of food, whilst the fish is cleansed of ectoparasites that were feeding from it. Both organisms benefit.

Other mutualistic associations are much closer. All legumes (plants such as peas and beans that produce seeds in pods) form a mutualistic association with nitrogen-fixing bacteria (see page 54). These bacteria make additional nitrogen available to the plants (in the form of ammonia), whilst they receive a supply of organic molecules, such as sugars, from the plant. With the additional nitrogen, the plants can make more amino acids, which can be converted to protein, and so the plants grow better. The bacteria can multiply more quickly using the organic materials supplied by the plant. Both organisms benefit.

Predator–prey relationships

As we saw earlier, a population is a group of organisms of the same species that occupy the same habitat at any one time. However, the numbers of organisms in a population don't remain static, and any population must first establish itself. A species must **colonise** an area and multiply. If it is sufficiently well adapted, it will achieve this. If not, it will die out.

Predation is the continued feeding of one animal species on another. For example, lions are predators of antelope, and killer whales are predators of seals.

44

Factors that can limit the growth of a population include **predation**, disease, limited food supply and limited space for breeding.

An animal that hunts, kills and eats other animals is called a **predator**. The animal that is eaten by a predator is called the predator's **prey**.

The relationship between the numbers of a predator and its prey is a very close one. One of the best researched examples of predator–prey numbers involves the snowshoe hare (the prey) and the lynx (the predator) in Canada (Figure 3.9).

Figure 3.8 *The lynx (a) and its prey, the snowshoe hare (b).*

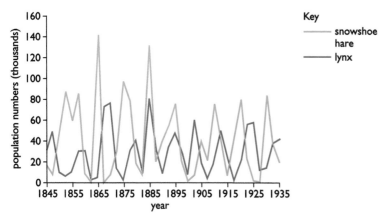

Figure 3.9 *The relationship between the numbers of lynx (b) and its prey, the snowshoe hare (b).*

The patterns in the graph are typical of predator–prey relationships in which the prey is the only food or the main food of the predator. They can be explained as follows:

1 As the numbers of prey increase, there is more food available for the predators.

2 The predators kill and eat more prey and so their numbers increase.

3 As more prey are now being eaten, their numbers start to decrease.

4 There is now *less* food for an *increased* number of predators.

5 The numbers of predators decrease as they compete for scarce food.

6 Fewer predators means that there will be fewer prey eaten and so the prey population recovers.

7 There are more prey and so there is more food available for the predators, and the cycle begins again.

Predators keep populations in check. They prevent dramatic surges in numbers. They also help the prey population by killing and eating mainly the weaker members. This means that only the healthier, fitter prey survive.

In the case of fur seals in the Antarctic, the effects of predation and food supply coincided. Sealing was banned and so predation of the seals by humans ceased. At the same time, the numbers of whales in the Antarctic decreased. Less plankton was eaten by the whales, so there was more for the fish. The numbers of fish increased, so there was more food for the seals. The numbers of seals surged. Later, as their numbers continued to increase, they had to compete for available food (fish) and their numbers stabilised, even decreasing in some years. These changes are shown in Figure 3.10.

> You may be asked to identify which is the predator and which is the prey from a graph like the one in Figure 3.9. Remember:
>
> - the numbers of prey will usually be higher than those of the predator as they are lower in the food chain
>
> - changes in the numbers of prey occur first.

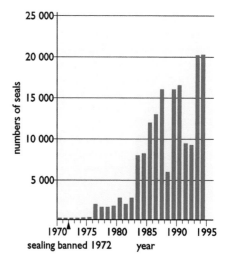

Figure 3.10 *Fur seal numbers at Signy Island in the Antarctic from 1970 to 1995.*

In an established natural ecosystem, nutrients are constantly being recycled and made available for a new generation of organisms. Populations remain more or less stable, showing natural fluctuations typical of predator–prey relationships. Complex food webs mean that fluctuations in numbers of one species do not necessarily have a dramatic effect on others. Sometimes, human influence upsets the balance in an ecosystem. Chapter 5 deals with some of the ways in which humans influence the environment.

Ecological pyramids

Ecological pyramids are diagrams that represent the relative amounts of organisms at each trophic level in a food chain. There are two main types:

- **pyramids of numbers**, which represent the numbers of organisms in each trophic level in a food chain, irrespective of their mass

- **pyramids of biomass**, which show the total mass of the organisms in each trophic level, irrespective of their numbers.

Consider these two food chains:

grass → grasshopper → frog → bird   hibiscus → aphid → ladybird → bird

The diagrams below show the pyramids of numbers and biomass for these two food chains.

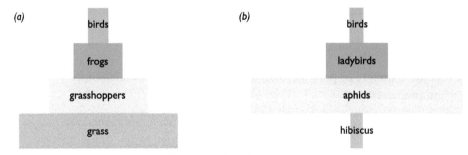

Figure 3.11 *Pyramids of numbers for the two food chains.*

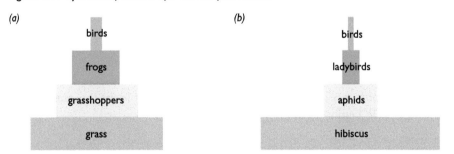

Figure 3.12 *Pyramids of biomass for the two food chains.*

The two pyramids for the 'grass' food chain look the same – the numbers at each trophic level decrease. The *total* biomass also decreases along the food chain – the mass of *all* the grass plants in a large field would be more than that of *all* the grasshoppers, which would be more than that of *all* the frogs, and so on.

The two pyramids for the 'hibiscus' food chain look different because of the size of the hibiscus shrubs. Each hibiscus plant can support many thousands of aphids, so the numbers *increase* from the first to second trophic levels.

But, each ladybird will need to eat many aphids and each bird will need to eat many ladybirds, so the numbers *decrease* at the third and fourth trophic levels. However, the total biomass *decreases* at each trophic level – the biomass of one hibiscus tree is much greater than that of the thousands of aphids it supports. The total biomass of all these aphids is greater than that of the ladybirds, which is greater than that of the birds.

Suppose the birds in the second food chain are parasitised by nematode worms. The food chain now becomes:

hibiscus → aphid → ladybird → bird → nematode worm

The pyramid of numbers now takes on a very strange appearance (Figure 3.13) because of the large numbers of parasites on each bird. The pyramid of biomass, however, has a true pyramid shape because the total biomass of the nematode worms must be less than that of the birds they parasitise.

Why do diagrams of feeding relationships give a pyramid shape?

The explanation is relatively straightforward (Figure 3.14). When a goat eats grass, not all of the materials in the grass plant end up as goat! There are losses:

- Some parts of the grass are not eaten (the roots for example).

- Some parts are not digested and so are not absorbed – even though goats have a very efficient digestive system.

- Some of the materials absorbed form excretory products.

- Many of the materials are respired to release energy, with the loss of carbon dioxide and water.

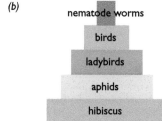

(a)

(b)

Figure 3.13 *(a) A pyramid of numbers and (b) a pyramid of biomass for the parasitised food chain.*

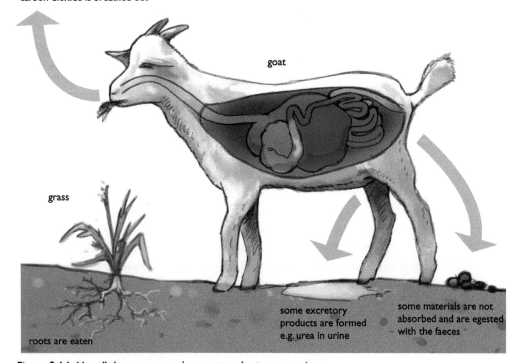

carbon dioxide is breathed out

goat

grass

roots are eaten

some excretory products are formed e.g. urea in urine

some materials are not absorbed and are egested with the faeces

Figure 3.14 *Not all the grass eaten by a goat ends up as goat tissue.*

In fact, only a small fraction of the materials in the grass ends up in new cells in the goat. Similar losses are repeated at each stage in the food chain, so smaller and smaller amounts of biomass are available for growth at successive trophic levels. The shape of pyramids of biomass reflects this.

Feeding is a way of transferring energy between organisms. Another way of modelling ecosystems looks at the energy flow between the various trophic levels.

The flow of energy through ecosystems

This approach focuses less on individual organisms and food chains and rather more on energy transfer between trophic levels (producers, consumers and decomposers) in the whole ecosystem. There are a number of key ideas that you should understand at the outset.

- Photosynthesis 'fixes' sunlight energy into chemicals such as glucose and starch.

- Respiration releases energy from organic compounds such as glucose.

- Almost all other biological processes (e.g. muscle contraction, growth, reproduction, excretion and active transport) use the energy released in respiration.

- If the energy released in respiration is used to produce new cells (general body cells in growth and sex cells in reproduction), then the energy remains 'fixed' in molecules in that organism. It can be passed on to the next trophic level through feeding.

- If the energy released in respiration is used for other processes then it will, once used, eventually escape as heat from the organism. Energy is therefore lost from food chains and webs at each trophic level.

This can be shown in an **energy flow diagram**. Figure 3.15 shows the main ways in which energy is transferred in an ecosystem. It also gives the amounts of energy transferred between the trophic levels of a grassland ecosystem.

As you can see, only about 10% of the energy entering a trophic level is passed on to the next trophic level. This explains why not many food chains have more than five trophic levels. Think of the food chain:

$$A \rightarrow B \rightarrow C \rightarrow D \rightarrow E$$

If we use the idea that only about 10% of the energy entering a trophic level is passed on to the next level, then, of the original 100% reaching A (a producer), 10% passes to B, 1% (10% of 10%) passes to C, 0.1% passes to D and only 0.01% passes to E. There just isn't enough energy left for another trophic level. In certain parts of the world some marine food chains have six trophic levels because of the huge amount of light energy reaching the surface waters.

Cross-curricular links:

physics, types of energy; chemistry, exothermic and endothermic reactions

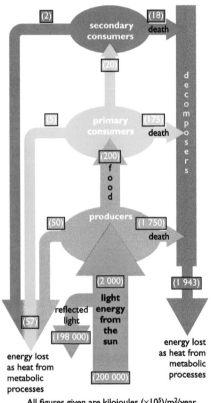

All figures given are kilojoules ($\times 10^5$)/m^2/year.

Figure 3.15 *The main ways in which energy is transferred in an ecosystem. The amounts of energy transferred through 1 m^2 of a grassland ecosystem per year are shown in brackets.*

Chapter summary

In this chapter you have learnt that:

- food chains illustrate the feeding relationships between producers and consumers

- food webs illustrate how many different food chains interact in an ecosystem and show the interdependence of the organisms in the food web

- decomposers are important in the process of decay, which releases nutrients from dead organisms

- in a parasite–host relationship, the parasite derives benefit whilst the host is harmed

- in a commensal relationship, only one organism benefits, but the other is not obviously harmed in any way

- in a mutualistic relationship, both organisms derive benefit from the association

- pyramids of numbers show the relative numbers of organisms at each stage of a food chain

- pyramids of biomass show the relative total mass of the organisms at each stage of a food chain

- energy flow diagrams show how energy is passed from one trophic level to another in an ecosystem or food chain

- energy is lost from a food chain when organisms:

  - respire – some energy is lost as heat

  - excrete – some energy is contained in the excretory products themselves

  - egest waste foods – these also contain some energy.

Questions

1. Which one of the following does **not** result in energy being lost from a food chain?

   A   respiration
   B   feeding
   C   excretion
   D   egestion

2. Commensalism is an association in which:

   A   one organism benefits and the other is harmed
   B   both organisms benefit
   C   both organisms are harmed
   D   one organism benefits and the other is not harmed and does not benefit

3. In predator–prey relationships, the numbers of the predator are usually:

   A   smaller than those of the prey and changes in their numbers occur before changes in the numbers of the prey
   B   larger than those of the prey and changes in their numbers occur before changes in the numbers of the prey
   C   larger than those of the prey and changes in their numbers occur after changes in the numbers of the prey
   D   smaller than those of the prey and changes in their numbers occur after changes in the numbers of the prey

4. Part (a) of the diagram shows a woodland food web. Part (b) shows a pyramid of numbers and a pyramid of biomass for a small part of this wood.

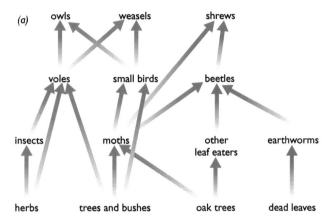

(a)

(b) pyramid of numbers
(numbers per 0.1 hectare)

pyramid of biomass
(grams per square metre)

a) Write out *two* food chains (from the food web) containing four organisms, both involving moths. *(2)*

b) Name *one* organism in the food web which is both a primary consumer and a secondary consumer. *(1)*

c) Suggest how a reduction in the amount of dead leaves may lead to a reduction in the numbers of voles. *(3)*

d) Explain how a reduction in the number of voles might lead to:

i) a reduction in the numbers of weasels

ii) an increase in the numbers of weasels

iii) the numbers of weasels remaining static. *(3)*

e) In part (b) of the diagram, explain why level Y is such a different width in the two pyramids. *(3)*

5. Read the following description of the ecosystem of a mangrove swamp.

Pieces of dead leaves (detritus) from mangrove plants in the water are fed on by a range of crabs, shrimps and worms. These, in turn, are fed on by young butterfly fish, angelfish, tarpon, snappers and barracuda. Mature snappers and tarpon are caught by fishermen as the fish move out from the swamps to the open seas.

a) Use the description to construct a food web of the mangrove swamp ecosystem. *(4)*

b) Write out *two* food chains, each containing four organisms from this food web. Label each organism in each food chain as producer, primary consumer,

secondary consumer or tertiary consumer. *(2)*

6. A marine food chain is shown below:

plankton → small crustacean → krill → seal → killer whale

a) Which organism is the producer and which is the secondary consumer? *(2)*

b) What term best describes the killer whale? *(1)*

c) Suggest why five trophic levels are possible in this case, when many food chains only have three or four. *(2)*

7. In natural ecosystems, there is competition between members of the same species as well as between different species.

a) Explain how competition between members of the same species helps to control population growth. *(3)*

b) Crop plants must often compete with weeds for resources. Farmers often control weeds by spraying herbicides (weedkillers).

i) Name two factors that the crop plants and weeds may compete for and explain the importance of each. *(4)*

ii) Farmers usually prefer to spray herbicides on weeds early in the growing season. Suggest why. *(2)*

c) Two species of the flour beetle *Tribolium* compete with each other for flour. Both are parasitised by a protoctist. The graphs below show the changes in numbers of the two species over 900 days when the parasite is absent and when it is present.

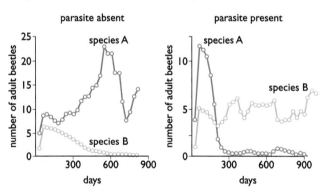

i) Which species is the most successful when the parasite is absent? Justify your answer. *(2)*

ii) What is the effect of the parasite on the relative success of the two beetles? Suggest an explanation for your answer. *(4)*

8. Explain the following statements:

a) Food chains begin with plants.

b) As the trophic levels increase, the number of organisms decreases. *(4)*

9. State **two** types of relationship that may be present within a food web. Give an example of each relationship. *(4)*

Chapter 4: Cycles in the Environment

When you have completed this chapter, you will be able to:

- describe how carbon and nitrogen are cycled through an ecosystem
- explain the role of decomposers and other bacteria in these nutrient cycles
- explain how biodegradable materials return nutrients to the environment
- discuss the importance of recycling
- identify difficulties encountered in recycling.

Cycling nutrients through ecosystems

The chemicals that make up our bodies have all been around before – probably many times! You may have in your body some carbon atoms that were part of the carbon dioxide molecules breathed out by Winston Churchill while making one of his famous speeches in the second World War, or by Sir Garfield Sobers when he made 365 runs not out for the West Indies cricket team against Pakistan in 1958. This constant recycling of substances is all part of the cycle of life, death and decay.

The carbon cycle

Carbon is a component of all major biological molecules. Carbohydrates, lipids, proteins, nucleic acids, vitamins and many other molecules all contain carbon. The following processes are important in cycling carbon through ecosystems:

- Photosynthesis 'fixes' carbon atoms from carbon dioxide into **organic compounds**.
- Feeding and assimilation pass carbon atoms already in organic compounds along food chains.
- Respiration produces inorganic carbon dioxide from organic compounds (mainly carbohydrates) as they are broken down to release energy.

Organic compounds all contain carbon and hydrogen. Starch and glucose are organic molecules, but carbon dioxide (CO_2) isn't.

- Fossilisation – sometimes living things do not decay fully when they die due to the conditions in the soil (decay is prevented if it is too acidic) and so form fossil fuels (coal, oil, natural gas and peat).
- Combustion releases carbon dioxide into the atmosphere when fossil fuels are burned.

Figures 4.1 and 4.2 show the role of these processes in the carbon cycle in different ways.

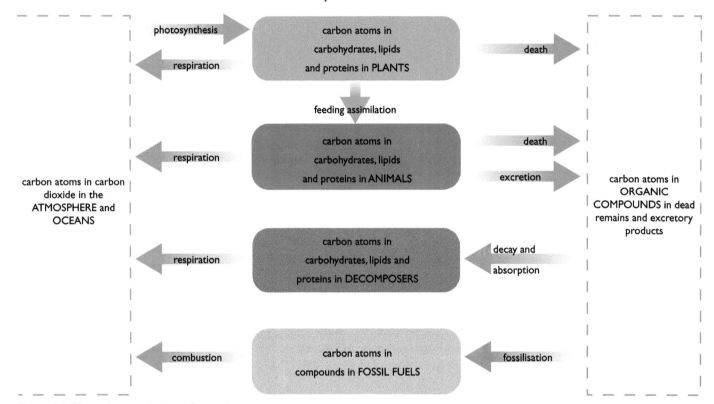

Figure 4.1 *The main stages in the carbon cycle.*

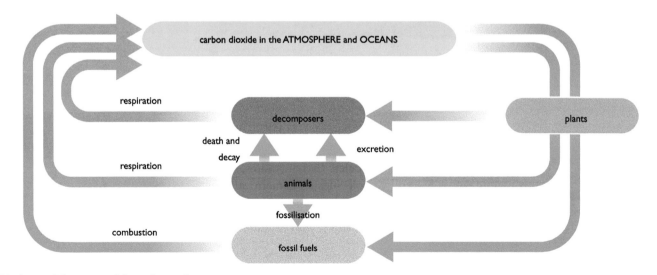

Figure 4.2 *A typical illustration of the carbon cycle.*

Activity 1: Investigating carbon dioxide exchange in an ecosystem

You will need:

- four boiling tubes
- four bungs (stoppers)
- a measuring cylinder
- a test-tube rack
- pondweed
- water snails or other water animals
- a lamp
- hydrogen carbonate indicator solution

Carry out the following:

1. Number the boiling tubes 1–4.

2. Construct a table like the one below.

Tube	Colour at start	Colour at end	Change (darker/ paler)	Carbon dioxide added/removed
1				
2				
3				
4				

Hydrogen carbonate indicator solution is normally a light orange colour. If extra carbon dioxide is added, it becomes paler – usually a yellow colour. If carbon dioxide is taken away from it, it becomes darker – usually pink or purple.

3. Measure 25 cm³ of hydrogen carbonate indicator solution into each boiling tube.

4. Place pondweed only in tube 1. Place the water animals only in tube 2. Place both pondweed and the water animals in tube 3 (use the same amount of each as you used in tubes 1 and 2). Add nothing to tube 4.

5. Note the colour of the hydrogen carbonate indicator solution in each tube at the start of the investigation.

6. Place the rack containing the tubes in a safe place, and leave illuminated by the lamp for 12–24 hours.

7. Note any colour changes in the table.

8. Write up your investigation and try to explain the colour changes in the hydrogen carbonate indicator solution in each tube.

The nitrogen cycle

Nitrogen is a key element in many biological compounds. It is present in proteins, amino acids, most vitamins, DNA, RNA and adenosine triphosphate (ATP). Like the carbon cycle, the nitrogen cycle involves feeding, assimilation, death and decay. Photosynthesis and respiration are not directly involved in the nitrogen cycle as these processes fix and release carbon, not nitrogen.

Some nitrogen-fixing bacteria are free-living in the soil. Others form associations with the roots of legumes (legumes are plants that produce seeds in a pod, like peas and beans). They form little bumps or 'root nodules' (Figure 4.3).

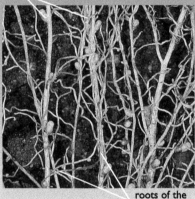

these nodules contain millions of nitrogen-fixing bacteria

roots of the clover plant

Figure 4.3 *Root nodules on a clover plant.*

This is an example of **mutualism**. The bacteria receive a supply of organic nutrients from the plants and the plants receive a supply of ammonia from the bacteria. They can use the ammonia to form amino acids in the same way as they use nitrates.

To remember if an organic compound contains nitrogen, check to see if the letter **N** (symbol for nitrogen) is present.

Protei**N**s, ami**N**o acids and **DNA** all contain nitrogen. Carbohydrates and fats don't – they have no **N**.

The following processes are important in cycling nitrogen through ecosystems:

- Feeding and assimilation pass nitrogen atoms already in organic compounds along food chains.
- Decomposition (putrefaction) by decomposers produces ammonia from the nitrogen in compounds like proteins, DNA and vitamins.
- The ammonia is oxidised first to nitrite and then to nitrate by **nitrifying bacteria**. This overall process is called **nitrification**.
- Plant roots can absorb the nitrates. They are combined with carbohydrates (from photosynthesis) to form amino acids and then proteins, as well as other nitrogen-containing compounds.

This represents the basic nitrogen cycle, but other bacteria carry out processes that affect the amount of nitrate in the soil that is available to plants:

- **Denitrifying bacteria** use nitrates as an energy source and convert them into nitrogen gas. **Denitrification** *reduces* the amount of nitrate in the soil.
- **Nitrogen-fixing bacteria** convert nitrogen gas in the soil into ammonia. This can then be oxidised by nitrifying bacteria to nitrate. **Nitrogen fixation** *increases* the amount of nitrate available to plants.

In addition to all the processes described so far, lightning converts nitrogen gas in the air into various oxides of nitrogen. These dissolve in rainwater and enter the soil to be converted into nitrates by nitrifying bacteria. Figure 4.4 shows the role of these processes in the nitrogen cycle.

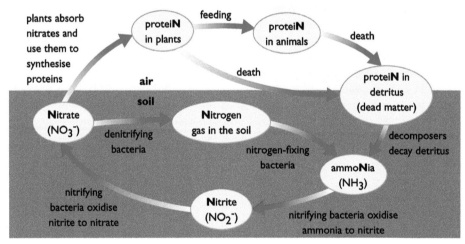

Figure 4.4 *The main stages in the nitrogen cycle.*

9. Which biological process uses carbon dioxide to make organic molecules?
10. Which biological process releases carbon dioxide into the environment?
11. Describe the importance of decomposers in the carbon cycle and the nitrogen cycle.
12. What is the importance of nitrogen-fixing bacteria in the nitrogen cycle?

Recycling

Just as substances found naturally in the environment are recycled, so are substances that are human-made. Recycling involves the reuse of materials that people discard, such as glass, paper, plastic and even metals, for example iron, copper, silver and gold.

The importance of recycling

Environmental pollution, the need to reduce manufacturing costs, and the need to use natural resources sustainably are just some of the reasons why recycling is important.

Reducing waste and pollution

Whenever items are reused, it reduces the amount of waste produced. In turn, the volume of waste that has to be collected and sent to landfills also decreases. Some people reduce waste by making compost at home using biodegrable materials, such as vegetable peels. The compost is later used in the garden, to enrich the soil. On a larger scale, some sugar factories use the waste from the sugarcane to make fertiliser, to make bagasse board, or to create biogas to fuel the factories' boilers.

Figure 4.5 *A family constructing a compost heap in their garden.*

Substances that can be decomposed, such as the remains of dead plants and animals, are considered to be **biodegradable**. Their compounds are recycled in nature. Substances that cannot be broken down by decomposers, for example plastics and metals, are considered to be **non-biodegradable**. Because of this, only a limited range of items can be recycled.

Waste can be classified as solid or liquid. Solid waste is normally burnt, buried or taken to landfills. These practices are harmful to the environment as they can cause air pollution and create breeding sites for pests and microorganisms. If more solid waste were recycled, the volume of waste in landfills would decrease.

Figure 4.6 *In areas where refuse collection services are poor, some people burn their refuse. The smoke from the fires contributes to air pollution.*

Table 4.1 identifies how different materials are recycled by certain Caribbean countries.

Materials	Sources	Processing method	Recycling centres
cardboard	Barbados, Guyana, Jamaica, Trinidad and Tobago	baled	Guyana (limited), Continental USA, Venezuela
glass	Barbados, Grenada, Guyana, Trinidad and Tobago	collected in steel bins and then ground	Trinidad and Tobago
paper	Trinidad and Tobago	baled or compacted in super sacks	Puerto Rico, Continental USA
HDPE (high-density polyethylene) plastic	Trinidad and Tobago	ground and then packed in super sacks	Canada
vehicle batteries	Antigua, Barbados, Trinidad and Tobago	drained and then shrink-wrapped	Venezuela
waste oil	Eastern Caribbean	stored in steel drums and then re-refined	Trinidad and Tobago
green waste	Antigua, Barbados, Barbuda, Belize, Bermuda, Dominica, Guyana, Jamaica, Montserrat, St Kitts and Nevis, St Lucia, St Vincent and the Grenadines, Trinidad and Tobago	made into compost	landfills, households

Table 4.1: *Recovery and recycling initiatives in the Caribbean.*

Some communities pay for sewerage, and the wastewater is sent to a water treatment plant before being released into the environment. Some hotels and factories also treat their effluent before it is released. Depending on the level of water treatment, the solid part of the waste may be used as fertiliser and the liquid part as irrigation water for lawns. For example, some lawn sprinklers at the University of the West Indies' Mona campus use recycled water.

Figure 4.7 *Soapberry wastewater treatment plant in St Catherine, Jamaica. It was designed to treat sewage from areas in four parishes on the island.*

Reducing costs

It costs less money and uses less energy to make products from recycled materials than it does to make them from new materials. For example, glass bottles can be collected and returned to beverage factories, where they can be cleaned and refilled.

Reducing the use of natural resources

Trees, oil and bauxite are all natural resources; they are used to make paper, fuel and aluminium, respectively. When these materials are recycled, forests, oil and ore reserves are conserved. So, recycling helps to preserve the availability of natural resourcers for future generations, and it helps to protect the environment.

Limitations in recycling

Although recycling has many benefits, it also presents some challenges:

- Breaking down a manufactured material to a point where it can be reused can take longer than it took to make the material originally.

- Some chemicals used in breaking down manufactured materials are harmful to the environment.

- Collecting, transporting and sorting waste is expensive. Because of this, few recycling companies operate in Caribbean countries and a lot of waste is still sent to landfills.

- City infrastructure is damaged or lost because metal items are stolen and then sold on to scrap-metal dealers. Typical items include metal railings from bridges and cables used in telecommunications networks.

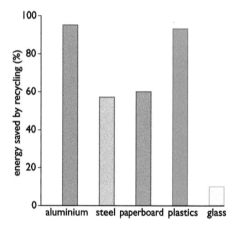

Figure 4.8 *Energy savings from recycling some materials.*

Ways to help reduce waste

- Create a compost heap for biodegradeable materials.
- Choose to buy products that have less packaging.
- Refill plastic water bottles instead of buying new bottles of water.
- Instead of throwing away old clothes, books and magazines, donante them to charity.
- Reuse shopping bags and gift wrapping.

Chapter summary

In this chapter you have learnt that:

- carbon is cycled through ecosystems by a combination of photosynthesis, feeding and digestion, respiration, death and decay, and fossilisation
- nitrogen is cycled through ecosystems by feeding and digestion, death and decay, excretion, and the action of nitrifying and nitrogen-fixing bacteria
- decomposers are important in the process of decay, which releases nutrients from dead organisms
- recycling can save on costs and energy, reduce pollution, and preserve natural resources and the environment
- high processing costs, toxic chemicals and theft of metals limit the process of recycling.

Questions

1. In the nitrogen cycle, nitrifying bacteria convert:

   A nitrates to ammonia
   B ammonia to nitrates
   C nitrogen gas to nitrates
   D nitrates to nitrogen gas

2. The diagram below shows the carbon cycle. Which processes take place at I and II?

	I	II
A	respiration	combustion
B	photosynthesis	combustion
C	respiration	photosynthesis
D	decomposition	photosynthesis

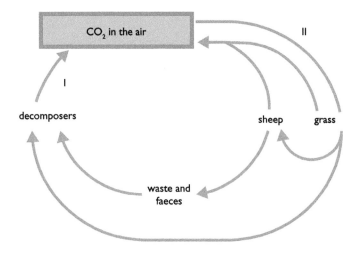

3. The diagram shows part of the nitrogen cycle.

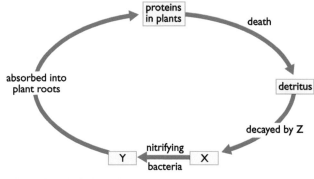

   a) What do X, Y and Z represent? (3)

   b) Name the process by which plant roots absorb nitrates. (1)

   c) i) What are nitrogen-fixing bacteria? (1)

      ii) Describe the type of association shown between nitrogen-fixing bacteria and legumes. (2)

   d) Give two ways, that are not shown in the diagram, in which animals can return nitrogen to the soil. (2)

4. Carbon is cycled through ecosystems by the actions of plants, animals and decomposers. Humans influence the cycling of carbon more than other animals.

   a) Explain the importance of plants in the cycling of carbon through ecosystems. (2)

   b) Describe two human activities that have significant effects on the global cycling of carbon. (4)

   c) The graph shows the activity of decomposers acting on the bodies of dead animals under different conditions.

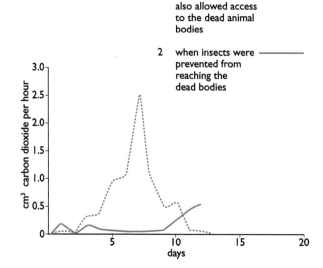

1 when insects were also allowed access to the dead animal bodies

2 when insects were prevented from reaching the dead bodies

   i) Why was carbon dioxide production used as a measure of the activity of the decomposers? (2)

   ii) Describe and explain the changes in decomposer activity when insects were also allowed access to the dead bodies curve (1). (3)

   iii) Describe two differences between curves 1 and 2. Suggest an explanation for the differences you describe. (4)

5. The diagram shows the carbon cycle.

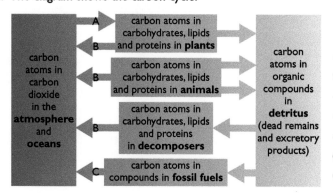

a) Name the processes occurring at A, B and C. *(3)*

b) What has been the consequence of an increase in process C over the past 200 years? *(3)*

c) What is the importance of the decomposers in the carbon cycle? *(2)*

6. The element nitrogen is found in soils as nitrate ions (NO_3^-) and as ammonium ions (NH_4^+).

a) *i)* Explain how ammonium ions come to be in soils. *(3)*

ii) Explain, as fully as you can, how ammonium ions are converted to nitrate ions in soils. *(3)*

b) The graph shows the growth of three different species of plant at different pHs using either ammonium ions or nitrogen ions as the main source of nitrogen.

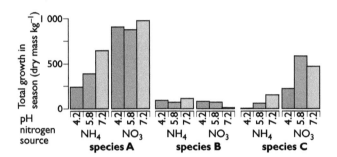

i) Which of the three species would not survive in an alkaline soil in which nitrate was the only source of nitrogen? Explain your answer. *(3)*

ii) Which of the three species is best adapted to live in soils with a pH of between 5.8 and 7.2? Explain your answer. *(3)*

iii) Which of the three species would survive best in a soil with little oxygen? Explain your answer. *(3)*

7. Explain two benefits of recycling materials. *(2)*

8. Explain two difficulties of recycling materials. *(2)*

Chapter 5: The Impact of Human Activity on the Environment

When you have completed this chapter, you will be able to:

- describe what a resource is and state examples
- distinguish between renewable and non-renewable resources
- define pollution
- be aware of some of the ways in which air, water and land can be polluted
- understand that pollution of the environment is a direct consequence of our own actions
- appreciate that individual action can have an impact on levels of pollution and conservation.

Natural resources

For millions of years, living things have relied on natural resources. A natural resource is any substance found in nature that people are able to use. Natural resources have been used to create tools, build cities and make living more comfortable. Our Earth supplies all the materials needed to support life: gases for breathing, water in rivers and seas, and sources of food, fuel and shelter.

Natural resources are either renewable or non-renewable.

Renewable resources

A natural resource that can be replaced in nature at about the same rate at which it is used is a renewable resource. Examples include sunlight, plants and animals.

Non-renewable resources

A natural resource is formed in nature over extremely long periods of time (often millions of years). So, natural resources are consumed faster than they can be replaced in nature. This means that the availability of all natural resources is finite. Examples of non-renewable resources include oil, natural gas and coal.

Table 5.1 describes the impact of human activities on natural resources.

Resource	Uses	Impact of human activities
water	• cooking, washing, recreation • hydro-electricity • transportation of ships	• various forms of pollution reduce water quality
plants	• making paper, furniture, dyes	• forest clearance destroys habitats • soil may be left bare, which leads to soil erosion and flooding
animals	• food for other animals • waste from animals form biomass fuel	• pollution, habitat destruction and hunting reduce animal numbers
minerals and rocks, e.g. bauxite, gypsum, marble, gold	• materials for use in buildings, household products, jewellery	• excessive use leads to decrease in availability globally
fossil fuels, e.g. oil, natural gas	• fuel for vehicles, generating electricity, making plastics and wax	• little conservation of fossil fuels leads to rapid decrease in their availability globally

Table 5.1: *The impact of human activities on natural resources.*

Pollution – the consequences of our actions

Pollution means releasing substances into the environment in amounts that cause harmful effects, and which natural biological processes cannot easily remove. A key feature is the *amount*. Small amounts of sulphur dioxide and carbon dioxide would easily be absorbed by the environment and, over time, made harmless. It is the sheer mass of the pollutants that poses the problem.

Modern agricultural practices, building and other industries, as well as individual actions, release many pollutants into the environment. We pollute the air, land and water with a range of chemicals and heat.

Air pollution

We pollute the air with many gases. The main ones are carbon dioxide, carbon monoxide, sulphur dioxide, nitrogen oxides, methane and CFCs (chlorofluorocarbons).

Carbon dioxide

The levels of carbon dioxide have been rising for several hundred years. Over the last 100 years alone, the level of carbon dioxide in the atmosphere has increased by nearly 30%. This recent rise has been due mainly to the increased burning of fossil fuels, including petrol and diesel in vehicle engines. It has been made worse by cutting down large areas of tropical rainforest. These extensive forests have been called 'the lungs of the Earth' because they absorb such vast quantities of carbon dioxide and produce equally large amounts of oxygen. Extensive deforestation means that less carbon dioxide is being absorbed. Figure 5.1 shows changes in the level of carbon dioxide in the atmosphere from 1960 to 1990.

Sulphur dioxide and carbon dioxide are known as 'greenhouse' gases. Other greenhouse gases include methane (CH_4), nitrous oxide (N_2O) and chlorofluorocarbons (CFCs).

If there were no greenhouse gases and *no* global warming, the Earth would be the same temperature as the Moon. Life, as we know it, would be impossible.

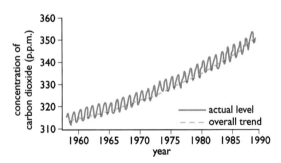

Figure 5.1 *The changes in levels of CO2 at Mauna Loa, Hawaii from 1960 to 1990.*

In any one year, there is a peak and a trough in the levels of carbon dioxide. This is shown more clearly in Figure 5.2.

The increased levels of carbon dioxide contribute to **global warming**. Carbon dioxide is just one of the so-called 'greenhouse gases' that form a layer in the Earth's atmosphere.

Short-wave radiation from the Sun strikes the planet. Some is absorbed and some is reflected as longer wave radiation. The greenhouse gases absorb, then re-emit towards the Earth, some of this long-wave radiation, which would otherwise escape into space. This is the '**greenhouse effect**' and it is a major factor in global warming (Figure 5.3).

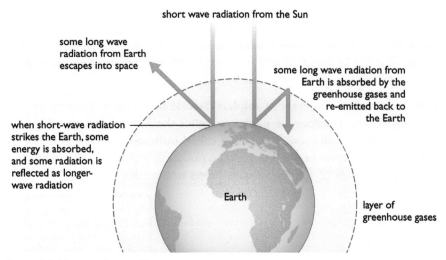

Figure 5.3 *The greenhouse effect.*

A rise in the Earth's temperature of only a few degrees would have many effects:

- Polar ice caps would melt and sea levels would rise.

- A change in the major ocean currents would result in warm water being redirected into previously cooler areas.

- A change in global rainfall patterns could result. With all the extra water in the seas, there would be more evaporation from the surface and so more rainfall in most areas.

- It could change the nature of many ecosystems. If species could not migrate quickly enough to a new, appropriate habitat, or adapt quickly enough to the changed conditions in their current habitat, they could become extinct.

- Changes in agricultural practices would be necessary as some pests become more abundant. Higher temperatures might allow some pests to complete their life cycles more quickly.

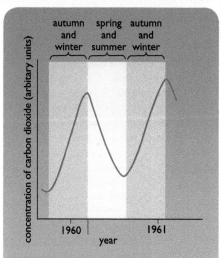

Figure 5.2 *Seasonal fluctuations in carbon dioxide levels.*

In the autumn and winter, trees lose their leaves. They photosynthesise much less and so absorb little carbon dioxide. They still respire, which produces carbon dioxide, so in the winter months, they give out carbon dioxide and the level in the atmosphere rises. In the spring and summer, with new leaves, the trees photosynthesise faster than they respire. As a result, they absorb carbon dioxide from the atmosphere and the level decreases. However, because there are fewer trees overall, it doesn't quite get back to the low level of the previous summer.

Global warming and the Caribbean

Scientists predict increased temperatures, a rise in sea levels and lower rainfall as some of the consequences of global warming. In the Caribbean, these three areas of change would affect:

- tourism – beaches would be lost either by flooding or erosion as the coral reefs would have died

- agriculture – lack of rainfall to naturally irrigate plants, creating severe periods of drought, which would reduce crop growth

- economy – less money would be generated if tourism and agriculture decline. Caribbean countries would have to incur great expense erecting structures to protect coastal areas from flooding.

Carbon monoxide

When substances containing carbon are burned in a limited supply of oxygen, carbon monoxide (CO) is formed. This happens when petrol and diesel are burned in vehicle engines. Exhaust gases contain significant amounts of carbon monoxide. It is a dangerous pollutant as it is colourless, odourless and tasteless, and can cause death by asphyxiation. Haemoglobin binds more strongly with carbon monoxide than with oxygen. If a person inhales carbon monoxide for a period of time, more and more haemoglobin becomes bound to carbon monoxide and so cannot bind with oxygen. The person may lose consciousness and, eventually, may die as a result of a lack of oxygen.

Sulphur dioxide

Sulphur dioxide (SO_2) is an important pollutant as it is a major constituent of **acid rain**. It is formed when fossil fuels are burned, and it can be carried hundreds of miles in the atmosphere before finally combining with rainwater to form acid rain.

Some lichens are more tolerant of sulphur dioxide than others. Patterns of lichen growth can, therefore, be used to monitor the level of pollution by sulphur dioxide. The different lichens are called **indicator species** as they 'indicate' different levels of sulphur dioxide pollution (see Figure 5.4, which shows different levels of sulphur dioxide levels in the UK).

In Europe there are now strict laws controlling the permitted levels of carbon monoxide in the exhaust gases produced by newly designed engines. These levels are lower than those allowed in the M.O.T. test of road-worthiness of vehicles three or more years old.

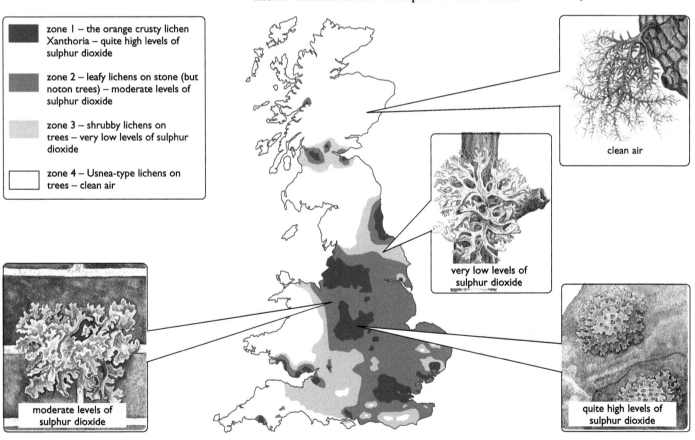

zone 1 – the orange crusty lichen Xanthoria – quite high levels of sulphur dioxide

zone 2 – leafy lichens on stone (but not on trees) – moderate levels of sulphur dioxide

zone 3 – shrubby lichens on trees – very low levels of sulphur dioxide

zone 4 – Usnea-type lichens on trees – clean air

clean air

very low levels of sulphur dioxide

moderate levels of sulphur dioxide

quite high levels of sulphur dioxide

Figure 5.4 *Lichens are sensitive to pollution levels.*

You can make a rough estimate of how much sulphur dioxide is in the air around you by observing which lichens grow on trees and walls. If the only lichens that grow are flat, and pressed close to the surface, this indicates a high level of sulphur dioxide. If the lichens are 'shrubby' in appearance and grow well clear of the surface, there is likely to be less sulphur dioxide polluting the air.

Nitrogen oxides

Nitrogen oxides (NO_x) are also constituents of acid rain. They are formed when petrol and diesel are burned in vehicle engines.

Acid rain

Rain normally has a pH of about 5.5 – it is slightly acidic due to the carbon dioxide dissolved in it. Both sulphur dioxide and nitrogen oxides dissolve in rainwater to form a mixture of acids, including sulphuric acid and nitric acid. As a result, the rainwater is more acidic with a much lower pH than normal rain (Figure 5.5).

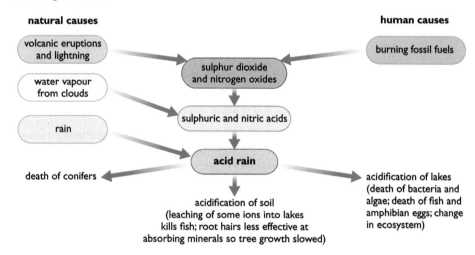

Figure 5.5 *The formation of acid rain and its effects on living organisms.*

Methane

Methane (CH_4) is an organic gas. It is produced when microorganisms ferment larger organic molecules to release energy. The most significant locations of these microorganisms are:

- decomposition of waste in landfill sites by microorganisms
- fermentation by microorganisms in the rumen of cattle and other ruminants
- fermentation by bacteria in rice paddy fields.

Methane is a greenhouse gas, with effects similar to carbon dioxide. Although there is less methane in the atmosphere than carbon dioxide, each molecule has a bigger greenhouse effect.

Freshwater pollution

The four main pollutants of fresh water are nitrates from fertilisers, organic waste, detergents and solid waste.

Besides its effects on living things, acid rain causes damage to stonework by reacting with carbonates and other ions in the stone (Figure 5.8).

Figure 5.6 *This stone lion has been dissolved by acid rain.*

Herds of cattle can produce up to 40 dm³ of methane per animal per hour. This adds up to a lot of methane being belched and farted into the atmosphere!

Figure 5.7 *Some detergents contain nitrates and phosphates which increase nutrient levels in rivers, causing algal blooms*

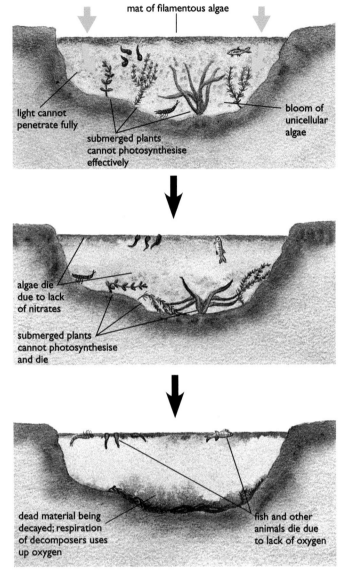

mat of filamentous algae

light cannot penetrate fully

submerged plants cannot photosynthesise effectively

bloom of unicellular algae

algae die due to lack of nitrates

submerged plants cannot photosynthesise and die

dead material being decayed; respiration of decomposers uses up oxygen

fish and other animals die due to lack of oxygen

Figure 5.8 *Stages in eutrophication in a pond.*

Nitrates from fertilisers

Farmers add inorganic fertilisers to soils to replace mineral ions lost when crops are removed. The ions in these fertilisers (particularly nitrates) are very soluble. As a result, they are easily **leached** (carried out with water) from the soils and can enter waterways. The level of nitrates (and other ions) can rise rapidly in these lakes and rivers. This increase in mineral ions is called **eutrophication**, which is a natural process in nearly all waterways (Figure 5.8). What is *not* natural is the speed with which it happens due to leaching of ions in fertilisers from soils. Rapid eutrophication can have disastrous consequences for a waterway.

1   As nitrate levels rise, algae reproduce rapidly. They use the nitrates to make extra proteins for growth, just like the crop plants for which the nitrates were intended.

2   The algae form an **algal bloom** – a kind of algal pea soup if the algae are unicellular, or a mass of filaments if the algae are filamentous.

3   The algae prevent light from penetrating further into the water.

4   Submerged plants cannot photosynthesise and so die.

5   The algae also die as they run out of nitrates.

6   Bacteria decay the dead plants and algae (releasing more nitrates and allowing the cycle to start again).

7   The bacteria reproduce (due to the large amount of dead matter) and their respiration uses up more and more oxygen.

8   The water may become totally **anoxic** (without oxygen) and all life in the water will die.

The problems can be more severe in hot weather because the nitrates can become more concentrated as the heat evaporates water. All the processes are speeded up due to increased enzyme activity. The problems are less severe in moving water because the nitrates are rapidly diluted and the water is continually being re-oxygenated.

Rapid eutrophication is less likely when farmers use organic fertilisers (like manure). The organic nitrogen-containing compounds in manure are less soluble and so are leached less quickly from the soil. However, water can sometimes be polluted with large amounts of organic matter, for example when untreated sewage is released into waterways. Bacteria and fungi decay this and use up oxygen as they respire. The water becomes anoxic in the same way as in eutrophication. Fish and other animals die due to a lack of oxygen.

The level of organic water pollution can be monitored by the 'indicator species' present. Figure 5.9 shows some of these species.

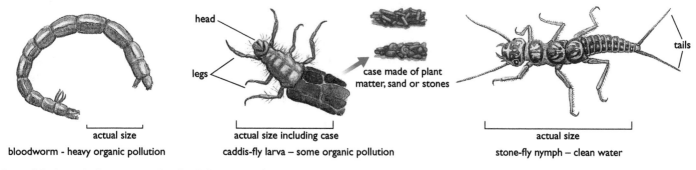

head
legs
case made of plant matter, sand or stones
tails
actual size
bloodworm - heavy organic pollution
actual size including case
caddis-fly larva – some organic pollution
actual size
stone-fly nymph – clean water

Figure 5.9 *Some freshwater animals will only live in very clean water, while others can survive in very polluted areas.*

When organic matter pollutes moving water, the point of the polluting outlet becomes very low in oxygen as bacteria decompose the organic material. Only those species adapted to such conditions can survive. As the water moves away from the outlet, it becomes oxygenated again as it mixes with the air at the surface. The increase in oxygen levels allows more species of 'clean water' animals to survive. Figure 5.10 shows the changes in oxygen content, numbers of clean-water animals and bloodworms at, and just downstream from, a sewage outlet into a river.

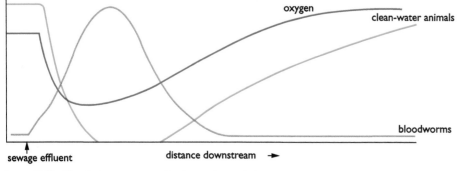

oxygen
clean-water animals
bloodworms
sewage effluent
distance downstream ➞

Figure 5.10 *The changes in oxygen levels, numbers of clean-water animals and numbers of blood worms around a sewage outlet into a river.*

Marine pollution

The world's oceans are becoming increasingly polluted. Because of their size, they have always been thought of as a virtually unlimited dumping ground for all sorts of waste materials. The theory is that the waste will become so dilute in the oceans that it will be harmless. Now we are realising that this is not true and that there is a limit to the amount of waste that can safely be dumped in the sea. Marine pollution includes the examples listed below.

Pollution by oil

Oil pollution has, in the past, been a minor problem as small amounts of oil from ships' engines have polluted the sea beds and caused small-scale damage to these ecosystems. However, with the development of supertankers carrying vast quantities of oil, the problems have become much more serious.

There have been several incidents in recent years where supertankers have been damaged in a number of ways, resulting in the release of vast quantities

Figure 5.11 *An oil slick forming as crude oil escapes from a sinking tanker, the MV Braer, off the coast of the Shetland Islands in 1993.*

Figure 5.12 *This seabird is covered in crude oil from a damaged tanker.*

of crude oil. This oil floats, forming huge 'slicks' and sometimes can be driven onto beaches, killing much of the plant and animal life there. Oil also leaks from offshore oil-drilling rigs and pipelines that bring the oil to land from the oceans.

Oil slicks are sometimes 'dispersed' by using detergents. This breaks the oil into smaller droplets so that it disperses more easily. However, the detergents also break up the natural oils in the skin and feathers of many animals as well as the lipids in cell membranes. In addition, breaking the oil into smaller droplets makes it easier for animals to absorb the oil and so increases its toxicity.

Pollution by solid waste

Some countries have built gullies in urban areas to serve as waterways to reduce flooding. In such areas, large volumes of water are channelled away from roads and houses and emptied into the sea. Unfortunately, some people dump solid waste into these gullies. While certain sections of the gullies allow for solid waste to be removed, items that are not removed enter the sea (Figure 5.13).

Solid waste that is allowed to enter the sea is then carried by ocean currents to other areas. Some solid waste may float into a wetland and cause physical damage to the flora (plant life). The flora protect the wetland itself and provide habitat for animals which live on branches, tree trunks and roots. So damage done to the flora will decrease the biodiversity of the wetland.

> Polluted water:
>
> - destroys habitats, which in turn affects ecosystems
> - decreases the aesthetic appeal of a country and reduces its attractiveness to tourists. Visitors on holiday do not wish to risk swimming in water that may cause skin or ear infections.

Figure 5.13 *Polluted gullies cause blockages which create breeding sites for pests and increase the chances of flooding in an area*

Pollution by organic matter

Organic matter from untreated sewage pollutes the coastal waters of many countries, where it has the same effects as organic pollution of fresh water.

Pollution by heavy metals

Many oceans have become contaminated with heavy metals, such as mercury. These metals become concentrated in the tissues of animals along the food chain. It is estimated that it might take up to 100 years for the mercury to work its way out of an ecosystem, once present, so all that can be done is to ensure that levels of pollution do not increase from present levels.

Other forms of pollution

We pollute our air with poisonous gases, and our water with organic pollutants and nitrates. We also pollute our environment in a number of other ways.

Thermal pollution

Water is used as a coolant in power stations and in other industrial plants. As water removes the heat, it becomes warmer and its ability to dissolve oxygen decreases. This can affect the number of animals able to survive in the water. Besides the effects of decreasing oxygen content, the direct effect of changing temperature may also kill animals and plants. Many animals cannot regulate their body temperature (as mammals and birds can). Their temperature changes with that of the environment. There have been cases reported in North America of river water being heated to over 80°C and being completely lifeless. This is thermal pollution of water at its worst.

Pesticide pollution

Farmers frequently use pesticides to control many different types of pests. Many of these pesticides have no serious side effects, but some are persistent – they are not degraded (decomposed) easily. Traces of some herbicides (weedkillers) can remain on or in the crops that were sprayed, with the risk that they could then be eaten with the food. A single, small 'dose' of herbicide is unlikely to cause any harm, but if the dose is repeated many times, then the amount may accumulate and begin to have more serious consequences. Pesticides are sometimes stored in fatty tissue where the amount builds up over a period of time. This is called **bioaccumulation**.

Sometimes the effects of bioaccumulation can be *magnified* as a pesticide is passed along a food chain. The best documented cases of this happening involve the insecticide **DDT**. DDT is an extremely effective insecticide and was widely used after the Second World War. It prevented millions of deaths from malaria in some areas by significantly reducing the mosquito population. However, DDT was passed along food chains and accumulated in harmful amounts in the top carnivores. This happened because DDT is extremely persistent (a single application can take over 20 years to degrade) and is fat soluble, so is easily stored in living tissue. The food chain in Figure 5.14 shows the extent to which DDT can be accumulated. The amounts in brackets are parts per million of DDT.

> An ideal pesticide should:
>
> - control the pest effectively
> - be biodegradable, so that no toxic products are left in the soil or on crops
> - be specific, so that only the pest is killed
> - not accumulate in organisms
> - be safe to transport and store
> - be easy and safe to apply.

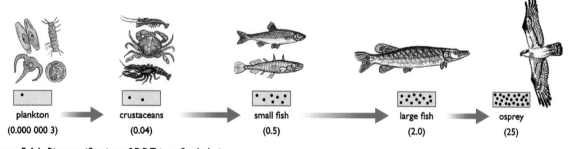

Figure 5.14 *Biomagnification of DDT in a food chain.*

plankton (0.000 000 3) → crustaceans (0.04) → small fish (0.5) → large fish (2.0) → osprey (25)

This increase in concentration along a food chain is called **biomagnification**. It happens because each organism in the chain eats many of the preceding organisms and accumulates the DDT (bioaccumulation). The effects of bioaccumulation are therefore magnified at each stage in the food chain.

The use of DDT has been restricted since 1972 because of its ecological effects.

> 1. Name four gases that pollute the atmosphere. For each gas you name, explain how the gas is produced and describe one harmful effect of the gas.
> 2. What is acid rain? Describe how it is formed and give some of its effects.
> 3. List four ways in which fresh water or marine water can be polluted. For each method you describe, explain how it is brought about and describe the main consequences of this form of pollution.
> 4. What are indicator species? Give two examples.
> 5. Explain what is meant by bioaccumulation and biomagnification.

Conservation – what can we do about it?

Some people think that conservation means leaving the Earth alone and not using any of its resources. This is not the case. Conservation simply means what the name implies – conserving whatever is present now for future generations to be able to use.

Many agricultural and industrial practices have resulted in the depletion of habitats for wildlife. We have a 'duty of care' for the wildlife of the Earth and should try to ensure that when we take resources from the Earth, we do not needlessly destroy the habitat of another species. Many species are now classified as 'endangered species' and could easily become extinct if their populations do not recover.

Figure 5.15 *The blue iguana of the Cayman Islands is an endangered species.*

As a result of commercial whaling, several species of whales have populations that are only a fraction of what they used to be. Destruction of the habitat of the Bengal tiger has reduced the population to the point where extinction is very likely. Felling tropical rainforest does not just remove trees, but destroys the habitats of a whole range of organisms.

Many organisms, including some as yet undiscovered, will become extinct if the tropical rainforests are lost. We will also lose valuable biological knowledge and potential medical products.

What can we do to help? Many people believe that little can be done about the environmental problems that face us, or that only governments working together on a global scale can have any impact. Neither of these is true. Individuals can do a great deal to preserve and conserve our natural environment.

Table 5.2 shows some possible courses of action to reduce the effects of pollution.

Figure 5.16 *The Bengal tiger is an endangered species.*

Problem	Possible individual action	Possible government/industrial action
global warming (greenhouse gases, CO_2, CH_4)	use as little energy as possible – less then needs to be generated; reduce use of private transport as far as possible	international agreements to set acceptable levels of CO_2 emissions; reduce deforestation: encourage sustainable felling and replanting schemes; encourage use of more recycled metals
acid rain (SO_2, NO_x)	reduce use of private transport as far as possible; reduce use of electricity where possible	legislation to enforce desulphurisaton of emissions from power stations; encourage use of 'cleaner' fuels such as natural gas (methane) and low-sulphur petrol
ozone depletion (overuse of CFCs)	reduce use of any aerosols containing CFCs	international legislation to restrict the use of CFCs
organic pollution of water	individual farmers to ensure safe storage and use of organic fertiliser (manure)	legislation to enforce treatment of sewage before discharge into waterways
nitrate pollution of water	individual farmers to make more use of 'organic' farming practices – use of crop rotations and organic manure; we can encourage organic farming by buying more organic produce	increase monitoring of waterways; legislation to limit levels of nitrates in water
pesticide pollution of soils	individual farmers to reduce use of pesticides and use biological control and crop rotation to limit pest build up	encourage development of 'safer' pesticides

Table 5.2: *Problems associated with pollution and possible solutions.*

Saving energy and recycling

We can all do something to help conserve the resources of our planet for future generations:

- Use as many recycled products as is appropriate. This reduces the energy consumption needed to extract metals from ores and to manufacture new products, so reducing the pollution that goes with energy generation.

- Turn off electrical appliances when they are not needed – again this reduces energy consumption.

- Have cars serviced regularly so that the engine is running efficiently and producing the minimum levels of pollutants.

- Try to make fewer journeys alone in the car – share transport or use public transport where convenient.

- Don't have endless conversations on your mobile 'phone. The batteries in it are expensive to produce – in energy terms as well as financially!

- Try to encourage wildlife in your garden, if you have one. Construct a small wildlife pond to provide a habitat for frogs and newts.

- Use compost on the garden rather than inorganic fertilisers.

We need to alter our way of thinking, so that conserving wildlife, materials and energy becomes a way of life, not just a special event.

6. What is conservation?
7. How can individuals help to reduce (a) global warming, (b) acid rain, (c) pesticide pollution of soils?
8. Explain two benefits of recycling materials.
9. Give two ways in which individuals can help to conserve resources.

Chapter summary

In this chapter you have learnt that:

- resources are either renewable or non-renewable

- non-renewable resources in the Caribbean include bauxite, oil and natural gas

- the air has been polluted by the products of the combustion of fossil fuels such as oil, coal and petrol

- the consequences of air pollution include acid rain and the greenhouse effect

- waterways have been polluted by fertilisers, pesticides, organic matter, oil and heavy metals

- it is possible for individual action as well as government intervention to have an impact on levels of pollution

- conservation ensures that resources can be used today, while still being available for future generations to use.

Questions

1. Which of the following is a non-renewable resource?

   A crude oil
   B natural gas
   C wood
   D coal

2. Eutrophication, acid rain and oil spills are some problems associated with:

   A conservation
   B fossil fuels
   C pollution
   D recycling

3. The diagram shows three areas (A, B and C) which contribute to pollution of the river. State two sources of pollution from each site.

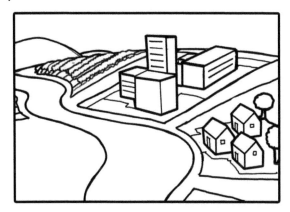

4. The table gives information about the pollutants produced in extracting aluminium from its ore (bauxite) and in recycling aluminium.

Pollutants	Amount (g per kg aluminium produced)	
Air	**Extraction from bauxite**	**Recycling aluminium**
sulphur dioxide	88 600	886
nitrogen oxides	139 000	6 760
carbon monoxide	34 600	2 440
Water		
dissolved solids	18 600	575
suspended solids	1 600	175

a) Calculate the percentage reduction in sulphur dioxide pollution by recycling aluminium. (2)

b) Explain how extraction of aluminium from bauxite may contribute to the acidification of water hundreds of miles from the extraction plant. (3)

c) Suggest *two* reasons why there may be little plant life in water near an extraction plant. (4)

5. The graph shows the changing concentrations of carbon dioxide at Mauna Loa, Hawaii, over a number of years.

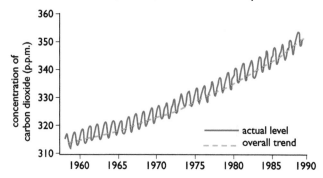

a) Describe the overall trend shown by the graph. (2)

b) Explain the trend described in (a). (3)

c) In any one year, the level of atmospheric carbon dioxide shows a peak and a trough. Explain why. (5)

6. The diagram shows how the greenhouse effect is thought to operate.

short wave radiation from the Sun

some long wave radiation from Earth escapes into space

some long-wave radiation from Earth is absorbed by the greenhouse gases and re-emitted back to the Earth

when short-wave radiation strikes the Earth, some energy is absorbed and the radiation is re-emitted as longer wave radiation

Earth

layer of greenhouse gases

a) Name *two* greenhouse gases. (2)

b) Explain *one* benefit to the Earth of the greenhouse effect. (2)

c) Suggest why global warming may lead to malaria becoming more common in Europe. (6)

7. The diagram shows the profile of the ground on a farm either side of a pond.

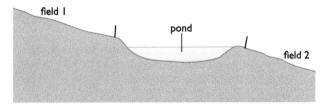

The farmer applies nitrate fertiliser to the two fields in alternate years. When he applies the fertiliser to Field 1, the pond often develops an algal bloom. This does not happen when fertiliser is applied to Field 2.

a) Explain why an algal bloom develops when he applies the fertiliser to Field 1. *(3)*

b) Explain why no algal bloom develops when he applies the fertiliser to Field 2. *(3)*

c) Explain why the algal bloom is more pronounced in hot weather. *(4)*

8. Insecticides are used by farmers to control the populations of insect pests. New insecticides are continually being developed.

a) A new insecticide was trialled over three years to test its effectiveness in controlling an insect pest of potato plants. Three different concentrations of the insecticide were tested. Some results are shown in the table.

Concentration of insecticide	Percentage of insect pest killed each year		
	Year 1	Year 2	Year 3
1 (weakest)	95	72	18
2 (intermediate)	98	90	43
3 (strongest)	99	91	47

    i) Describe, and suggest an explanation for, the change in the effectiveness of the insecticide over the three years. *(3)*

    ii) Which concentration would a farmer be most likely to choose to apply to potato crops? Explain your answer. *(3)*

b) The trials also showed that there was no significant bioaccumulation of the insecticide.

    i) What is bioaccumulation? *(1)*

    ii) Give an example of bioaccumulation of an insecticide and describe its consequences. *(2)*

    iii) Explain why it is particularly important that there is no bioaccumulation of this insecticide. *(1)*

Chapter 6: Population

When you have completed this chapter, you will be able to:

- define a population
- explain how the size of a population can change over time
- list factors that can affect a population, and explain how each factor exerts its effect
- describe and account for changes in the human population in recent times
- appreciate that there is no single, easy solution to the problems of overpopulation.

Populations

A population is a group of organisms of the same species that occupy the same habitat at any one time. However, the numbers of organisms in a population fluctuate, and any species must first establish itself in an area. If the species is sufficiently well adapted, it will multiply and eventually colonise the area. If it is not sufficiently well adapted, the species will die out.

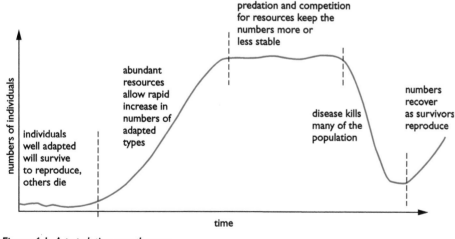

predation and competition for resources keep the numbers more or less stable

abundant resources allow rapid increase in numbers of adapted types

individuals well adapted will survive to reproduce, others die

disease kills many of the population

numbers recover as survivors reproduce

numbers of individuals

time

Figure 6.1 *A population growth curve.*

A population cannot increase in numbers indefinitely. Eventually, something will limit its growth and numbers will not increase further unless conditions change. These changes can be shown in a population growth curve, as in Figure 6.1.

Factors that can limit the growth of a population include predation, disease, limited food supply and limited space for breeding.

As a population grows, the available food must be shared between more and more organisms. They must *compete* for the food. Eventually, there will not be enough food for any more organisms. Those best able to obtain food will be more likely to survive, and others will die. The numbers in the population cannot increase beyond this level.

Plants compete for light. Overcrowding of plants means that some cannot get enough light to photosynthesise effectively. Plants that can grow quickly and can get their leaves into a position to obtain sufficient light will survive at the expense of others.

Disease can drastically reduce a population if few individuals are immune. Numbers recover as those with immunity who survived begin to reproduce.

Predation can also limit the numbers of a population. The relationship between the numbers of a predator and its prey is a very close one. Predator–prey relationships are illustrated in Chapter 3.

Predators that are **invasive species** also limit the numbers within a population.

An 'invasive species' is a plant or animal that is non-native (alien) to an ecosystem, and whose introduction is likely to cause damage to the environment, human health or the economy in that ecosystem. Once an alien species becomes established, it is extremely difficult to control its spread.

venomous spines protect the fish from predation

The spine can cause a painful sting to humans.

Figure 6.2 *Pterois volitans* – the red lionfish.

The lionfish (Figure 6.2) is a species invasive to the Caribbean Sea. It is a type of fish that has elongated and venomous dorsal and anal fin spines. It is related to the scorpion fish. It is a predator which can consume economically important fish (both juveniles and adults) and crustaceans in large numbers. This may reduce the number of fish in a given country, as well as biodiversity. In the Caribbean Sea, the lionfish has no natural predators, so its population may continue to increase in this ecosystem.

Organisms **migrate** in search of food and to increase their chances of survival. The numbers in the population could decrease as the organisms may not return.

Many Caribbean people migrate from one Caribbean country to another, or to Canada, the United Kingdom or the United States of America. This reduces the number of people in the Caribbean's population.

Disasters can reduce the size of a population as they often cause loss of habitat. Organisms that survive the disaster may nevertheless die due to lack of food, or they may migrate. Disasters are classified as being either natural, for example hurricanes, earthquakes and volcanic eruptions, or human-made, for example oil spills and fires.

Figure 6.3 *Flooding, following very high levels of rainfall*

Figure 6.4 *Volcanic eruptions in Montserrat.*

Figure 6.5 *A forest fire.*

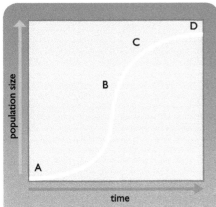

Figure 6.6 *The typical population growth curve (sigmoid growth curve).*

A lag phase: a set number of organisms migrate to a new area; a period of little growth

B log phase: organisms reproduce, so numbers increase greatly

C transitional phase: growth rate begins to slow down as competition for D resources increases

D stationary phase: little to no growth; births and deaths help to level off the numbers

The human population

For many thousands of years, the human population remained more or less constant and was subject to the same sort of checks as many other populations. However, the past 3 000 years has seen a dramatic increase in the human population; an increase that has got faster as time has gone by. The increase has been largely due to:

- improved food supplies
- better understanding of how disease is caused
- improved sanitation.

The increase in the population is often greatest in developing countries. As a result, many resources are in short supply in these countries. There is often a shortage of:

- food – a larger number of people must share the available food
- space – a larger number of people must compete for space to live, work and rest
- energy – there is less energy available for heating, lighting, machinery and industry in general
- water – the increased number of people must share often limited amounts of water for drinking, washing, cooking, etc.

In addition to the problems created by population increase, the distribution of global resources is also unbalanced, as Table 6.1 shows.

Resource	Industrialised countries	Developing countries
population	25%	
energy use	80%	
world industry	86% (60% in the five most industrialised countries)	14% (0.2% in the 44 least industrialised countries)
fossil fuel use	70%	30%
water used per person	350–1 000 litres per day	20–40 litres per day
protein intake per day (recommended intake is 65 g per person per day)	90 g per person	28 g per person
% of workforce working in agriculture	8	70
earnings	500 million people earn more than $25 000 per year	2 billion people earn less than $600 per year

Table 6.1: *The unbalanced distribution of global resources.*

Many people advocate a programme of birth control to limit the population as a solution to the problem (China has introduced a programme of incentives to couples who limit their families to two children). But this is not the only way of ensuring that fewer people die of starvation and malnutrition.

Our pattern of eating often does nothing to help the world food problem. One acre of arable land can support nearly ten times as many people if they eat the plants instead of eating other animals that eat the plants grown (Figure 6.8 on page 79).

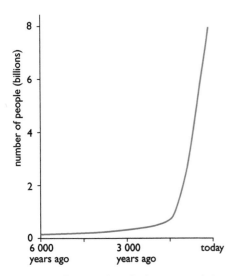

Figure 6.7 *The growth in the human population.*

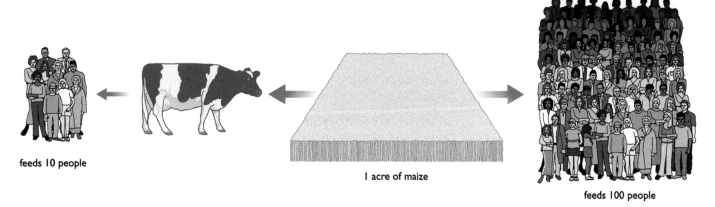

feeds 10 people

1 acre of maize

feeds 100 people

Figure 6.8 *A field of plants can feed more people if we eat the plants instead of eating other animals that eat the plants.*

As the population has grown, people have often become concentrated in relatively small areas. The parish of St Andrew in Jamaica houses about 573 369 people. Unless hygiene is maintained, the potential for the spread of disease is obvious. A large number of people in a small space means disease-causing organisms can easily be transferred from one person to another. In the 14th century, when the need for sanitation and hygiene was poorly understood, the Black Death (bubonic plague) killed 25 million people in Europe. Most of the deaths were in cities where there were large concentrations of people living in areas with poor sanitation.

In developing countries, where hygiene is often of a much lower standard, disease still often spreads rapidly. Also, malnutrition often means that people who become ill are less able to resist the infection.

Periodically, serious new diseases develop and have the potential to reduce the population as few individuals are immune to the new disease-causing organism. The SARS virus is the most recent example of such a disease. AIDS is a relatively new disease that is still killing millions throughout the world, but particularly in Africa. Every few years a new influenza ('flu) virus appears that is more virulent (able to infect and cause disease) than usual and the number of deaths from 'flu increases significantly. Sometimes, exceptionally virulent forms of 'flu appear. In 1918, Spanish 'flu killed millions across the world. The Asian 'flu of 1958 and Hong Kong 'flu of 1968 were less serious, but still killed many thousands of people.

Figure 6.9 *A discussion of population and resource distribution.*

1. What is a population?
2. List four factors that can influence the size of a population.
3. Explain, as fully as you can, why the human population has increased dramatically in recent times.

Chapter summary

In this chapter you have learnt that:

- a population is a group of organisms of the same species living in the same area

- disasters, predation, competition, migration and diseases are factors that affect population growth

- the human population has increased rapidly in recent times because we have a better understanding of the nature of disease, better food supply and improved sanitation

- problems associated with overpopulation can be solved only by measures that address birth control and resource control.

Questions

1. A population is best defined as:

A the total number of organisms of all species in an ecosystem at any one time

B the total number of organisms of one species in an ecosystem at any one time

C the total number of organisms of one species in an ecosystem over one year

D the total number of organisms of all species in an ecosystem over one year

2. Which statement best describes the log phase in population growth?

A organisms die, so the numbers decrease greatly

B organisms reproduce, so the numbers increase greatly

C organism numbers remain constant

D organisms migrate, so the numbers increase

3. Which factor limits the growth of a population?

A available space

B good food supply

C no predation

D diseases

4. Describe two modern-day practices that have led to increase of the human population. (4)

5. Identify and explain two methods that a family may use to control the number of people in a population. (4)

6. The diagram shows a population growth curve for some deer, which were allowed to colonise a small island where there were no natural predators.

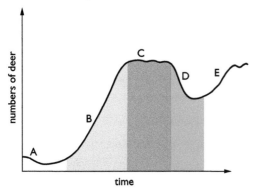

Suggest explanations for:

a) the initial decrease in population (A) (2)

b) the rapid increase in population (B) (2)

c) the steady population (C) (2)

d) the rapid decrease in population (D)§ (2)

e) the recovery in numbers (E). (2)

7. The human population has increased rapidly in the past century. Better understanding of disease, an improved food supply and improved sanitation have been important factors in bringing about this increase.

a) Explain the importance of each of the factors listed in bringing about the recent rapid increase in the human population. (6)

b) Explain why both resource management and birth control are important in managing the problem of human population growth. (4)

c) The table gives some details about energy, food and water consumption in four countries.

Country	Yearly energy usage per person (MJ)	Yearly water usage per person (dm³ × 10³)	Food eaten per person per day (kJ)
USA	295	2 162	14 054
UK	147	507	8 628
India	9	612	900
Nigeria	5	44	780

i) Plot a bar chart to compare the levels of water usage in each country. (3)

ii) Describe and suggest an explanation for the pattern of energy usage in the four countries. (3)

iii) Suggest what might be done to make the food usage per person more equal. (4)

Chapter 7: Cell Structure and Function

When you have completed this chapter, you will be able to:

- understand the structures of generalised plant, animal and bacterial cells
- describe the structure and function of the cell wall, cell membrane, nucleus, chromosomes, cytoplasm, vacuoles, mitochondria and chloroplasts
- explain the differences between plant and animal cells
- use a hand lens and microscope correctly
- explain the importance of cell specialisation in multicellular organisms.

All living organisms are composed of units called **cells**. The simplest organisms are made from single cells (Figure 7.1), but more complex plants and animals, like ourselves, are multicellular, composed of millions of cells. In a **multicellular** organism there are many different types of cells, with different structures. They are specialised so that they can carry out particular functions in the animal or plant. Despite all the differences, there are basic features that are the same in all cells.

(a) Amoeba

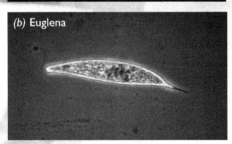
(b) Euglena

Figure 7.1 *Many simple organisms have 'bodies' made from single cells.*

The single-celled microbes shown in Figure 7.1 are known as **protists** – they belong to the kingdom **Protista**. Protists are sometimes described as the 'dustbin kingdom' because they are a mixed group of organisms, mostly single-celled, which are neither plants, animals, nor fungi. Some protists look like animal cells, such as *Amoeba* (Figure 7.1a), which lives in pond water. Other protists have chloroplasts, like plant cells, such as the organism *Euglena* (Figure 7.1b). Compare these protists with the animal and plant cells on page 83.

Life processes

There are seven life processes which are common to most living things. Organisms:

- **obtain nutrition** – either by making food, as in plants, or eating other organisms, as animals do
- **excrete** – get rid of toxic waste products
- **move** – by the action of muscles in animals, and slow growth movements in plants
- **grow** – increase in size and mass, using materials from their food
- **respire** – get energy from their food
- **respond to stimuli** – are sensitive to changes in their surroundings
- **reproduce** – produce offspring.

Cell structure

For over 160 years scientists have known that animals and plants are made from cells. All cells contain some common parts, such as the nucleus, cytoplasm and cell membrane. Some cells have structures missing, for instance red blood cells lack a nucleus, which is unusual. A biology textbook usually shows diagrams of 'typical' plant and animal cells. In fact there is really no such thing as a 'typical' cell. Humans, for example, are composed of hundreds of different kinds of cells, from nerve cells to blood cells, and from skin cells to liver cells. What we really mean by a 'typical' cell is a general diagram that shows all the features that you might find in most cells, without their being too specialised. Figure 7.2 shows the features you would expect to see in many animal and plant cells. However, not all these are present in all cells – the parts of a plant which are not green do not have chloroplasts, for example.

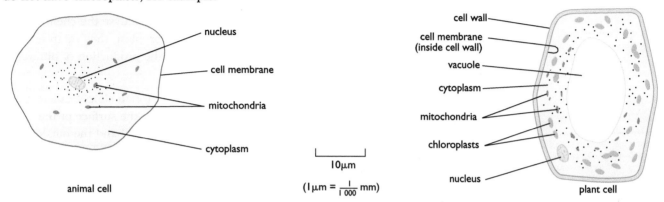

Figure 7.2 *The structure of a 'typical' animal cell and plant cell.*

The living material that makes up a cell is called **cytoplasm**. Cytoplasm is 70% water, and it has a texture rather like sloppy jelly, in other words somewhere between a solid and a liquid. Unlike a jelly, it is not made of one substance but is a complex material made of many different structures. You can't see many of these structures under an ordinary light microscope. An electron microscope has a much higher magnification, and can show the details of these structures, which are called **organelles** (Figure 7.3).

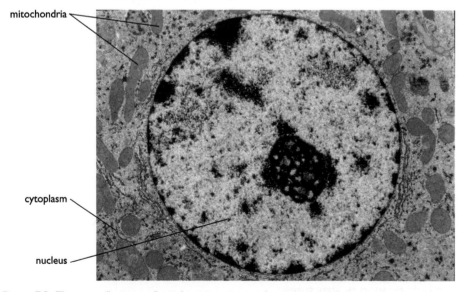

Figure 7.3 *The organelles in a cell can be seen using an electron microscope.*

The largest organelle in a cell is the **nucleus**. Nearly all cells have a nucleus. The few types that don't are usually dead (e.g. the xylem vessels in a stem, Chapter 15) or don't live for very long (e.g. mature red blood cells, Chapter 14). The nucleus controls the activities of the cell. It contains **chromosomes** (46 in human cells), which carry the genetic material, or **genes**. Genes are made of a chemical called deoxyribonucleic acid (DNA). You will find out much more about

The scale on Figure 7.2 shows a length of 10 μm (10 micrometres). A micrometre is one millionth of a metre (m) or one thousandth of a millimetre (mm):

  1 m = 1 000 mm
  1 mm = 1 000 μm
  1 m = 1 000 000 μm

1. Use the scale to estimate the length of one of the nuclei shown in Figure 7.2.
2. The diameter of the animal cell drawing in Figure 7.2 is about 45 mm. The real width of the cell is 45 μm. Use the following equation to calculate the magnification of the cell. Remember that the measurements must be in the same units.

$$\text{magnification} = \frac{\text{drawing width}}{\text{real width}}$$

Cross-curricular links:
chemistry, DNA, proteins, chlorophyll,
carbohydrates

genes and inheritance later in the book. Genes control the activities in the cell by determining which proteins the cell can make. One very important group of proteins found in cells is **enzymes**. Enzymes control chemical reactions that take place in the cytoplasm.

Enzymes are biological **catalysts**. They speed up the rate of reactions inside the cell. The structure of a cell, and everything that a cell does, is controlled by enzymes. For example, the reactions that build up proteins in a muscle cell are catalysed by enzymes. In a liver cell, enzymes catalyse the chemical pathways leading to the production of urea. In a chloroplast, they catalyse the reactions of photosynthesis. Different genes control the production of different enzymes, which in turn control different reactions.

All cells are surrounded by a **cell surface membrane** (often simply called the cell membrane). This is a thin layer, like a 'skin', on the surface of the cell. It forms a boundary between the cytoplasm of the cell and the outside. However, it is not a complete barrier. Some chemicals can pass into the cell and others can pass out (the membrane is **permeable** to them). In fact the cell membrane *controls* which substances pass in either direction. We say that it is **differentially** permeable.

One organelle that is found in the cytoplasm of nearly all living cells is the **mitochondrion** (plural **mitochondria**). There are many mitochondria in cells that need a lot of energy, such as muscle or nerve cells. This gives us a clue to the role of mitochondria. They carry out some of the reactions of **respiration** (see Chapter 13) to release energy that the cell can use. In fact most of the energy from respiration is released in the mitochondria.

All of the structures we have described so far are found in both animal and plant cells. However, some structures are only found in plant cells. There are three in particular – the cell wall, a permanent vacuole and chloroplasts.

The **cell wall** is a layer of non-living material that is found outside the cell membrane of plant cells. It is made mainly of a carbohydrate called **cellulose**, although other chemicals may be added to the wall in some cells. Cellulose is a tough material that helps the cell keep its shape. This is why plant cells have a fairly fixed shape. Animal cells, which lack a cell wall, tend to be more variable in shape. Plant cells absorb water, producing internal pressure which pushes against other cells of the plant, giving them support. Without a cell wall to withstand these pressures, this method of support would be impossible. The cell wall has large holes in it, so it is not a barrier to water or dissolved substances. In other words it is **freely permeable**.

Mature (fully grown) plant cells often have a large central space surrounded by a membrane, called a **vacuole**. This vacuole is a permanent feature of the cell. It is filled with a watery liquid called **cell sap**, a store of dissolved sugars, mineral ions and other solutes. Animal cells can have small vacuoles, but they are only temporary structures.

Cells of the green parts of plants, especially the leaves, have another very important organelle, the **chloroplast**. Chloroplasts absorb light energy to make food in the process of photosynthesis (see Chapter 9). The chloroplasts are green because they contain a green pigment called **chlorophyll**. Cells from the parts of a plant that are not green, such as the flowers, roots and woody stems, have no chloroplasts.

Figure 7.4 shows some animal and plant cells seen through a light microscope.

Both animal and plant cells have:

- cytoplasm
- a nucleus
- a cell surface membrane
- mitochondria.

The differences between plant and animal cells are listed in Table 7.1.

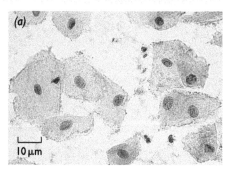

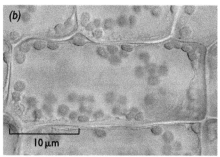

Figure 7.4 *(a) Cells from the lining of a human cheek. (b) Cells from the photosynthetic tissue of a leaf.*

Plant cells	Animal cells
have a cellulose cell wall surrounding the cell membrane	have no cell wall
may contain chloroplasts with chlorophyll	have no chloroplasts or chlorophyll
may contain a large permanent central vacuole filled with cell sap	have only small temporary vacuoles
have a regular, fixed shape	often are irregular or variable in shape
often contain starch granules	never contain starch (but they may contain a similar storage compound called glycogen)

Table 7.1 *A comparison of animal and plant cells.*

3. What is the main component of cytoplasm?
4. Where in a cell are the chromosomes found?
5. How do genes control the activities of a cell?
6. Enzymes are biological catalysts. What does this mean?
7. What happens inside mitochondria?
8. Explain the difference between the terms freely permeable and differentially permeable.

Bacteria

Bacteria are small single-celled organisms. However, bacterial cells are much smaller than those of animals, plants or protists, and they have a much simpler structure. To give you some idea of their size, a typical animal cell might be 10 μm to 50 μm in diameter. Compared with this, a typical bacterium is only 1 μm to 5 μm in length, and its volume can be thousands of times less than the larger cell (Figure 7.5).

There are three basic shapes of bacteria: spheres, rods and spirals, but they all have a similar internal structure (Figure 7.6 on page 86).

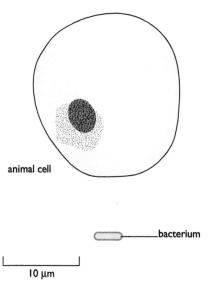

animal cell

bacterium

10 μm

Figure 7.5 *A bacterium is much smaller than an animal cell.*

(a) Some different bacterial shapes

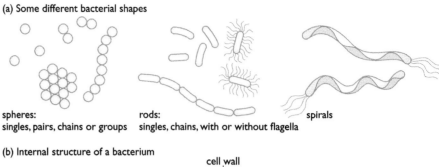

spheres:
singles, pairs, chains or groups

rods:
singles, chains, with or without flagella

spirals

(b) Internal structure of a bacterium

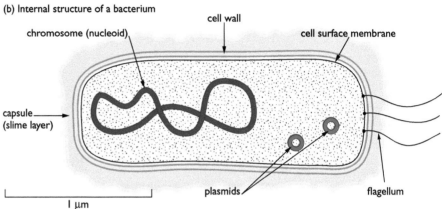

chromosome (nucleoid)

cell wall

cell surface membrane

capsule (slime layer)

plasmids

flagellum

1 μm

Figure 7.6 *Structure of a bacterium.*

All bacteria are surrounded by a cell wall, which protects the bacterium and keeps the shape of the cell. Whereas the cell wall of a plant cell is made of cellulose, and cell walls of fungi are made of chitin, bacterial cell walls contain neither of these two substances. Instead, they are composed of complex chemicals made of polysaccharides and proteins. Some species have another layer outside this wall, called a **capsule** or **slime layer**. Both give the bacterium extra protection. Underneath the cell wall is the cell membrane, as in other cells. The middle of the cell is made of cytoplasm. One major difference between a bacterial cell and the more complex cells of animals and plants is that the bacterium has no nucleus. Instead, its genetic material (DNA) is in a **single chromosome**, loose in the cytoplasm, forming a circular loop.

Some bacteria can swim, and are propelled through water by corkscrew-like movements of structures called **flagella** (a single one of these is called a flagell*um*). However, many bacteria do not have flagella and cannot move by themselves. Other structures present in the cytoplasm include the **plasmids**. These are small circular rings of DNA, carrying some of the bacterium's genes. Not all bacteria contain plasmids, although about three-quarters of all known species do. Plasmids have very important uses in **genetic engineering** (see Chapter 25).

Some bacteria contain a form of chlorophyll in their cytoplasm and can carry out photosynthesis. However, most bacteria feed off other living or dead organisms. Along with the fungi, many bacteria are important **decomposers** (see Chapter 17), recycling dead organisms and waste products in the soil and elsewhere. Some bacteria are used by humans to make food, such as *Lactobacillus bulgaricus*, a rod-shaped species used in the production of yoghurt from milk (Figure 7.7). Other species are **pathogens**, which means that they cause disease (Figure 7.8).

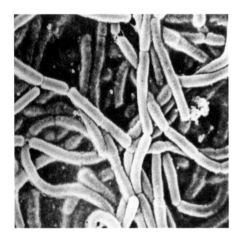

Figure 7.7 *The bacterium Lactobacillus bulgaricus, used in the production of yoghurt.*

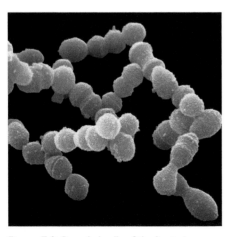

Figure 7.8 *Rounded cells of the bacterium Pneumococcus, the cause of pneumonia.*

The light microscope

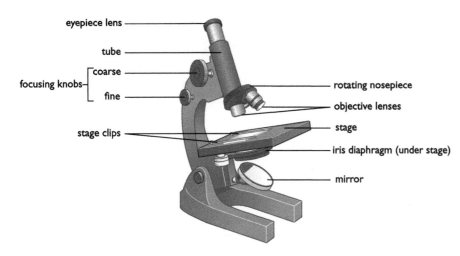

Figure 7.9 *A compound microscope.*

To magnify very small objects such as cells, a compound microscope is used. These instruments can magnify anything up to ×1 500, although school microscopes usually have a maximum magnification of ×400 or less. A typical school microscope is shown in Figure 7.9.

It is called a 'compound' microscope because it has two sets of lenses in series. At the top is the eyepiece lens, and at the bottom an objective lens. There is usually a choice of objective lenses on a rotating nosepiece to allow for different magnifications of the specimen. The specimen is placed on a glass slide and covered with a smaller square of thin glass called a cover slip. The slide is then placed on the stage of the microscope and the specimen illuminated by light from a lamp. Some microscopes have a built-in lamp. Others, like the one in Figure 7.9, reflect the light off a mirror.

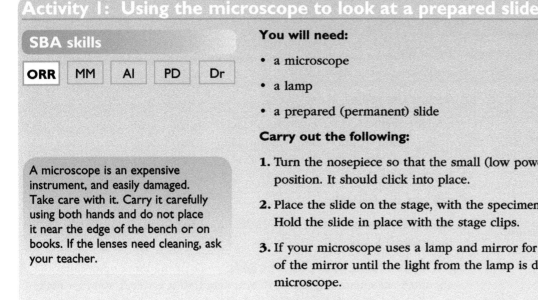

Activity 1: Using the microscope to look at a prepared slide

SBA skills

ORR	MM	AI	PD	Dr

A microscope is an expensive instrument, and easily damaged. Take care with it. Carry it carefully using both hands and do not place it near the edge of the bench or on books. If the lenses need cleaning, ask your teacher.

You will need:

- a microscope
- a lamp
- a prepared (permanent) slide

Carry out the following:

1. Turn the nosepiece so that the small (low power) objective lens is in position. It should click into place.

2. Place the slide on the stage, with the specimen over the hole in the stage. Hold the slide in place with the stage clips.

3. If your microscope uses a lamp and mirror for illumination, alter the angle of the mirror until the light from the lamp is directed up through the microscope.

4. Look through the microscope. You should see an illuminated area called the field of view. Adjust the iris diaphragm so the light is bright but not dazzling.

Activity 1: Using the microscope to look at a prepared slide (continued)

5. Look at the microscope from the side. Turn the coarse focusing knob so that the objective lens is as low down as possible without touching the slide.

6. Now look through the microscope again. Slowly turn the coarse focusing knob in the *opposite direction* to the way you turned it in step 5, so that the lens moves away from the slide. The specimen will gradually appear in focus. Use the fine focusing knob to get the focus as sharp as possible.

7. Adjust the iris diaphragm again. You will find that when you shut the diaphragm down, so that the hole for the light to pass through is made smaller, you will see the specimen with much better contrast.

8. *Without touching the focusing knobs*, rotate the nosepiece so that the high power objective clicks into place. With a good microscope, the specimen should still be in focus. It may need slight focusing with the fine adjustment, but take care not to let the lens touch the slide.

9. Readjust the iris diaphragm if necessary.

You will have noticed that it is more difficult to focus on a specimen under high magnification. You should always start with the low power objective lens. In the next two activities you will make some temporary slides of living plant cells.

Activity 2: Making a slide of onion cells

SBA skills

ORR	MM	AI	PD	Dr

One of the easiest slide preparations is of the cells from an onion bulb epidermis. An onion is a plant storage organ, consisting of many swollen leaves containing sugars and other nutrients. The leaves are covered with a single layer of cells called an epidermis. This tissue does not carry out photosynthesis, so the cells have no chloroplasts. The cells are large, so it is easy to see the nuclei, cell walls and cytoplasm.

You will need:

- a microscope
- a slide and cover slip
- forceps and scalpel
- a board or tile
- tissues
- iodine solution

Carry out the following:

1. Place the onion on a tile and cut it into four segments. Separate one of the inner fleshy leaves.

2. Use a pair of forceps to peel off a small piece of the inner epidermis, about 0.5 cm square (Figure 7.10). You may find it easier if you first make a square cut in the leaf with the scalpel.

3. Place the epidermis onto a slide and add a drop of iodine solution. This helps to show up the cell walls and nuclei.

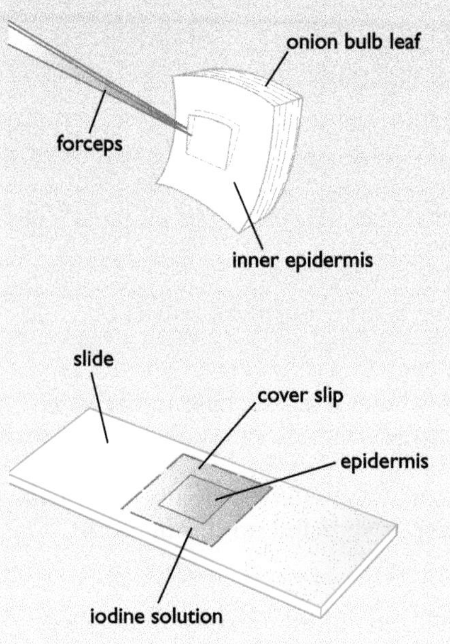

Figure 7.10 *Making a slide of onion epidermis.*

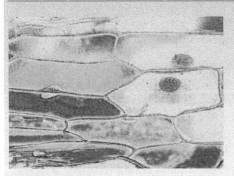

Figure 7.11 *Onion epidermis cells seen through a compound microscope.*

4. Carefully lower a cover slip onto the specimen. Avoid getting any air bubbles on the slide.

5. Clean up your slide with a tissue and observe the specimen under the microscope. Use a low power objective lens first, so that it is easy to focus on the cells. Now switch to a higher power to view a few cells in detail.

6. Make a labelled drawing of three or four cells.

Under high power, the epidermis cells appear as large, elongated box-shaped structures (Figure 7.11). The cytoplasm of these cells forms a thin layer underneath the cell wall, and usually a nucleus can be seen in the cytoplasm. The cytoplasm surrounds a large sap vacuole, but this is transparent and not easily seen.

Activity 3: Making a slide of leaf cells

SBA skills

| ORR | MM | AI | PD | Dr |

To see plant cells with many chloroplasts, you need to look at a green, photosynthesising part of a plant.

You will need:

- a microscope
- a slide and cover slip
- a piece of Canadian pondweed (*Elodea*)
- scissors
- a dropping pipette
- a beaker of water

Carry out the following:

1. Take a single leaf from a piece of Canadian pondweed.

2. Using sharp scissors, cut off a tiny piece of the leaf (about 2 mm square) and place it on a clean microscope slide.

3. Add a drop of water and a cover slip.

4. Remove any excess water from the slide with a tissue and place it on the stage of the microscope. Focus on the tissue under low power. Now switch to high power to study the cells in detail.

5. Draw and label a few cells.

Under high power, the cells will show up clearly, each surrounded by a cell wall. They contain many green chloroplasts (Figure 7.4b). Here you are not using a stain, so the nuclei of these plant cells are more difficult to see. If you carefully watch one cell for a few minutes, you may be able to see the chloroplasts moving around in the living cytoplasm.

6. Explain why the pondweed cells contain chloroplasts but the onion cells do not.

Cell division and differentiation

Multicellular organisms like animals and plants begin life as a single fertilised egg cell, called a **zygote**. This divides into two cells, then four, then eight, and so on, until the adult body contains countless millions of cells (Figure 7.12).

This type of cell division is called **mitosis** and is under the control of the genes. You can read a full account of mitosis in Chapter 24, but it is worthwhile considering an outline of the process now. Firstly, the chromosomes in the nucleus are copied, then the nucleus splits into two, so that the genetic information is shared equally between the two 'daughter' cells. The cytoplasm then divides (or in plant cells a new cell wall develops) forming two smaller cells.

These then take in food substances to supply energy and building materials so that they can grow to full size. The process is repeated, but as the developing **embryo** grows, cells become specialised to carry out particular roles. This specialisation is also under the control of the genes, and is called **differentiation**. Different kinds of cells develop depending on where they are located in the embryo, for example a nerve cell in the spinal cord, or an epidermal cell in the outer layer of the skin.

Large organisms, such as a human or a flowering plant, are made up from countless millions of cells. Most of these have a structure which is adapted for a particular function. We say that cells with different functions show **division of labour**. For example, muscle cells are adapted for contraction, so they can cause movement. A sperm cell has a tail for swimming, and so on (Figure 7.13). Each type of cell has a structure which is specialised to carry out a certain role, while still carrying out the activities that are characteristic of living things. What is hard to understand about the process of differentiation, is that through mitosis all the cells of the body have the *same* genes. How is it that some genes are 'switched on' and others are 'switched off' to produce different cells? The answer to this question is very complicated, and scientists are only just beginning to work it out.

Cells, tissues and organs

Cells with a similar function are grouped together as **tissues**. For example, the muscle of your arm contains millions of similar muscle cells, all specialised for one function – contraction to move the arm bones. This is muscle tissue. However, a muscle also contains other tissues, such as blood, nervous tissue and epithelium (lining tissue). A collection of several tissues carrying out a particular function is called an **organ**. The main organs of the human body are shown in Figure 7.14 on page 91. Plants also have tissues and organs. Leaves, roots, stems and flowers are all plant organs.

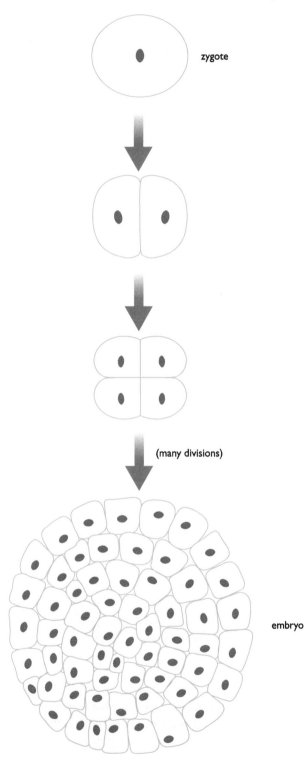

zygote

(many divisions)

embryo

Figure 7.12 *Animals and plants grow by cell division.*

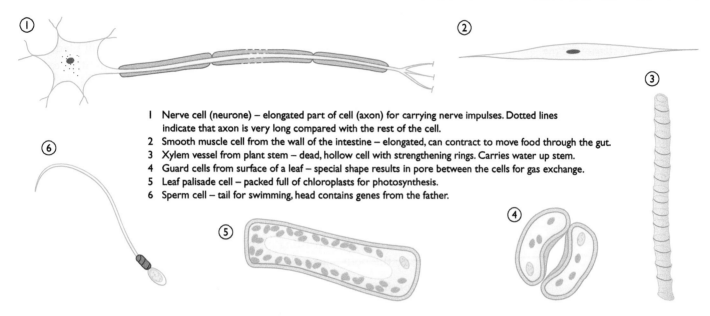

1 Nerve cell (neurone) – elongated part of cell (axon) for carrying nerve impulses. Dotted lines indicate that axon is very long compared with the rest of the cell.
2 Smooth muscle cell from the wall of the intestine – elongated, can contract to move food through the gut.
3 Xylem vessel from plant stem – dead, hollow cell with strengthening rings. Carries water up stem.
4 Guard cells from surface of a leaf – special shape results in pore between the cells for gas exchange.
5 Leaf palisade cell – packed full of chloroplasts for photosynthesis.
6 Sperm cell – tail for swimming, head contains genes from the father.

Figure 7.13 *Some cells with very specialised functions. They are not drawn to the same scale.*

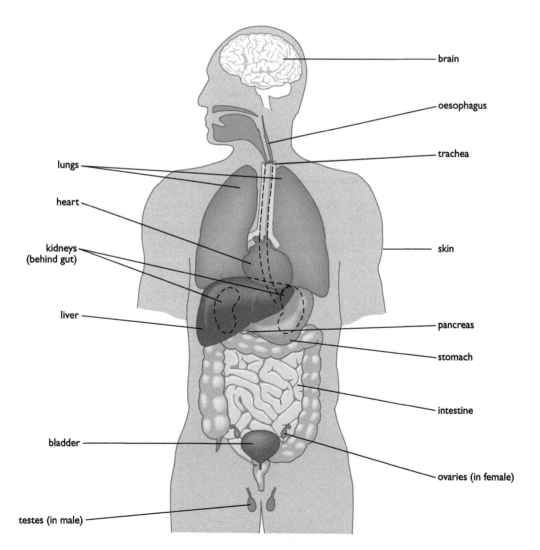

Figure 7.14 *Some of the main organs of the human body.*

In animals, jobs are usually carried out by several different organs working together. This is called an **organ system**. For example, the digestive system consists of the gut, along with glands such as the pancreas and gall bladder. The function of the whole system is to digest food and absorb the digested products into the blood. There are seven main systems in the human body:

- the **digestive** system

- the **respiratory** system – the lungs, which exchange oxygen and carbon dioxide

- the **circulatory** system – the heart and blood vessels, which transport materials around the body

- the **excretory** system, including the kidneys, which filter toxic waste materials from the blood

- the **nervous** system, consisting of the brain, spinal cord and nerves, which coordinate the body's actions

- the **endocrine** system – glands secreting hormones, which act as chemical messengers

- the **reproductive** system, producing sperm in males and eggs in females, and allowing the development of the embryo.

Plants also have cells organised into tissues and organs. Leaves, roots, stems and flowers are all plant organs. A leaf contains several types of tissue, each adapted for roles such as protection, photosynthesis, gas exchange, transport and support. Figure 7.15 shows a section through part of a leaf. You can see a number of tissues in this diagram, for example:

- the epidermis reduces water loss from the leaf and protects it against attack by harmful microbes

- palisade mesophyll is the main site of photosynthesis

- spongy mesophyll forms the main gas exchange surface

- xylem transports water and minerals to the leaf cells during the process of transpiration

- phloem carries away the products of photosynthesis during translocation.

You will learn about these functions and investigate plant tissues more fully in Chapters 9 and 13.

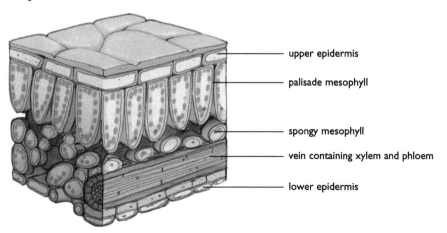

Figure 7.15 *A 3D view of a section through a leaf.*

upper epidermis

palisade mesophyll

spongy mesophyll

vein containing xylem and phloem

lower epidermis

Magnifying larger specimens

Biologists sometimes need to magnify specimens that are too large to put under the microscope. For this purpose, a hand lens can be used. This is a convex lens with a magnification of about 10 times (×10). This means that by using a hand lens you can see an object ten times bigger than it really is. The correct way to use a hand lens is to hold it 2–3 cm from your eye and bring the object up towards the lens until it is in focus. Make sure that the object is well lit from the side (Figure 7.16).

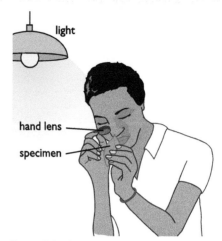

Figure 7.16 *Using a hand lens.*

Activity 4: Making a scale drawing of a leaf

SBA skills

ORR	MM	AI	PD	Dr

A leaf is a plant organ made up of a number of different tissues, with cells adapted for a particular function. Here you can use a hand lens to help you make a scale drawing of the external features of a leaf.

You will need:

- a hand lens
- a leaf (with stalk)

Carry out the following:

1. Observe the external features of one leaf by eye and also by using a hand lens. Identify the leaf stalk (petiole), leaf blade (lamina), midrib and the branching veins. Using the lens, you may be able to see other features, depending on the plant species used.

2. Make an enlarged scale drawing of the leaf (in other words in the correct proportions). Label any features that you can see.

3. Measure the length of the leaf and add a scale bar to your drawing (see Figure 7.2 on page 83).

4. The magnification of a drawing is given by the equation:

$$\text{magnification} = \frac{\text{drawing length}}{\text{real length}}$$

Calculate the magnification of your drawing using the above equation.

Chapter summary

In this chapter you have learnt that:

- organisms are composed of units called cells

- cells of multicellular organisms have specialised structures which are adaptations to allow them to carry out their function

- the seven life processes common to most living things are nutrition, excretion, movement, growth, respiration, reproduction and response to stimuli

- the living material in a cell is called cytoplasm. It consists of a watery solution of different solutes and contains various structures called organelles

- both plant and animal cells have a nucleus, which contains chromosomes

- chromosomes are made up of genes, which are composed of DNA

- genes control the activities in a cell by determining which proteins the cell can make

- an important group of proteins are enzymes, which are biological catalysts

- plant and animal cells contain mitochondria, which release energy from respiration

- plant and animal cells have a cell membrane, which controls the entry and exit of substances into or out of the cell (it is selectively permeable)

- plant cells have a cellulose cell wall, which maintains the shape of the cell, and a large permanent vacuole, which contains cell sap

- plant cells that are able to photosynthesise contain chloroplasts, where light energy is absorbed by chlorophyll

- bacteria are unicellular organisms that are much smaller than plant or animal cells

- bacterial cells have a simple structure, with no nucleus and few organelles

- the magnification of a drawing is given by:

$$\text{magnification} = \frac{\text{drawing width}}{\text{real width}}$$

- a fertilised egg cell (zygote) divides by mitosis to produce an embryo, under the control of the genes. As the embryo grows, cells differentiate, and their structure becomes specialised to carry out different functions

- cells with similar functions are grouped together as tissues

- a collection of different tissues carrying out a particular function is called an organ

- several different organs working together are collectively known as an organ system.

Questions

1. Which of the following is a structure that is present in the nucleus of the cell and made up of genes?

   A deoxyribonucleic acid (DNA)
   B mitochondrion
   C chromosome
   D organelle

2. Which of the following comparisons of plant and animal cells is **not** true?

	Plant cells	Animal cells
A	have permanent vacuoles	have temporary vacuoles
B	have cellulose cell walls	do not have cellulose cell walls
C	have chloroplasts	do not have chloroplasts
D	do not have cell membranes	have cell membranes

3. Which of the following descriptions is true?

   A The cell wall is freely permeable and the cell membrane is selectively permeable.
   B The cell wall is selectively permeable and the cell membrane is freely permeable.
   C Both the cell wall and the cell membrane are freely permeable.
   D Both the cell wall and the cell membrane are selectively permeable.

4. Which of the following is the **best** definition of 'differentiation'?

   A The organisation of the body into cells, tissues and organs.
   B The process by which the structure of a cell becomes specialised for a particular function.
   C A type of cell division resulting in the growth of an embryo.
   D The adaptation of a cell for its function.

5. Which of the following is a tissue?

   A pancreas
   B skin
   C testis
   D blood

The diagram below shows some structures present in a plant cell. Use this diagram to answer questions 6 and 7.

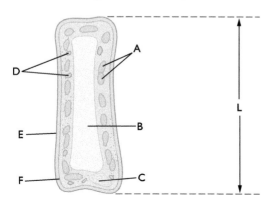

6. a) Copy the diagram and label structures A to F. (6)

   b) Briefly (about one sentence each) explain the function of each of these structures. (6)

7. a) Which of the structures shown in the diagram would be found in:

   i)   a cell from a human muscle

   ii)   a palisade cell from a leaf

   iii) a cell from the root of a plant? (3)

   b) Explain fully why your answers to (ii) and (iii) above are different. (2)

   c) The real length of the cell, shown as L, is 68 μm. Measure the length of the cell in the diagram and calculate its magnification. Show all your working. (4)

8. Explain, using two sentences only, the meaning of each of the following terms.

   a) mitosis (2)
   b) division of labour (2)
   c) enzyme (2)
   d) organ system (2)

9. Copy and complete the following account.

   Plants have cell walls made of _____ . They store carbohydrates as the insoluble compound called _____ or sometimes as the sugar _____ . Plants make these substances as a result of the process called _____ . Animals, on the other hand, store carbohydrates as the compound _____ . Both animal and plant cells have nuclei, but the cells of bacteria lack a true nucleus, having their DNA in a circular chromosome. They sometimes also contain small rings of DNA called _____, which are used in genetic engineering. Bacteria and fungi break down organic matter in the soil. They are known as _____ . Some bacteria are pathogens, which means that they _____ . (8)

10. a) With the aid of a clearly labelled diagram, describe the structure of a plant cell that is capable of carrying out photosynthesis. (7)

    b) Explain the function of *four* of the structures you have labelled in the diagram. (4)

    c) Write down *three* structural differences between the cell you have drawn and a 'typical' animal cell. (3)

11. **a)** Describe how you would prepare a specimen of onion bulb epidermis for viewing through a light microscope. Suggest a suitable stain to use with this tissue, and explain why it is sometimes necessary to use stains in a slide preparation. *(6)*

b) Why does onion bulb epidermis lack chloroplasts? *(2)*

c) Explain the difference between a tissue, an organ and an organ system, giving an example of each. *(6)*

12. These three organelles are found in cells: nucleus, chloroplast and mitochondrion.

a) Which of the above organelles would be found in:

i)   a cell from a human muscle *(1)*

ii)   a palisade cell from a leaf *(1)*

iii)   a cell from the root of a plant? *(1)*

b) Explain fully why the answers to ii) and iii) above are different. *(1)*

c) What is the function of each organelle? *(3)*

d) Draw a diagram of a plant cell. Label all of the parts. Alongside each label, write the function of that part. *(5)*

e) Write down *three* differences between the cell you have drawn and a 'typical' animal cell. *(3)*

13. In multicellular organisms, cells are organised into tissues, organs and organ systems.

The diagram shows a section through an artery and a capillary.

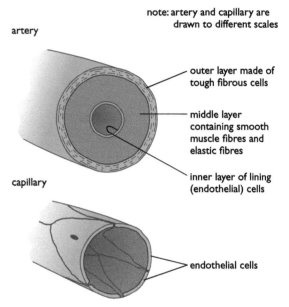

note: artery and capillary are drawn to different scales

artery

outer layer made of tough fibrous cells

middle layer containing smooth muscle fibres and elastic fibres

inner layer of lining (endothelial) cells

capillary

endothelial cells

Explain why an artery can be considered to be an organ whereas a capillary cannot. *(2)*

Chapter 8: Movement of Substances in Cells

When you have completed this chapter, you will be able to:

- explain the processes of diffusion and osmosis
- discuss the importance of diffusion, osmosis and active transport in living systems.

Cells need to be able to take in certain substances from their surroundings, such as glucose and oxygen, and get rid of others, such as carbon dioxide and water. As you have seen, the cell surface membrane is selective about which chemicals can pass in and out. There are three main ways that molecules and ions can move through the membrane. They are diffusion, osmosis and active transport.

Diffusion

Many substances can pass through the cell membrane by **diffusion**. Diffusion happens when a substance is more concentrated in one place than another. For example, if a cell is making carbon dioxide by respiration, the concentration of carbon dioxide inside the cell will be higher than outside the cell. This difference in concentration is called a **concentration gradient**. The molecules of carbon dioxide are constantly moving about because of their kinetic energy. The cell membrane is permeable to carbon dioxide, so these molecules can move in either direction through it. Because there is a higher concentration of carbon dioxide molecules inside the cell than outside it, over time more molecules will move from inside the cell to outside it than move in the other direction. We say that there is a net movement of the molecules from inside to outside (Figure 8.1).

Cross-curricular links:
physics, diffusion; chemistry, particulate nature of matter

> Diffusion is the net movement of particles (molecules or ions) from a region of high concentration to a region of low concentration, i.e. down a concentration gradient.

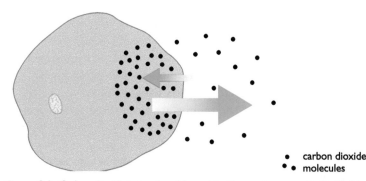

carbon dioxide
molecules

Figure 8.1 *Carbon dioxide is produced by respiration, so its concentration builds up inside the cell. Although the carbon dioxide molecules diffuse in both directions across the cell membrane, the overall (net) movement is out of the cell, down the concentration gradient.*

The opposite happens with oxygen. Respiration uses up oxygen, so there is a concentration gradient of oxygen from outside to inside the cell. There is therefore a net movement of oxygen *into* the cell by diffusion.

Notice that the only reason for the movement of the particles is their own kinetic energy. This means that diffusion is a relatively slow process. If the temperature is raised, the particles will have more kinetic energy, so they will diffuse faster. Another factor that affects the rate of diffusion is the steepness of the concentration gradient. If the gradient is steeper, the rate of diffusion will be greater.

Diffusion happens throughout the bodies of animals and plants, exchanging materials between the cells and their surroundings. However, it is not fast enough to transport materials over long distances. For this, larger organisms have evolved transport systems, such as the human blood system, or the phloem and xylem tissues of a plant. These are known as **mass flow** systems. They rely on an external force being used to move the fluid in bulk. In the blood system this driving force is provided by the pumping action of the heart muscle. In xylem, mass flow is driven by evaporation of water at the leaves. You will read about these processes in later chapters.

Because diffusion is a slow process, organs which exchange materials by diffusion are adapted by having a very large surface area in proportion to their volume. In animals, two examples are the alveoli (air sacs) of the lungs and the villi of the small intestine. The alveoli allow the exchange of oxygen and carbon dioxide to take place between the air and the blood, during breathing. The villi of the small intestine provide a large surface area for the absorption of digested food. In plants, exchange surfaces are also adapted by having a large surface area, such as the spongy mesophyll cells of the leaf, which exchange gases during respiration and photosynthesis.

> A protist such as *Amoeba* is a small unicellular organism, so there is only a short distance between the inside and outside of the cell. As a result diffusion is fast enough for moving oxygen and carbon dioxide across the organism's surface membrane.

1. Explain the meaning of diffusion.
2. Explain the effect of temperature and concentration gradient on the rate of diffusion.
3. Give three examples of situations in which diffusion takes place in animals or plants.

Active transport

> Active transport is the movement of particles against a concentration gradient, using energy from respiration.

Sometimes a cell needs to take in a substance when there is very little of that substance outside the cell, in other words *against* a concentration gradient. It can do this by a process called **active transport**. The cell uses energy from respiration to take up the particles, rather like a pump uses energy to move a liquid from one place to another. In fact biologists usually speak of a cell 'pumping' ions or molecules in or out. The pumps are large protein molecules located in the cell membrane. An example of a place where this happens is in the human small intestine, where some glucose in the gut is absorbed into the cells lining the intestine by active transport. The roots of plants also take up certain mineral ions in this way.

Activity 1: Demonstrating diffusion in a jelly

SBA skills

| ORR | MM | AI | PD | Dr |

Agar jelly has a consistency similar to the cytoplasm of a cell. Like cytoplasm, it has a high water content. In this activity you can use agar as a model to show how substances can diffuse through a cell.

You will need:

- a Petri dish containing a 2-cm depth of agar jelly that has been dyed with potassium permanganate solution

- a sharp scalpel and forceps

- a tile or board (to use as a chopping block)

- a stopwatch or clock

- a 250-cm³ beaker

- a ruler

- dilute hydrochloric acid

(When hydrochloric acid comes into contact with the potassium permanganate, the purple colour of the permanganate will disappear.)

Carry out the following:

1. Cut out some cubes of the agar jelly with the following side lengths: 2 cm, 1 cm, and 0.5 cm.

2. Place about 100 cm³ of dilute hydrochloric acid in the beaker.

3. At the same time, drop each of the cubes of agar into the acid (Figure 8.2) and note the time.

4. Record the time taken for each cube to turn completely colourless.

5. Write up this investigation. In your conclusion, explain the differences in the times taken for the cubes to turn colourless.

6. If the three cubes represent cells of different sizes, which cell would have the most difficulty obtaining substances by diffusion?

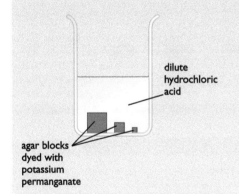

dilute hydrochloric acid

agar blocks dyed with potassium permanganate

Figure 8.2 *Apparatus for investigating diffusion in a jelly.*

Activity 2: Comparing diffusion with mass flow in a gas

SBA skills

| ORR | MM | AI | PD | Dr |

You will need:

- two open-ended glass tubes (each about 30 cm long)

- two bungs to fit the tubes

- red litmus paper

- forceps

- cotton wool

- a stopwatch or clock

- ammonia solution (ammonia vapour will turn red litmus blue)

- distilled water

Carry out the following:

1. Wet about eight small pieces of red litmus paper with distilled water.

2. Using forceps, place the pieces of damp litmus at intervals along each of the tubes.

3. At one end of each tube, place a piece of cotton wool which has had a few drops of ammonia solution added to it. (WARNING: Ammonia has an irritating smell. Do not breathe in the fumes.)

4. Place a bung in each end of the first tube (Figure 8.3) and place it on the table, making sure it does not roll off. Note the time.

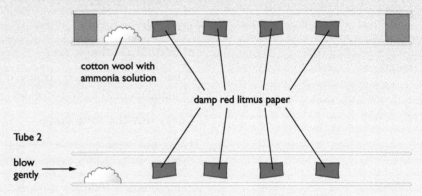

Figure 8.3 *Apparatus for investigating diffusion of ammonia gas.*

5. Leave the second tube open at each end. Gently blow down the tube as shown in Figure 8.3.

6. Note what happens in the second tube, and record how long it takes for changes to be completed in the first tube.

7. Write up this investigation. Make sure that you describe fully what happens to the litmus paper in each tube. In your conclusion, explain how the ammonia moves in each tube. In which tube does it move by diffusion and in which one by mass flow?

> 4. Explain why large organisms cannot rely entirely on diffusion to move materials around their bodies.

Osmosis

Osmosis is the name of a process by which water moves into and out of cells. To be able to understand how water moves through a plant, you need to understand the mechanism of osmosis. Osmosis happens when a material called a **partially permeable membrane** separates two solutions. One artificial partially permeable membrane is called Visking tubing. This is used in kidney

Cross-curricular links:
physics, osmosis; chemistry, particulate nature of matter

machines, which are used to treat patients with kidney failure. Visking tubing has microscopic holes in it, which let small molecules like water pass through (it is *permeable* to them) but is not permeable to some larger molecules, such as the sugar sucrose. This is why it is called 'partially' permeable. You can show the effects of osmosis by filling a Visking tubing 'sausage' with concentrated sucrose solution, attaching it to a capillary tube and placing the Visking tubing in a beaker of water (Figure 8.4).

The level in the capillary tube rises as water moves from the beaker to the inside of the Visking tubing. This movement is due to osmosis. You can understand what's happening if you imagine a highly magnified view of the Visking tubing separating the two liquids (Figure 8.5).

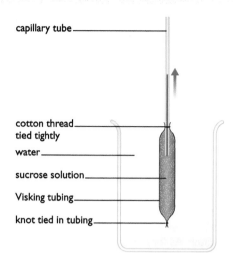

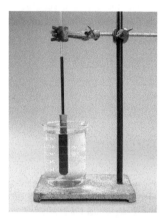

Figure 8.4 *Water enters the Visking tubing 'sausage' by osmosis. This causes the level of liquid in the capillary tube to rise. In the photograph, the contents of the Visking tubing have had a red dye added to make it easier to see the movement of the liquid.*

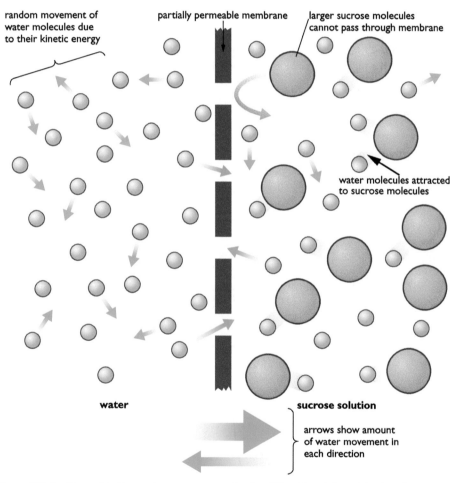

Figure 8.5 *In this model of osmosis, more water molecules diffuse from left to right than from right to left.*

The sucrose molecules are too big to pass through the holes in the partially permeable membrane. The water molecules can pass through the membrane in either direction, but those on the right are attracted to the sugar molecules. This slows them down and means that they are less free to move – they have less kinetic energy. As a result of this, more water molecules diffuse from left to right than from right to left. In other words, there is a greater diffusion of water molecules from the more dilute solution (in this case pure water) to the more concentrated solution.

So, one definition of osmosis is:

> Osmosis is the net diffusion of water across a partially permeable membrane, from a solution with a higher concentration of water molecules to one with a lower concentration of water molecules.

In other words, osmosis can be regarded as a special case of diffusion.

How 'free' the water molecules are to move is called the **water potential**. The molecules in pure water can move the most freely, so pure water has the highest water potential. The more concentrated a solution is, the lower its water potential. In the model in Figure 8.5, water moves from a high to a low water potential. This is the law which applies whenever water moves by osmosis. We can bring these ideas together in an alternative definition of osmosis:

> Osmosis is the net diffusion of water across a partially permeable membrane, from a solution with a high water potential to one with a lower water potential.

Osmosis in plant cells

So far we have only dealt with osmosis through Visking tubing. However, there are partially permeable membranes in cells too. The cell surface membranes of both animal and plant cells are partially permeable, and so is the inner membrane around the plant cell's sap vacuole (Figure 8.6).

It is important to realise that neither of the two solutions has to be pure water. As long as there is a difference in their concentrations (and their water potentials), and they are separated by a partially permeable membrane, osmosis can still take place.

In Chapter 7, we described the cell membrane as being differentially permeable. Here it is called partially permeable. They are not quite the same thing. 'Differentially' permeable means that the membrane actually controls what substances enter or leave the cell. Visking tubing is not alive, so it cannot be differentially permeable. The living cell membrane is both partially and differentially permeable.

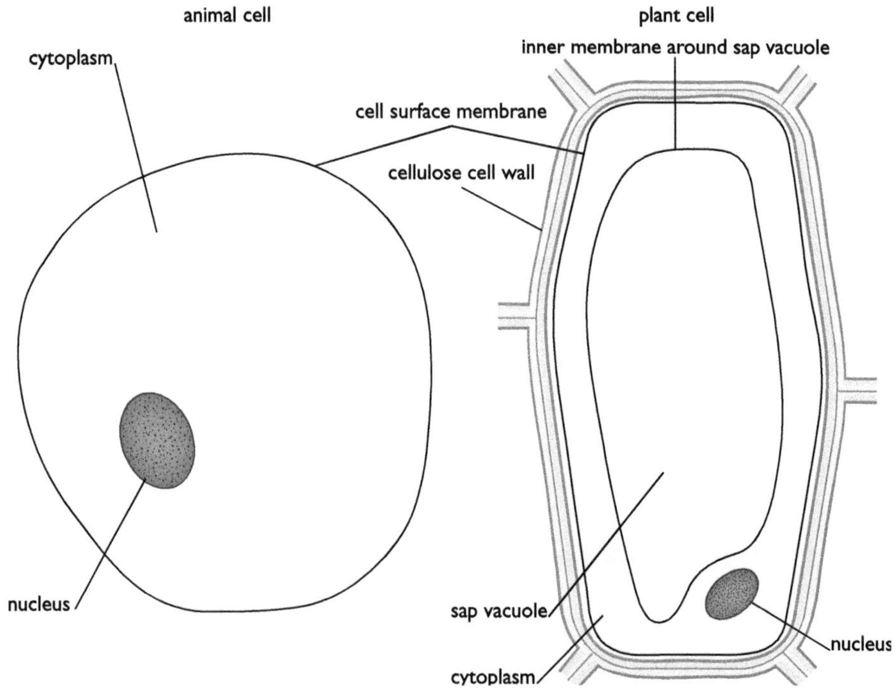

Figure 8.6 *Membranes in animal and plant cells.*

The cell *wall* has large holes in it, making it fully permeable to water and solutes. Only the cell *membranes* are partially permeable barriers that allow osmosis to take place.

Around the plant cell is the tough cellulose cell wall. This outer structure keeps the shape of the cell, and can resist changes in pressure inside the cell. This is very important, and critical in explaining the way that plants are supported. The cell contents, including the sap vacuole, contain many dissolved solutes, such as sugars and ions.

If a plant cell is put into pure water or a dilute solution, the contents of the cell have a lower water potential than the external solution, so the cell will absorb water by osmosis (Figure 8.7). The cell then swells up and the cytoplasm pushes against the cell wall. A plant cell that has developed an internal pressure like this is called **turgid**.

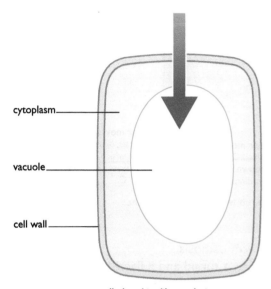
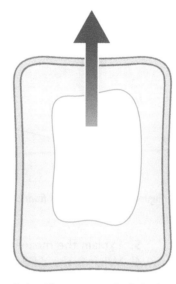
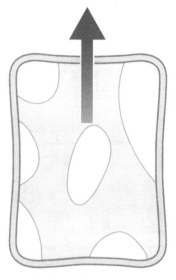

cytoplasm

vacuole

cell wall

cell placed in dilute solution, or water, absorbs water by osmosis and becomes turgid

cell placed in concentrated solution loses water by osmosis and becomes flaccid

excessive loss of water by osmosis causes the cell to become plasmolysed

Figure 8.7 *The effects of osmosis on plant cells.*

On the other hand, if a cell is placed in a concentrated sucrose solution that has a lower water potential than the cell contents, it will *lose* water by osmosis. The cell decreases in volume and the cytoplasm no longer pushes against the cell wall. In this state, the cell is called **flaccid**. Eventually the cell contents shrink so much that the membrane and cytoplasm split away from the cell wall and gaps appear between the wall and the membrane. A cell like this is called **plasmolysed**. You can see plasmolysis happening in the plant cells shown in Figure 8.8. The space between the cell wall and the cell surface membrane will now be filled with the sucrose solution.

Turgor (the state a plant is in when its cells are turgid) is very important to plants. The pressure inside the cells pushes neighbouring cells against each other, like a box full of inflated balloons. This supports the non-woody parts of the plant, such as young stems and leaves, and holds stems upright, so the leaves can carry out photosynthesis properly. Turgor is also important in the functioning of stomata. If a plant loses too much water from its cells so that they become flaccid, this makes the plant **wilt**. You can see this in a pot plant which has been left for too long without water. The leaves droop and collapse. In fact this is a protective action. It cuts down water loss by reducing the exposed surface area of the leaves and closing the stomata.

Inside the plant, water moves from cell to cell by osmosis. If a cell has a higher water potential than the cell next to it, water will move from the first cell to the second. In turn, this will dilute the contents of the second cell, so that it has a higher water potential than the next cell. In this way, water can move across a plant tissue, down a gradient of water potential (Figure 8.9, page 104).

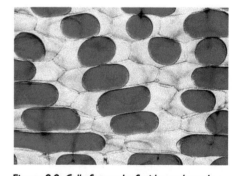

Figure 8.8 *Cells from a leaf with a coloured epidermis, showing plasmolysis. The cell membranes can be seen pulling away from the cell walls.*

Osmosis also happens in animal cells, but there is much less water movement. This is because animal cells do not have a strong cell wall around them, and can't resist the changes in internal pressure resulting from large movements of water. For example, if red blood cells are put into water, they will swell up and burst. If the same cells are put into a concentrated salt solution, they lose water by osmosis and shrink, producing cells with crinkly edges (Figure 8.10).

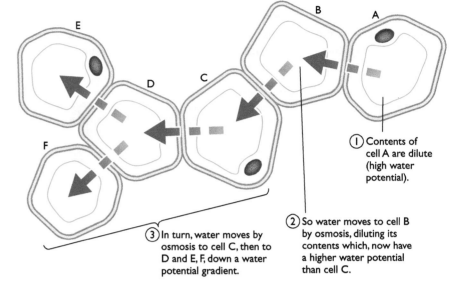

① Contents of cell A are dilute (high water potential).

② So water moves to cell B by osmosis, diluting its contents which, now have a higher water potential than cell C.

③ In turn, water moves by osmosis to cell C, then to D and E, F, down a water potential gradient.

Figure 8.9 *Water moves from cell to cell down a water potential gradient.*

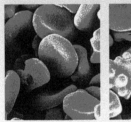

Figure 8.10 *Compare the red blood cells on the right, which were placed in a 3% salt solution, with the normal cells on the left. Blood plasma has a concentration equal to a 0.85% salt solution.*

5. Explain the meaning of the term osmosis.
6. What is the difference between a turgid and a flaccid cell?
7. Explain why plant cells can withstand far greater movements of water in or out of the cell by osmosis than animal cells can.

Activity 3: Investigating the effects of osmosis in onion epidermis cells

SBA skills

ORR	MM	AI	PD	Dr

An alternative species that you can use is the plant *Rhoeo discolor*. The cells of the lower epidermis of the leaves of this plant are purple, and show up the effects of osmosis well. Tearing a leaf will provide a specimen of lower epidermis that can be mounted on a slide.

You will need:

- an onion (use a red onion if possible, as the contents of the cells are coloured and easier to see)

- a microscope

- two microscope slides and cover slips

- concentrated (molar) sucrose solution

- filter paper

- tissues

- dropping pipettes

- forceps and scalpel

- a tile or board

- two small beakers

Carry out the following:

1. Place a drop of sucrose solution on one microscope slide, and a drop of tap water on the other. Use a different pipette to transfer each liquid.

2. Remove two small squares of inner epidermis from one of the outer fleshy leaves of the onion, as shown in Figure 7.11, page 89. Transfer one square to the tap water and the other to the sucrose solution. This should be done as quickly as possible, so that the cells do not dry out.

3. Add one drop of the correct solution to the top of each specimen, followed by a cover slip (Figure 8.11). Clean up your slide, blotting any excess liquid with tissue or filter paper.

4. Examine each slide through the microscope for several minutes. Note any differences between the cells on each slide, comparing your findings with Figures 8.7 and 8.8.

5. Make accurate labelled drawings of two or three representative cells from each slide.

6. Replace the sucrose solution of the second slide with tap water. You can do this quite easily without removing the cover slip. Place some water on one side of the cover slip and draw it across the slide using filter paper (Figure 8.11). You may have to do this for a while to ensure that all the sucrose solution is replaced by water.

7. Examine the cells again and draw a couple of cells that have been placed in water after being in sucrose solution.

8. Write up your investigation and explain your findings in terms of osmosis.

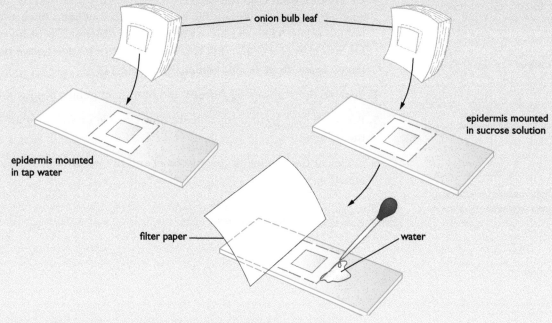

Figure 8.11 *Procedure for observing onion bulb epidermis in sucrose solution and water.*

Activity 4: Investigating the effects of osmosis on potato tuber tissue

ORR	MM	AI	PD	Dr

A potato tuber is a plant storage organ. It is a convenient tissue to use to investigate the effects of osmosis on the mass of the tissue.

You will need:

- a potato
- a chopping board or tile
- three boiling tubes
- forceps
- two small beakers
- a sharp knife or scalpel
- a ruler
- a balance reading to 0.1 g
- filter paper
- concentrated (molar) sucrose solution
- a marker pen or wax pencil

Carry out the following:

1. Half fill one boiling tube with tap water and a second with the sucrose solution. Leave the third empty. Label the tubes.

2. Cut chips of potato measuring 5 cm × 1 cm × 1 cm. Make these measurements as accurate as possible so that the three chips are the same size. You should be able to measure to ±1 mm. Make sure that no skin is left on the potato tissue.

3. Gently blot each chip to remove excess moisture and find the mass of each by weighing them on the balance. Place one chip in each of the three boiling tubes. Their masses will be slightly different, so make sure you know which is which – this is best recorded in a table (as on the next page). Leave them for 30 minutes (Figure 8.12).

4. Remove the chips using forceps and blot them gently, then re-weigh them.

5. Feel each chip in turn to compare how flexible or stiff they are. Note the differences.

6. Calculate the change in mass (+ or –) and the percentage change, from the equation:

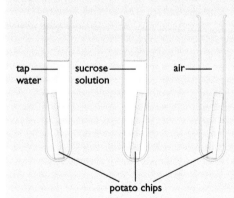

tap water   sucrose solution   air

potato chips

Figure 8.12 Apparatus for investigating the effects of osmosis on potato tuber tissue.

$$\% \text{ change} = \frac{\text{change in mass}}{\text{starting mass}} \times 100$$

7. Record your results in a table like this:

Tube	Starting mass (g)	Final mass (g)	Change in mass (g)	% change	Condition (flexible/stiff)
water					
sucrose solution					
nothing (air)					

8. Write up your investigation. In your conclusion, explain your results using your knowledge of osmosis. How large were the percentage changes in mass of the chips in the two liquids, compared with the one in air? Can you explain the final 'condition' of the chips, using terms such as 'flaccid' and 'turgid'? Can you think of any criticisms of this experiment? For example, does using one chip per tube yield reliable evidence?

9. You can extend this experiment into a more complete investigation. You should be able to make a prediction about what would happen to chips placed in solutions that were intermediate in concentration between water and the molar sucrose solution. Plan a method to test your hypothesis. You should include a description of the apparatus and materials to be used, descriptions of procedures and a statement of the expected results.

You don't have to use potato. You can try other vegetables such as carrots. You can also measure the length of the chips at the end of the experiment, and compare this with their starting length, instead of weighing them. However, weighing is more accurate and gives a bigger percentage change.

8. Explain what happens to green mango fruit or cucumber when we add salt to them for making a 'chow'.
9. When people buy heads of lettuce at the market, why do they place them in water afterwards?
10. Vendors at the market sprinkle water on their green vegetables. How does this keep them fresh?

Chapter summary

In this chapter you have learnt that:

- diffusion is the net movement of molecules or ions down a concentration gradient, due to their kinetic energy

- many substances move in or out of cells by diffusion

- mass flow moves substances more quickly than diffusion. It needs an external force, such as the contraction of the heart in the blood system, to move materials in bulk

- active transport is the movement of substances against a concentration gradient, using energy from respiration

- osmosis happens when two solutions of different concentrations are separated by a partially permeable membrane, such as Visking tubing or the selectively permeable cell membrane

- in osmosis, water moves across a partially permeable membrane from a more dilute solution to a more concentrated one, down a gradient of water potential

- if plant cells absorb water by osmosis, they become turgid; if they lose water by osmosis they become flaccid; excessive loss of water results in the cells becoming plasmolysed

- turgor is important in keeping the non-woody parts of a plant upright; loss of turgor results in wilting

- osmosis is important in moving water from cell to cell inside a plant; the water moves down a gradient of water potential

- plant cells are protected against damage by osmotic water movements because they have a strong cellulose cell wall

- there is less water movement in animal cells because they lack a cell wall and can't resist large changes in internal pressure

- osmosis can be studied in plant tissues by placing the tissues in solutions of different concentration; the cells can be studied under the microscope, or the tissues can be weighed or measured to show the effects of osmosis.

Questions

1. Movement of molecules against a concentration gradient, using energy from respiration' is called:

   A osmosis
   B active transport
   C diffusion
   D mass flow

   Use the following information to answer questions 2–4.

   A plant cell was placed in a concentrated salt solution. Its appearance through the microscope is shown in the diagram.

   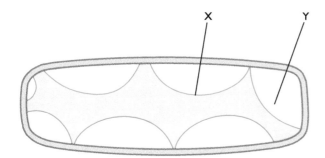

2. The condition of the cell is best described as:

   A flaccid
   B wilted
   C turgid
   D plasmolysed

3. What part of the cell is shown by the label 'X'?

   A cell membrane
   B cytoplasm
   C cellulose cell wall
   D sap vacuole

4. The space labelled 'Y' contains:

   A water
   B sap
   C salt solution
   D cytoplasm

5. A potato chip with a mass of 5 g was left in a concentrated sucrose solution for one hour. After it was removed and blotted, its mass had decreased to 4.5 g. What was its percentage loss in mass?

   A 0.5%
   B 4.5%
   C 5.0%
   D 10.0%

6. The diagram shows a cell from the lining of a human kidney tubule. A major role of the cell is to absorb salt from the fluid flowing along the tubule and pass it into the blood, as shown by the arrow on the diagram.

   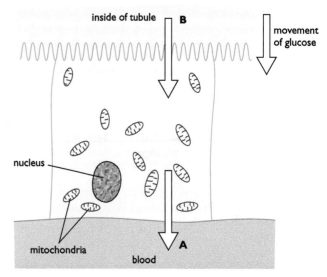

   a) What is the function of the mitochondria? *(1)*

   b) The tubule cell contains a large number of mitochlondria. They are needed for the cell to transport glucose across the cell membrane into the blood at 'A'. Suggest a method that the cell uses to do this and explain your answer. *(2)*

   c) The mitochondria are *not* needed to transport the glucose into the cell from the tubule at 'B'. Name the process by which the ions move across the membrane at 'B' and explain your answer. *(2)*

   d) The surface membrane of the tubule cell at 'B' is greatly folded. Explain how this adaptation helps the cell to carry out its function. *(2)*

7. Three 'chips' of about the same size and shape were cut from the same potato. Each was blotted, weighed and placed in a different sucrose solution (A, B or C). The chips were left in the solutions for one hour, then removed, blotted and re-weighed. Here are the results:

	Starting mass (g)	Final mass (g)	Change in mass (%)
solution A	7.4	6.5	−12.2
solution B	8.2	8.0	
solution C	7.7	8.5	+10.4

   a) Calculate the percentage change in mass for the chip in solution B. *(2)*

   b) Name the process that caused the chips to lose or gain mass. *(1)*

c) Which solution was likely to have been the most concentrated? *(1)*

d) Which solution had the highest water potential? *(1)*

e) Which solution had a water potential most similar to the water potential of the potato cells? *(1)*

f) The cell membrane is described as 'partially permeable'. Explain the meaning of this. *(2)*

8. Different particles move across cell membranes using different processes.

a) The table shows some ways in which active transport, osmosis and diffusion are similar and some ways in which they are different. Copy and complete the table with ticks and crosses. *(12)*

Feature	Active transport	Osmosis	Diffusion
particles move across using their own kinetic energy			
requires energy from respiration			
particles move down a concentration gradient			
process needs special carriers in the membrane			

b) The graph shows the results of an investigation into the rate of diffusion of sodium ions across the membranes of potato cells.

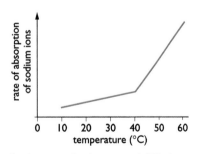

i) Explain the increase in the rate of diffusion up to 40°C. *(2)*

ii) Suggest why the rate of increase is much steeper at temperatures above 40°C. *(2)*

9. An experiment was carried out to find the effects of osmosis on blood cells. Three test tubes were filled with different solutions. 10 cm³ of water was placed in tube A, 10 cm³ of 0.85% salt solution in tube B and 10 cm³ of 3% salt solution in tube C. 1 cm³ of fresh blood was added to each tube. The tubes were shaken, and then a sample from each was observed using a microscope under high power.

The tubes were then placed in a centrifuge and spun around at high speed to separate any solid particles from solution. The results are shown in the diagram below.

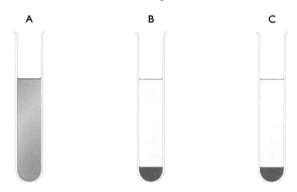

a) Which solution had a similar salt concentration to blood? *(1)*

b) Describe what you would expect to see when viewing the samples from all three tubes through the microscope. *(3)*

c) Explain the results shown in the diagram. *(6)*

d) When a patient has suffered from severe burns, damage to the skin results in a loss of water from the body. This condition can be treated by giving the patient a saline drip. This is a 0.85% salt solution which is fed into the patient's blood through a needle inserted into a vein. Explain why a 0.85% salt solution is used and not water. *(2)*

10. a) Explain the meaning of:

i) diffusion

ii) osmosis. *(4)*

b) With the aid of clearly labelled diagrams, explain the effect of placing onion epidermal cells in:

i) distilled water

ii) a concentrated sucrose (or salt) solution. *(4)*

c) Describe, giving full experimental details, how you could find the effect of different concentrations of sucrose solution on the mass of potato chips. Explain how you would calculate the percentage change in mass of a chip. *(7)*

Chapter 9: How Plants Obtain Nutrition

When you have completed this chapter, you will be able to:

* distinguish between autotrophic and heterotrophic nutrition
* describe how plants make food by the process of photosynthesis
* relate the structure of a leaf to its function in photosynthesis
* carry out experiments to show that light, chlorophyll and carbon dioxide are needed for photosynthesis
* explain how environmental factors affect the rate of photosynthesis
* discuss the importance of minerals in plant nutrition.

Life on Earth is dependent on green plants and the process of photosynthesis. Photosynthesis produces food for animals, supplies them with oxygen for respiration, and removes carbon dioxide from the air. Photosynthesis is the starting point for most food chains.

Autotrophic and heterotrophic nutrition

All living organisms need food for growth or repair of tissues, and as a source of energy. How an organism obtains its food is called its method of **nutrition**. This chapter deals with the way that plants make food. They do so by building up large complex organic molecules, such as starch, lipids and protein, from simple inorganic molecules such as water, carbon dioxide and minerals. This is the process known as **photosynthesis**. Obtaining food in this way is called **autotrophic** nutrition. The other type of nutrition is where an organism feeds on organic substances that have been made by other organisms. This is called **heterotrophic** nutrition. Animals and fungi both feed in this way. Humans, for example, eat plants, or they may eat other animals, which in turn have eaten plants. Most fungi feed on dead organic material, while a few species obtain their nutrition from other living organisms. Table 9.1 compares the two types of nutrition.

Autotrophic means self-feeding. Heterotrophic means different-feeding or feeding on other organisms.

Fungi or organisms which feed on dead organic material are often called **saprophytes** (they use **saprophytic** nutrition). However this is an old-fashioned term – 'phyte' means a plant, and dates back to when biologists thought fungi were plants. We now know they belong to a separate kingdom; so saprophytic has been updated to the modern term **saprobiotic**.

Autotrophic nutrition	Heterotrophic nutrition
builds up organic compounds from simple inorganic substances such as water, carbon dioxide and minerals	uses organic compounds such as carbohydrates, lipids and proteins from other organisms
carried out by green plants	carried out by animals and fungi

Table 9.1: *Autotrophic and heterotrophic nutrition compared.*

1. Give three examples of organic substances and two examples of inorganic molecules.
2. Explain the difference between autotrophic and heterotrophic nutrition.

Figure 9.1 *All these foods are made by plants and contain starch.*

Plants make starch

All the foods shown in Figure 9.1 are products of plants. Some, such as potatoes, rice and bread, form the staple diet of humans. They all contain *starch*, which is the main storage carbohydrate made by plants. Starch is a good way of storing carbohydrates because it is not soluble, is compact and can be broken down easily.

Activity 1: Testing leaves for starch

SBA skills

ORR	MM	AI	PD	Dr

You can test for starch in food by adding a few drops of red-brown iodine solution. If the food contains starch, a blue-black colour is produced.

Leaves that have been in sunlight also contain starch, but you can't test for it by adding iodine solution to a fresh leaf. The outer waxy surface of the leaf will not absorb the solution, and also, the green colour of the leaf would hide the colour change. To test for starch in a leaf, the outer waxy layer needs to be removed and the leaf decolourised. You can do this by placing the leaf in boiling ethanol. The steps in the method are shown in Figure 9.2 on page 113.

You will need:

- suitable potted plants, such as geraniums
- a large beaker
- a Bunsen burner
- a heatproof mat, tripod and gauze
- iodine solution
- forceps
- ethanol
- a white tile or Petri dish
- a bell jar
- a dish of potassium hydroxide solution

Carry out the following:

1. Set up a beaker of water on a tripod and gauze, and heat the water until it boils.

2. Remove a leaf from the plant, and holding it with forceps, kill it by placing it in the boiling water for 30 seconds (this stops all chemical reactions in the leaf).

FLAMMABLE

USE EYE PROTECTION

3. *Turn off the Bunsen burner*, place the leaf in a boiling tube containing ethanol and stand the boiling tube in a hot water bath. The tube containing ethanol *must not be heated directly, since ethanol is highly flammable*. The boiling point of ethanol (about 78°C) is lower than that of water (100°C) so the ethanol will boil for a few minutes until the water bath cools down. This is long enough to remove most of the chlorophyll from the leaf.

4. When the leaf has turned colourless or pale yellow, remove it and wash it with cold water to soften it.

5. Spread the leaf out on a white tile or Petri dish. Cover the leaf with a few drops of iodine solution and leave it for a few minutes, noting any colour change.

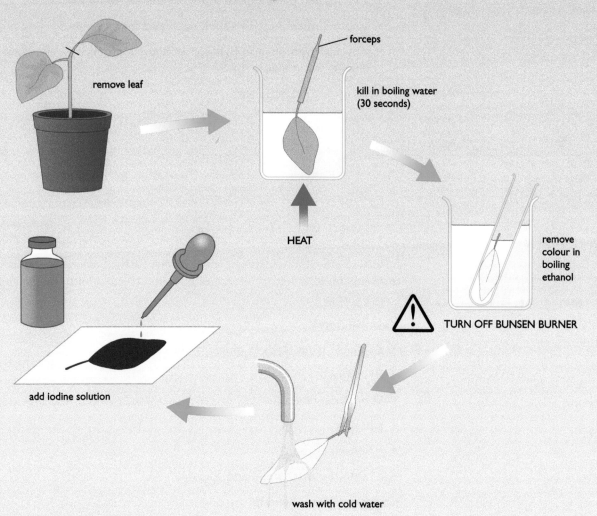

remove leaf

forceps

kill in boiling water (30 seconds)

HEAT

remove colour in boiling ethanol

TURN OFF BUNSEN BURNER

wash with cold water

add iodine solution

Figure 9.2 *How to test a leaf for starch.*

You can 'destarch' a plant by placing it in the dark for two or three days. The plant uses up the starch stores in its leaves.

When you try this method, you will see that the parts of the leaf that contain starch turn a very dark 'blue-black' colour as the iodine reacts with the starch. This will only work if the plant has had plenty of light for some time before the test.

Figure 9.3 *The leaf on the left was taken from a plant that was left under a bright light for 48 hours. The middle leaf is from a plant that was put in a dark cupboard for the same length of time. The third leaf is variegated, and only contains starch in the parts which were green.*

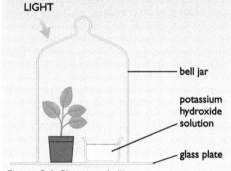

LIGHT

bell jar

potassium hydroxide solution

glass plate

Figure 9.4 *Plant in a bell jar.*

Starch is made only in the parts of leaves that are green. You can show this by testing a **variegated** leaf, which has green and white areas. The white regions, which lack the green pigment called chlorophyll, give a negative starch test. The results of starch tests on three leaves are shown in Figure 9.3.

Depriving a plant of light is not the only way you can prevent it making starch in its leaves. You can also place the plant in a closed container containing a beaker of potassium hydroxide solution (Figure 9.4). This substance absorbs carbon dioxide from the air around the plant. If the plant is kept under a bright light but with no carbon dioxide, it will be unable to make starch.

6. Repeat the starch test on leaves from the following plants:

   • a variegated plant kept in a bright light for 48 hours

   • a normal plant kept in a dark cupboard for 48 hours

   • a normal plant kept in light for 48 hours but deprived of carbon dioxide, as in Figure 9.4.

7. Write up your experiments.

SBA skills

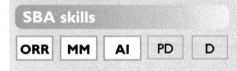

ORR | MM | AI | PD | D

Instead of starch, the leaves of some plants, such as onions and chives, store sugars. This activity describes how to test leaves of these plants for reducing sugars.

You will need:

• a suitable plant, such as onion or chives

• test-tubes

• Benedict's solution

• a beaker

• a Bunsen burner

• a heatproof mat, tripod and gauze

Carry out the following:

1. Cut the leaves into small pieces and place them in a small amount of water in a test-tube.

2. Add a few drops of Benedict's solution (enough to colour the water blue).

3. Boil the water by heating the test tube in a beaker over a Bunsen burner.

4. If a reducing sugar is present, a brick-red precipitate will form.

Where does the starch come from?

You have now found out three important facts about starch production by leaves:

- It uses carbon dioxide from the air.
- It needs light.
- It needs chlorophyll in the leaves.

As well as starch, there is another product of this process which is essential to the existence of most living things on the Earth – oxygen. When a plant is in the light, it makes oxygen gas. You can show this using an aquatic plant such as *Elodea* (Canadian pondweed). When a piece of this plant is placed in a test tube of water under a bright light, it produces a stream of small bubbles. If the bubbles are collected and their contents analysed, they are found to contain a high concentration of oxygen (Figure 9.5).

Starch is composed of long chains of glucose. A plant does not make starch directly, but first produces glucose, which is then joined together in chains to form starch molecules. A carbohydrate made of many sugar sub-units is called a **polysaccharide**. Glucose has the formula $C_6H_{12}O_6$. The carbon and oxygen atoms of the glucose come from the carbon dioxide gas in the air around the plant. The hydrogen atoms come from another molecule essential to living plants – water.

It would be very difficult in a school laboratory to show that a plant uses water to make starch. If you deprived a plant of water in the same way as you deprived it of carbon dioxide, it would soon wilt and die. However, scientists have proved that water is used in photosynthesis. They have done this by supplying the plant with water with 'labelled' atoms, for example using the 'heavy' isotope of oxygen (^{18}O). This isotope ends up in the oxygen gas produced by the plant. A summary of the sources of the atoms in the glucose and oxygen produced looks like this:

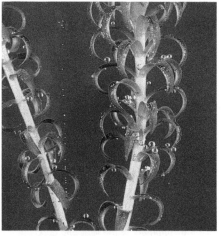

Figure 9.5 *The bubbles of gas released from this pondweed contain a higher concentration of oxygen than in atmospheric air.*

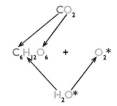

(*oxygen labelled with ^{18}O)

> Isotopes are forms of the same element with the same atomic number but different mass numbers (due to extra neutrons in the nucleus). Isotopes of some elements are radioactive and can be used as 'labels' to trace chemical pathways, others like ^{18}O are identified by their mass.

The photosynthesis reaction

Plants use the simple inorganic molecules carbon dioxide and water, in the presence of chlorophyll and light, to make glucose and oxygen. This process is called **photosynthesis**.

It is summarised by the equation:

$$\text{carbon dioxide + water} \xrightarrow[\text{chlorophyll}]{\text{light}} \text{glucose + oxygen}$$

or: $$6CO_2 + 6H_2O \longrightarrow C_6H_{12}O_6 + 6O_2$$

> The 'photo' in photosynthesis comes from the Greek word *photos*, meaning light, and a 'synthesis' reaction is one in which small molecules are built up into larger ones.

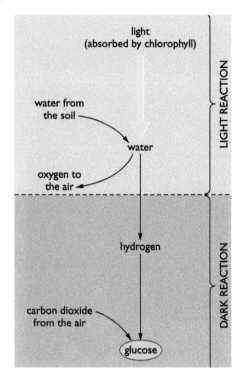

Figure 9.6 *Summary of the light and dark reactions of photosynthesis.*

Splitting of water in the light reaction is called **photolysis**. 'Photo' means light and 'lysis' means 'split'.

Did you know?
The fastest growing plant in the world is bamboo, a member of the grass family. Some large tropical species can grow up to 60 cm per day. The record is held by a bamboo plant in Japan, which grew 120 cm in 24 hours.

The role of the green pigment, chlorophyll, is to absorb the light energy needed for the reaction to take place. The products of the reaction (glucose and oxygen) contain more energy than the carbon dioxide and water (the reactants).

This simple equation is, in fact, a summary of a complex series of chemical reactions which take place in two stages (Figure 9.6).

Firstly, the light energy absorbed by the chlorophyll is used to split water molecules into hydrogen and oxygen. The waste product, oxygen, is given off as a gas. Because this process needs light, it is known as the **light reaction**.

Next, the hydrogen from the water is used to reduce the carbon dioxide to glucose. In this stage the carbon dioxide, by joining with the hydrogen, is **fixed**, or converted into an organic compound (the glucose). This stage does not need light, and is called the **dark reaction**.

Both these stages take place within the chloroplasts. The term 'dark' reaction can be a little confusing. The point is that the dark reaction does not need light, even though it takes place in daylight.

Glucose is the first organic substance made by plants, but as you have seen, they can turn it into starch, as well as many other organic molecules. We can summarise photosynthesis as 'the process by which plants make organic substances from inorganic ones'. The raw materials for the process are carbon dioxide and water. Carbon dioxide is obtained from the air around the leaves, and water is obtained from the soil, through the roots of the plant. The energy source to drive the reaction is sunlight, which is absorbed by the chlorophyll. We now need to look at the structure of leaves and how they are adapted for photosynthesis.

3. Write a word equation and a balanced symbol equation for photosynthesis.
4. What is chlorophyll and what is its function?
5. Explain what happens in the light and dark reactions of photosynthesis.

The structure of leaves

Most green parts of a plant can photosynthesise, but the leaves are the plant organs which are best adapted for this function. To be able to photosynthesise efficiently, leaves need to have a large surface area to absorb light, many chloroplasts containing the chlorophyll, a supply of water and carbon dioxide, and a system for carrying away the products of photosynthesis to other parts of the plant. They also need to release oxygen (and water vapour) from the leaf cells. Most leaves are thin, flat structures supported by a leaf stalk which can grow to allow the blade of the leaf to be angled to receive the maximum amount of sunlight (Figure 9.7, page 117).

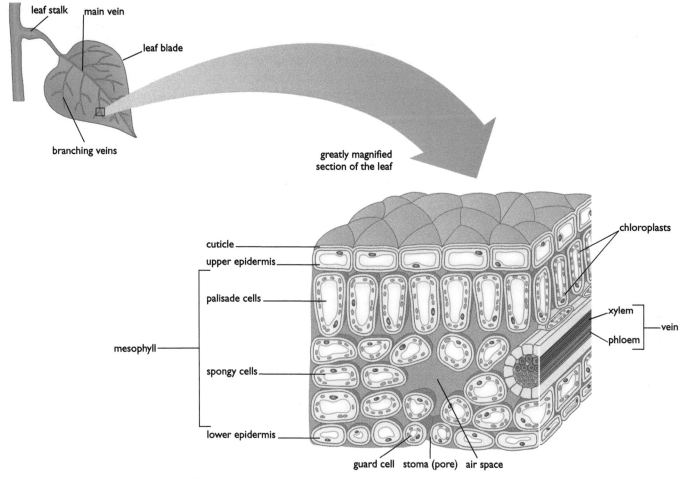

Figure 9.7 *External and internal features of a leaf.*

Inside the leaf are layers of cells with different functions:

- The two outer layers of cells (the upper and lower **epidermis**) have few chloroplasts and are covered by a thin layer of a waxy material, called the **cuticle**. This reduces water loss by evaporation, and acts as a barrier to the entry of disease-causing microorganisms such as bacteria and fungi.

- The lower epidermis has many pores called **stomata** (a single pore is a **stoma**). Usually the upper epidermis contains fewer or no stomata. The stomata allow carbon dioxide to diffuse into the leaf, in order to reach the photosynthetic tissues. They also allow oxygen and water vapour to diffuse out. Each stoma is formed as a gap between two highly specialised cells called **guard cells**, which can alter their shape to open or close the stoma (see Chapter 18).

- In the middle of the leaf are two layers of photosynthetic cells called the **mesophyll** ('mesophyll' just means 'middle of the leaf'). Just below the upper epidermis is the **palisade** layer. This is a tissue made of elongated cells, each containing hundreds of chloroplasts, and is the main site of photosynthesis. The palisade cells are close to the source of light, and the upper epidermis is relatively transparent, allowing light to pass through to the enormous numbers of chloroplasts which lie below.

A 'gas exchange' surface is a tissue that allows gases (usually oxygen, carbon dioxide and water vapour) to pass across it between the plant or animal and the outer environment. Gas exchange surfaces all have a large surface area in proportion to their volume, which allows large amounts of gases to diffuse across. Examples include the alveoli of the lungs, the gills of a fish and the spongy mesophyll of a leaf.

Starch is insoluble and so cannot be transported around the plant. The phloem carries only soluble substances such as sugars. These are converted into other compounds when they reach their destination.

- Below the palisade layer is a tissue made of more rounded, loosely packed cells with air spaces between them, called the **spongy** layer. These cells also photosynthesise, but have fewer chloroplasts than the palisade cells. They form the main **gas exchange surface** of the leaf, absorbing carbon dioxide, and releasing oxygen and water vapour. The air spaces allow these gases to diffuse in and out of the mesophyll.

- Water and mineral ions are supplied to the leaf by vessels in a tissue called the **xylem**. This forms a continuous transport system throughout the plant. Water is absorbed by the roots and passes up through the stem and through veins in the leaves in the **transpiration stream**. In the leaves, the water leaves the xylem and supplies the mesophyll cells.

- The products of photosynthesis, such as sugars, are carried away from the mesophyll cells by another transport system, the **phloem**. The phloem supplies all other parts of the plant, so that tissues and organs that can't make their own food receive the products of photosynthesis. The veins in the leaf contain both xylem and phloem tissue, and branch again and again to supply all parts of the leaf.

You can find out more about both plant transport systems in Chapter 18.

Activity 3: Drawing a leaf section

SBA skills

ORR	MM	AI	PD	Dr

You will need:

- a microscope
- a slide of a section through a leaf

Carry out the following:

1. Observe a prepared microscope slide of a section through a leaf under the microscope. Identify the various tissues you can see.

2. Make a drawing of the section. Do not try to draw all the cells, just three or four cells from each tissue.

3. Label your drawing and annotate it to show the functions and adaptations of the various tissues.

4. Add a title and the magnification to the drawing.

6. In a leaf, what are the functions of: (a) cuticle, (b) phloem, and (c) the spongy mesophyll cells?
7. How are palisade cells adapted for photosynthesis?

Photosynthesis and respiration

Through photosynthesis, plants supply animals with two of their essential needs – food and oxygen, as well as removing carbon dioxide from the air. However, remember that living cells, including plant cells, respire *all the time*, and they need oxygen for this. When the light intensity is high, a plant carries out photosynthesis at a much higher rate than it respires. So, in bright light, there is an overall uptake of carbon dioxide from the air around a plant's leaves, and a surplus production of oxygen that animals can use. A plant only produces more carbon dioxide than it uses in dim light. We can show this as a graph of carbon dioxide exchanged at different light intensities (Figure 9.8).

The concentration of carbon dioxide in the air around plants actually changes

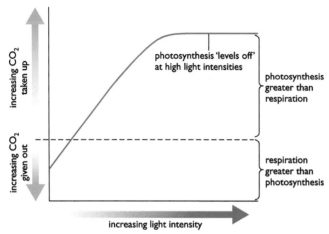

photosynthesis 'levels off' at high light intensities

photosynthesis greater than respiration

respiration greater than photosynthesis

increasing light intensity

The point at which the curve crosses the dashed line shows where photosynthesis is equal to respiration – there is no net gain or loss of CO_2.

Figure 9.8 *As the light intensity gets higher, photosynthesis speeds up, but eventually levels off in very bright light.*

throughout the day. Scientists have measured the level of carbon dioxide in the air in the middle of a field of long grass in summer. They found that the air contained the least amount of carbon dioxide in the afternoon, when photosynthesis was happening at its highest rate (Figure 9.9). At night when there was no photosynthesis, the level of carbon dioxide rose. This rise is due to less carbon dioxide being absorbed by the plants, while carbon dioxide was added to the air from the respiration of all organisms in the field.

Factors affecting the rate of photosynthesis

In Figure 9.8, you can see that when the light intensity increases, the rate of photosynthesis starts increasing as well, but eventually it reaches a maximum rate. What makes the rate 'level off' like this? It is because some other factor needed for photosynthesis is in short supply, so that increasing the light intensity does not affect the rate any more. Normally, the factor which 'holds up' photosynthesis is the concentration of carbon dioxide in the air. This is only about 0.03% to 0.04%, and the plant can only take up the carbon dioxide and fix it into carbohydrates at a certain rate. If the plant is put in a closed container with a higher than normal concentration of carbon dioxide, it will photosynthesise at a faster rate. If there is both a high light intensity and a high level of carbon dioxide, the temperature may limit the rate of photosynthesis, by limiting the rate of the chemical reactions in the leaf. A rise in temperature will then increase the rate. With normal levels of carbon dioxide, very low temperatures (close to 0 °C) slow the reactions, but high temperatures (above about 35 °C) also reduce photosynthesis by denaturing enzymes in the plant cells (see Chapter 11).

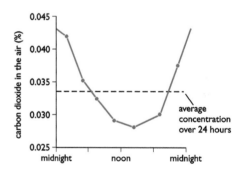

average concentration over 24 hours

Figure 9.9 *Photosynthesis affects the concentration of carbon dioxide in the air around plants. Over a 24-hour period, the concentration rises and falls as a result of the relative levels of photosynthesis and respiration.*

A limiting factor is the component of a reaction that is in shortest supply so that it prevents the rate of the reaction increasing, in other words sets a limit to it.

Knowledge of limiting factors is used in some glasshouses (greenhouses) to speed up the growth of crop plants such as tomatoes and lettuces. Extra carbon dioxide is added to the air around the plants, by using gas burners. The higher concentration of carbon dioxide, along with the high temperature in the glasshouse, increases the rate of photosynthesis and the growth of the leaves and fruits.

Light intensity, carbon dioxide concentration and temperature can all act as **limiting factors** in this way. This is easier to see as a graph (Figure 9.10).

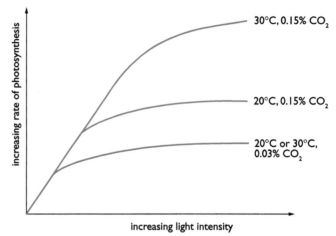

Figure 9.10 *Light intensity, carbon dioxide concentration and temperature can all act as limiting factors on the rate of photosynthesis.*

8. What is the average concentration of carbon dioxide in the air?
9. Explain what is meant by a limiting factor.
10. State three factors which can limit the rate of photosynthesis.

A plant's uses for glucose

As you have seen, some glucose that a plant makes is used in respiration to provide the plant's cells with energy. Some glucose is quickly converted into starch for storage. However, a plant is not made up of just glucose and starch, and must make all of its organic molecules, starting from glucose.

Glucose is a single sugar unit (a **monosaccharide**). Plant cells can convert it into other sugars, such as a monosaccharide called **fructose** (found in fruits) and the **disaccharide** (double sugar unit) **sucrose**, which is the main sugar carried in the phloem. It can also be changed into another polymer, the polysaccharide called **cellulose**, which forms plant cell walls.

All these compounds are carbohydrates. Plant cells can also convert glucose into lipids (fats and oils). Lipids are needed for the membranes of all cells, and are also an energy store in many seeds and fruits, such as peanuts, sunflower seeds and olives.

Carbohydrates and lipids both contain only three elements – carbon, hydrogen and oxygen – and so they can be inter-converted without the need for a supply of other elements. Proteins contain these elements too, but all amino acids (the building blocks of proteins) also contain nitrogen. This is obtained as nitrate ions from the soil through the plant's roots. Other compounds in plants contain other elements. For example, chlorophyll contains magnesium ions, which are also absorbed from water in the soil. Some of the products that a plant makes from glucose are summarised in Figure 9.11.

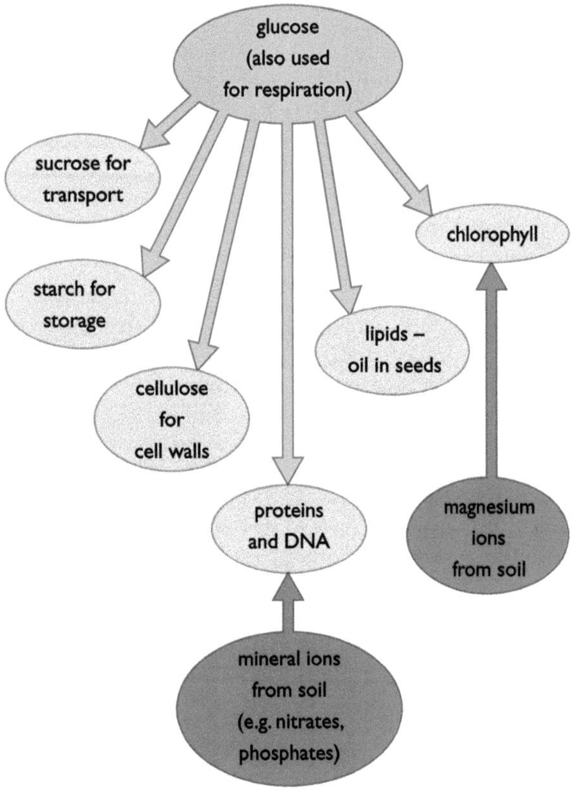

Figure 9.11 *Compounds that plant cells can make from glucose.*

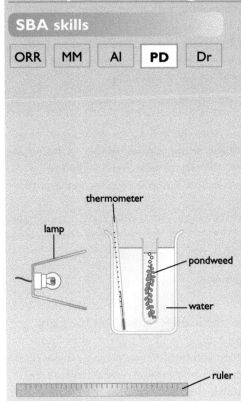

Figure 9.12 *Apparatus for Activity 4.*

You will need:

- a large beaker of water
- a boiling tube
- some pondweed (Canadian pondweed, *Elodea* is best)
- a ruler to measure up to 50 cm
- a stopwatch or clock
- a thermometer

You can measure the rate of photosynthesis of a plant by measuring how quickly it produces oxygen. With a land plant, this is difficult because the oxygen is released into the air, but with an aquatic plant bubbles of oxygen are released into the water around the plant (see Figure 9.5). If you count the bubbles formed per minute, this is a measure of the rate of photosynthesis of the plant. It is easiest to count the bubbles if the cut piece of weed is placed upside down in a test tube, as in Figure 9.12. You may have to weight it down by attaching a small paper clip to the bottom of the piece of weed.

You can change the light intensity by moving the lamp and altering the distance between the lamp and the pondweed. The beaker of water keeps the temperature of the plant constant. You can check this with the thermometer.

Carry out the following:

1. Design an experiment to find out if the rate of photosynthesis is affected by the light intensity. In your plan you should include:

   - a hypothesis – you should state what you think will happen as you alter the light intensity and why
   - a systematic way of changing the light intensity
   - how your experiment will be controlled so that nothing else is changed apart from the light intensity (Also think about what you will do about the background light in the laboratory?)
   - a way of ensuring that your results are reliable.

2. When you have completed your plan, you can discuss it with your teacher. If you are given permission, you can carry out the experiment.

3. How could you modify your plan to find the effect of changing the *temperature* on the rate of photosynthesis? What factors would you need to keep constant this time? What would be a suitable range of temperatures to try?

Glucose from photosynthesis is not just used as the *raw material* for the production of molecules such as starch, cellulose, lipids and proteins. Reactions like these, which synthesise large molecules from smaller ones, also need a source of energy. This energy is provided by the plant's *respiration* of glucose.

In fact, in addition to the ions listed in Knop's solution, plants need very small amounts of other mineral ions for healthy growth. Knop's culture solution only worked because the chemicals he used to make his solutions weren't very pure, and supplied enough of these additional ions by mistake!

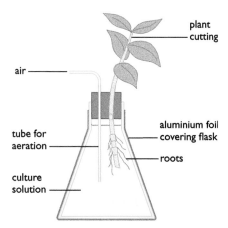

Figure 9.13 *A simple water culture method.*

Mineral nutrition

Nitrate ions are absorbed from the soil water, along with other minerals such as phosphate, potassium and magnesium ions. The element phosphorus is needed for the plant cells to make many important compounds, including DNA. Potassium ions are required for enzymes in respiration and photosynthesis to work, and magnesium forms a part of the chlorophyll molecule.

Water culture experiments

A plant takes only water and mineral ions from the soil for growth. Plants can be grown in soil-free cultures (water cultures) if the correct balance of minerals is added to the water. In the 19th century, the German biologist Wilhelm Knop invented one example of a culture solution. Knop's solution contains the following chemicals (per dm^3 of water):

0.8 g	calcium nitrate
0.2 g	magnesium sulfate
0.2 g	potassium nitrate
0.2 g	potassium dihydrogen phosphate
(trace)	iron(III) phosphate

Notice that these chemicals provide all of the main elements that the plant needs to make proteins, DNA and chlorophyll, as well as other compounds, from glucose. It is called a *complete* culture solution. If you were to make up a similar solution, but to replace, for example, magnesium sulfate with more calcium sulfate, this would produce a culture solution which was *deficient* (lacking) in magnesium. You could then grow plants in the complete and deficient solutions, and compare the results. There are several ways to grow the plants, such as using the apparatus shown in Figure 9.13, which is useful for plant cuttings. Seedlings can be grown by packing cotton wool around the seed, instead of using a rubber bung.

The plant is kept in bright light so that it can photosynthesise. The covering around the flask prevents algae from growing in the culture solution, and the aeration tube is used for short periods to supply the roots with oxygen for respiration of the root cells. Using methods like this, it soon becomes clear that mineral deficiencies result in poor plant growth. A shortage of a particular mineral results in particular symptoms in the plant, called a **mineral deficiency disease**. For example, lack of magnesium means that the plant won't be able to make chlorophyll, and so the leaves will turn yellow. Some of the mineral ions that a plant needs, their uses and the deficiency symptoms are shown in Table 9.2 on page 123. Compare the diagrams of the mineral deficient plants to that of the control plant in Figure 9.14 on the same page.

Mineral ion	Use	Deficiency symptoms
nitrate	making amino acids, proteins, chlorophyll, DNA and many other compounds	stunted growth of plant; older leaves turn yellow *A plant showing symptoms of nitrate deficiency.*
phosphate	making DNA and many other compounds; part of cell membranes	poor root growth; younger leaves turn purple *A plant showing symptoms of phosphate deficiency.*
potassium	needed for enzymes of respiration and photosynthesis to work	leaves turn yellow with dead spots *A plant showing symptoms of potassium deficiency.*
magnesium	part of chlorophyll molecule	leaves turn yellow *A plant showing symptoms of magnesium deficiency.*

Table 9.2: *Mineral ions needed by plants.*

Figure 9.14 *A healthy plant*

Some commercial crops such as lettuces can be grown without soil, in culture solutions. This is called **hydroponics**. The plants' roots grow in a long plastic tube which has culture solution passing through it (Figure 9.15). The composition of the solution can be carefully adjusted to ensure the plants grow well. Pests, which might live in soil, are also less of a problem.

Figure 9.15 *Lettuce plants grown by hydroponics.*

11. In what form is most carbohydrate transported through the phloem?
12. What mineral ions are needed by a plant in order for it to make DNA?
13. Deficiency in which mineral ion causes poor growth and yellowing of the older leaves?

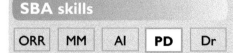

SBA skills

ORR	MM	AI	**PD**	Dr

You will need:

- suitable fast-growing seeds such as mustard, cress, red beans or bean sprouts

- clean, washed sand as a medium in which to plant the seeds

- containers to hold the sand, such as clean empty foil dishes or plastic trays

- distilled water

- a liquid fertiliser (e.g. the sort sold as a 'food' for houseplants)

- beakers

Carry out the following:

1. Design an experiment to find out if seedlings need minerals to grow normally. In your plan you should include:

   - a hypothesis – you should state what you think the difference will be between the growth of seedlings with and without minerals

   - a systematic way of changing the concentration of the minerals applied to the growing seedlings, and a way of measuring their growth (don't forget you can make observations as well as measurements)

   - how your experiment will be controlled so that nothing else is changed apart from the addition of the fertiliser

   - a way of ensuring that your results are reliable.

2. When you have finished your plan, you can show it to your teacher. If you are given permission, you can carry out the investigation.

Chapter summary

In this chapter you have learnt that:

- organisms such as green plants that produce their own food carry out autotrophic nutrition

- organisms such as animals and fungi, which get their food from other organisms, carry out heterotrophic nutrition

- plants produce their own food by photosynthesis. The raw materials of photosynthesis are carbon dioxide and water; the first products are glucose and oxygen

- the energy for the reaction is supplied by light, trapped by chlorophyll in the leaves

- the equation summarising photosynthesis is:

$$\text{carbon dioxide} + \text{water} \xrightarrow[\text{chlorophyll}]{\text{light}} \text{glucose} + \text{oxygen}$$
$$6CO_2 + 6H_2O \longrightarrow C_6H_{12}O_6 + 6O_2$$

- photosynthesis happens in two stages, called the light and dark reactions

- in the light reaction, light is absorbed by chlorophyll and the energy from the light is used to split water into hydrogen and oxygen. The oxygen is released into the air. The hydrogen is used to reduce carbon dioxide to carbohydrates

- the first carbohydrate formed is the sugar, glucose. Leaves and many other plant organs frequently store carbohydrates in the form of starch

- leaves are adapted to carry out photosynthesis efficiently. They have an epidermis that protects the leaf, mesophyll tissue which contains many chloroplasts, and the transport systems in phloem and xylem

- gases are exchanged through pores called stomata

- photosynthesis provides animals with food and oxygen, and uses up carbon dioxide from respiration

- in sunlight, photosynthesis exceeds respiration and causes a net reduction in atmospheric CO_2

- in darkness, respiration exceeds photosynthesis and the levels of CO_2 rise again

- as light intensity increases, so does the rate of photosynthesis, but above a certain intensity the rate levels off, due to another factor becoming limiting, such as temperature or CO_2 concentration

- a plant converts glucose into sucrose, starch, cellulose, lipids, DNA, proteins and many other compounds

- in order to make some compounds, a plant needs additional mineral ions from the soil. Chlorophyll synthesis requires magnesium and nitrate to make proteins.

Questions

1. The oxygen produced during photosynthesis comes from:

   A   water
   B   glucose
   C   carbon dioxide
   D   chlorophyll

   Use the following information to answer questions 2 and 3.

   A variegated leaf attached to a plant was set up in a test tube as shown in the following diagram.

   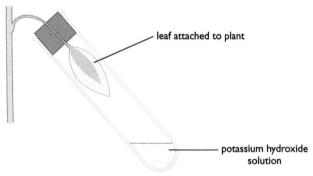

   The plant was left in bright light for 24 hours. The leaf was then removed and tested for starch.

2. Which of the following diagrams show the expected result?

   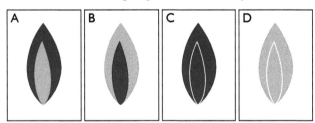

3. What is the purpose of the potassium hydroxide solution in the experiment?

   A   production of CO_2
   B   production of oxygen
   C   absorption of oxygen
   D   absorption of CO_2

4. At night, the concentration of carbon dioxide in the air in a forest is higher than average. The main reason for this is that:

   A   there is increased respiration by animals that are active at night
   B   plants are unable to absorb CO_2 because their stomata are shut
   C   photolysis stops, so no hydrogen is available to reduce CO_2
   D   plants begin to respire in the dark

5. Which of the following organic compounds can only be made by a plant if it has access to mineral ions?

   A   cellulose
   B   protein
   C   sucrose
   D   lipid

6. Which of the following is not normally a factor which limits the rate of photosynthesis?

   A   temperature
   B   light intensity
   C   oxygen concentration
   D   carbon dioxide concentration

7. A plant with variegated leaves had a piece of black paper attached to one leaf as shown in the diagram.

   The plant was kept under a bright light for 24 hours. The leaf was then removed, the paper taken off and the leaf was tested for starch.

   a)   Name the chemical used to test for starch, and describe the colour change if the test is positive.   (3)

   b)   Copy the leaf outline and shade in the areas which would contain starch.   (2)

   c)   Explain how you arrived at your answer to (b).   (2)

   d)   What is starch used for in a plant? How do the properties of starch make it suitable for this function? (3)

8. Copy and complete the following table to show the functions of different parts of a leaf. One has been done for you.   (4)

Part of leaf	Function
palisade mesophyll layer	main site of photosynthesis
spongy mesophyll layer	
stomata	
xylem	
phloem	

9. The graph shows the changes in the concentration of carbon dioxide in a field of long grass throughout a 24-hour period.

   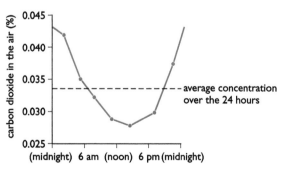

   a)   Explain why the levels of carbon dioxide are high at 2 am and low at noon.   (4)

b) What factor will limit the rate of photosynthesis at 4 am and at 2 pm? *(2)*

10. The table below shows some of the substances that can be made by plants. Give one use in the plant for each. The first has been done for you. *(5)*

Substance	Use
glucose	oxidised in respiration to give energy
sucrose	
starch	
cellulose	
protein	
lipid	

11. The apparatus shown in the diagram was used to grow a pea seedling in a water culture experiment.

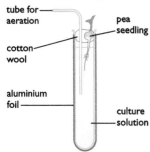

a) Explain the purpose of the aeration tube and the aluminium foil around the test-tube. *(4)*

b) After two weeks, the roots of the pea seedling had grown less than normal, although the leaves were well developed. What mineral ion is likely to be deficient in the culture solution? *(1)*

12. A piece of Canadian pondweed was placed upside down in a test-tube of water, as shown in the diagram. Light from a bench lamp was shone onto the weed, and bubbles of gas appeared at the cut end of the stem. The distance of the lamp from the weed was changed, and the number of bubbles produced per minute was recorded. The results are shown in the table below.

Distance of lamp (cm) (D)	Number of bubbles per minute
5	126
10	89
15	64
20	42
25	31
30	17
35	14
40	10

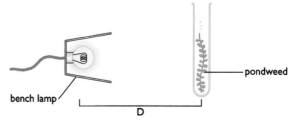

a) Plot a graph of the number of bubbles per minute against the distance of the weed from the lamp. *(6)*

b) Using your graph, predict the number of bubbles per minute that would be produced if the lamp was placed 17 cm from the weed. *(1)*

c) The student who carried out this experiment arrived at the following conclusion:

'The gas made by the weed is oxygen from photosynthesis, so the faster production of bubbles shows that the rate of photosynthesis is greater at higher light intensities.'

Write down three reasons why his conclusion could be criticised (Think about the bubbles, and whether the experiment was properly controlled). *(3)*

13. The diagram shows a cross-section through a leaf of a green plant with some cell details filled in.

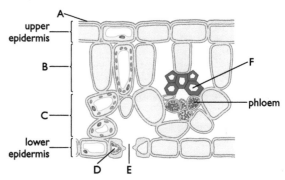

a) Copy the diagram and complete the labels A–F. *(6)*

b) What are the functions of:

i) structure A

ii) structure E

iii) structure F? *(6)*

c) Describe *two* ways in which tissue C differs in its structure from tissue B. *(2)*

14. **a)** Describe how carbohydrates are made by the process of photosynthesis in green plants. Include a summary equation in your answer. *(4)*

b) With the aid of a clearly labelled and annotated diagram, explain how a leaf is adapted to carry out the process of photosynthesis. *(6)*

c) Although the photosynthesis equation shows a sugar as a product, a leaf is not normally tested for sugar in order to show whether photosynthesis has taken place. Suggest a reason for this. *(1)*

d) State *three* ways that photosynthesis is important to animals. *(3)*

Chapter 10: Chemicals of Life

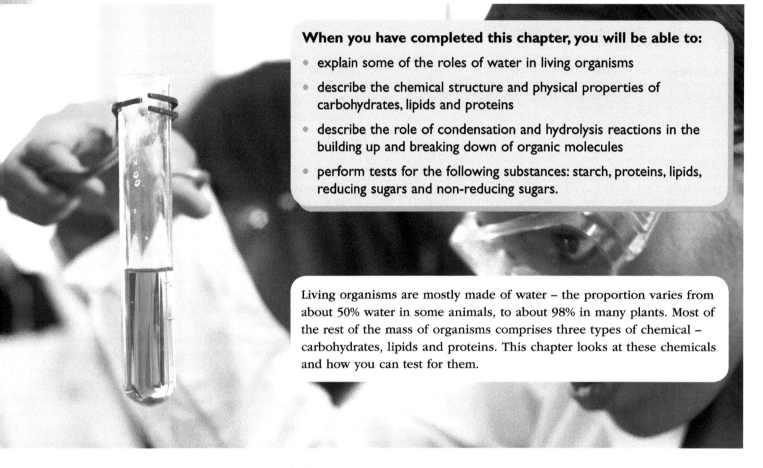

When you have completed this chapter, you will be able to:

- explain some of the roles of water in living organisms
- describe the chemical structure and physical properties of carbohydrates, lipids and proteins
- describe the role of condensation and hydrolysis reactions in the building up and breaking down of organic molecules
- perform tests for the following substances: starch, proteins, lipids, reducing sugars and non-reducing sugars.

Living organisms are mostly made of water – the proportion varies from about 50% water in some animals, to about 98% in many plants. Most of the rest of the mass of organisms comprises three types of chemical – carbohydrates, lipids and proteins. This chapter looks at these chemicals and how you can test for them.

Water

The reactions taking place inside cells are called the cell's **metabolism**. Metabolism is divided into two types of reactions. Some reactions break down large molecules (**catabolism**) while others build up large, complex molecules from simpler substances (**anabolism**). All of these processes can only take place in solution, which is one reason why water is essential to life. In addition, water forms the medium that transports solutes around the body of an organism, whether it is part of the blood plasma in animals, or the contents of the xylem and phloem tubes in plants. Water also forms an essential part of many reactions in cells. For example, it is one of the raw materials needed for photosynthesis, and a product of cell respiration. The mass of an adult human body is between 50% and 65% water.

Men contain more water than women. A man's body is 60 to 65% water, whereas a woman's is only 50 to 60%. This is because women have a higher percentage of body fat than men.

Organic molecules

The chemicals which make up the bodies of animals, plants or other organisms are known as organic molecules, whereas molecules that form part of the surroundings of an organism, such as carbon dioxide in the air or water in the soil, are called inorganic molecules. The human body is largely composed of water, but the solid material of the body is made up from various classes of substance, including carbohydrates, lipids and proteins. You need to know about the chemical nature of these materials, and how you can carry out simple chemical tests for them.

Carbohydrates

Carbohydrates make up less than 1% of the mass of the human body, but they have a very important role. They are the body's main fuel for supplying cells with energy. Cells release this energy by oxidising a sugar called glucose, in the process called cell respiration (see Chapter 13). Sugars are one class of carbohydrates.

Monosaccharides

The chemical formula of glucose is $C_6H_{12}O_6$ (Figure 10.1). Like all carbohydrates, glucose contains only the elements carbon, hydrogen and oxygen. The 'carbo' part of the name refers to carbon, and the 'hydrate' part refers to the fact that the hydrogen and oxygen atoms are in the ratio two to one, as in water (H_2O).

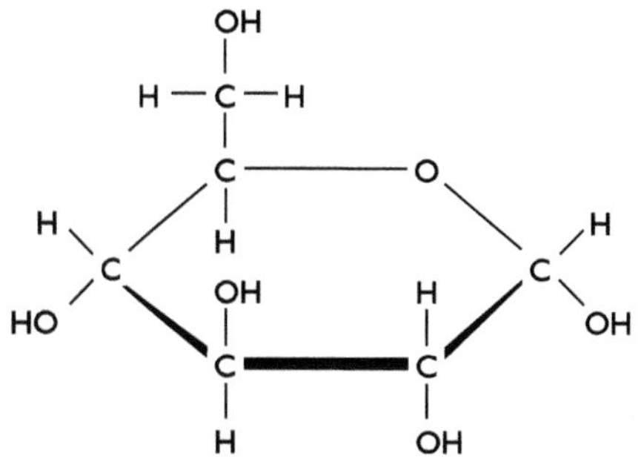

Glucose is found naturally in many sweet-tasting foods, such as fruits and vegetables. Other foods contain different sugars, such as the fruit sugar called **fructose**. Glucose and fructose are both **simple sugars**, each made of a single sugar molecule. We call them **monosaccharides**, where 'mono' means 'one' and a 'saccharide' is a sugar unit.

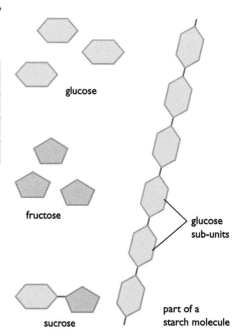

Figure 10.1 *The structure of the glucose molecule.*

Disaccharides

Monosaccharides can join together to form **complex sugars** made up from two sugar units. These are called disaccharides. Ordinary table sugar, the sort some people put in their tea or coffee, is a disaccharide called sucrose, made by linking together glucose with a fructose molecule (Figure 10.2).

Sucrose is the main sugar that is transported through plant stems. This is why we can extract it from the stems of sugarcane, which is a large grass-like plant. Other disaccharides include **maltose**, composed of two glucose units, and **lactose**, the sugar found in milk. Lactose is composed of glucose joined to a monosaccharide called **galactose** (Table 10.1).

Disaccharide	Common name	Component monosaccharides
sucrose	cane sugar	glucose + fructose
maltose	malt sugar	glucose + glucose
lactose	milk sugar	glucose + galactose

Table 10.1: *Some common disaccharides.*

Both monosaccharide and disaccharide sugars have two physical properties that you will probably know – they all taste sweet, and they are all soluble in water.

Polysaccharides

We get most of the carbohydrate in our diet not from sugars, but from starch. **Starch**, along with most other polysaccharides, is a large, insoluble molecule, and does not taste sweet. Because it does not dissolve in water, it is found as a storage carbohydrate in many plants, such as potato, rice, cassava, yam, eddoes and wheat. The staple diets of people from around the world are starchy foods like rice, potatoes, bread and pasta. Starch is made up of long chains of hundreds of glucose molecules joined together. In other words, it is a **polymer** of glucose, called a **polysaccharide** (Figure 10.2).

Figure 10.2 *Glucose and fructose are monosaccharides. A molecule of glucose joined to a molecule of fructose forms the disaccharide called sucrose. Starch is a polysaccharide of many glucose sub-units.*

Starch is only found in plant tissues, but animal cells sometimes contain a very similar carbohydrate called **glycogen**. This is also a polymer of glucose, and is found in tissues such as liver and muscle, where it can be converted into glucose when the body needs a source of energy.

As you will see, large insoluble carbohydrates such as starch and glycogen have to be broken down into simple soluble sugars during digestion, so that they can be absorbed into the blood (Chapter 12).

Another carbohydrate that is a polymer of glucose is cellulose, the material that makes up plant cell walls. We are not able to digest cellulose, because our gut doesn't make the enzyme needed to break down the cellulose molecule. This means that we are not able to use cellulose as a source of energy. However, it still has a vitally important function in our diet. It forms **dietary fibre** or **roughage**, which gives the muscles of the gut something to push against as the food is moved through the intestine. This keeps the gut contents moving, avoiding constipation and helping to prevent serious diseases of the intestine, such as colitis and bowel cancer.

1. What is meant by the term metabolism?
2. Name two monosaccharides.
3. Describe the structures of the disaccharides sucrose and lactose. What are the functions of these sugars?
4. In which cells or tissues would you find the polysaccharides starch, glycogen and cellulose?

Figure 10.3 *These foods are all rich in lipids.*

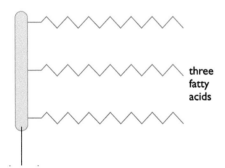

three fatty acids

Figure 10.4 *Lipids are made up of a molecule of glycerol joined to three fatty acids. The many different fatty acids form the variable part of the molecule.*

Lipids

Lipids are **fats** and **oils**. Most lipids are insoluble in water, and act as a store of energy in many animal and plant tissues. They contain the same three elements as carbohydrates – carbon, hydrogen and oxygen, but the proportion of oxygen in a lipid is much lower than in a carbohydrate. For example, beef and lamb both contain a lipid called tristearin, which has the formula $C_{51}H_{98}O_6$. This lipid is a solid at room temperature, but melts if you warm it up, making it a fat. On the other hand, plant lipids are usually liquid at room temperature, and so are called oils. Meat, butter, cheese, milk, eggs and oily fish are all rich in lipids, as well as foods fried in fat or oil, such as potato chips. Plant oils include many types that are used for cooking, such as olive oil, corn oil, coconut oil and soya oil, as well as products made from oils, such as margarine (Figure 10.3).

The chemical building blocks of lipids are two types of molecule called **glycerol** and **fatty acids**. Glycerol is an oily liquid. It is also known as glycerine, and is used in many types of make-up. In a lipid, a molecule of glycerol is joined to three fatty acid molecules (Figure 10.4). There are a large number of different fatty acid molecules, which gives us the many different kinds of lipid found in food. Both glycerol and fatty acids are water-soluble.

Although lipids are an essential part of our diet, too much lipid is unhealthy, especially a type called **saturated** lipids, as well as a compound called **cholesterol**. These substances have been linked to heart disease.

Saturated lipids are more common in food rich in animal fat, such as meat and dairy products. 'Saturated' is a word used in chemistry; it means that the fatty acids of the lipids contain no double bonds. Other lipids are **unsaturated**, which means that their fatty acids contain double bonds. These are more common in plant oils. There is evidence that unsaturated lipids are healthier for us than saturated ones. Cholesterol is a substance that the body gets from food such as eggs and meat, but we also make cholesterol in our livers. It is an essential part of all cells, but too much cholesterol can cause heart disease.

The percentage of lipid in the body can vary by large amounts. A slim, young male might contain only 10% lipid, while an adult overweight female may contain as much as 25% by mass. Conversely, males tend to have a higher protein and water content than women.

Proteins

The average protein content of the body is about 17%. All cells contain protein, so we need it for growth and repair of tissues. Many compounds in the body are made from protein, including enzymes, which are the biological catalysts that are essential for the metabolism of the cell.

Most foods contain some protein, but certain foods such as meat, fish, cheese and eggs are particularly rich in it. You will notice that these foods are animal products. Plant material generally contains less protein, but some foods, especially beans, peas and nuts, are richer in protein than others.

Like starch, proteins are also polymers, but whereas starch is made from a single molecular building block (glucose), proteins are made from combinations of 20 different sub-units called **amino acids**. All amino acids contain four chemical elements – carbon, hydrogen and oxygen (as in carbohydrates and fats) along with nitrogen. Two amino acids also contain sulphur. The amino acids are linked together in long chains (Figure 10.5). The chains are then folded up or twisted into spirals, with cross-links holding them together. In turn, these twists and spirals may be arranged to give the protein molecule a complex three-dimensional structure.

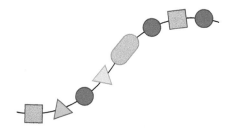

Figure 10.5 *A chain of amino acids forming part of a protein molecule. Each shape represents a different amino acid.*

This *shape* of a protein is very important in allowing it to carry out its function, and the *order* of amino acids in the protein decides its shape. Because there are 20 different amino acids, and they can be arranged in any order, the number of different protein structures that can be made is enormous. As a result, there are thousands of different kinds of proteins in organisms, from structural proteins such as collagen and keratin in skin and nails, to proteins with more specific functions, such as enzymes and haemoglobin.

Proteins are very large molecules. Most are insoluble, so except for a few small soluble proteins, they cannot be transported around the body. However, amino acids are soluble, and can be transported in the blood or in the contents of the phloem of a plant.

5. Briefly describe the structures of a lipid and a protein.
6. Which elements are present in a protein, but not in a carbohydrate or lipid?
7. How many different amino acids make up proteins?
8. Explain three functions of proteins in the body.

The chemical composition and properties of carbohydrates, lipids and proteins are summarised in Table 10.2.

	Solubility in water	Elements they contain	Sub-unit 'building blocks'
Carbohydrates monosaccharides disaccharides starch	soluble soluble insoluble	C, H, O C, H, O C, H, O	– monosaccharides monosaccharides
Lipids	insoluble	C, H, O	fatty acids and glycerol
Proteins	some soluble, some insoluble	C, H, O, N (two amino acids contain S)	amino acids

Table 10.2: *Components of some organic substances.*

Condensation and hydrolysis reactions

When a molecule of glucose joins with another, the disaccharide maltose is formed. The bond that forms between the two monosaccharides results from the loss of a hydrogen atom (H) from one of the glucose molecules and a hydroxyl (–OH) group from the other. These atoms join together, forming water (H_2O). Because water is formed, this is called a **condensation** reaction. More condensation reactions link further glucose units together during the formation of starch (Figure 10.6). When starch is broken down, as happens in digestion, water is *added* to the starch in the opposite way to a condensation reaction. This is called a **hydrolysis** reaction ('hydro' means water and 'lysis' means to split: the starch is 'split' using water).

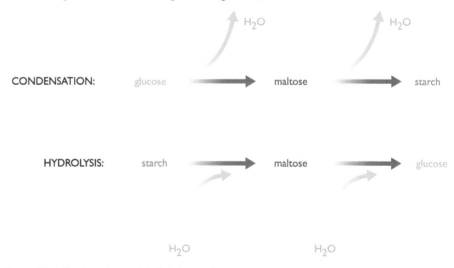

Figure 10.6 *Condensation and hydrolysis reactions.*

All disaccharides and polysaccharides are formed by condensation reactions between their component monosaccharides. Linking together fatty acids and glycerol to form lipids also occurs by condensation reactions, as does the joining together of amino acids to form proteins. When polysaccharides, lipids and proteins are broken down during digestion (Chapter 12), this happens by hydrolysis reactions.

Activity 1: Testing for some organic compounds

SBA skills

| ORR | MM | AI | PD | Dr |

Cross-curricular Links: chemistry, oxidation and reduction

You can carry out simple chemical tests for these organic compounds:

- starch
- reducing sugar
- non-reducing sugar
- lipid
- protein

A: Testing for a starch

You will need:

- iodine solution in a dropping bottle
- some powdered starch or a sample of food that you wish to test
- a white 'spotting' tile

Carry out the following:

1. Place a small amount of the food you wish to test on the white spotting tile.

2. Add a couple of drops of dilute iodine solution to the food.

3. The iodine reacts with the starch, forming a very dark blue, or blue-black colour (Figure 10.7a, page 135). Starch is insoluble, but this test will work on a solid sample of food, such as a potato, or a suspension of starch in water.

B: Testing for a reducing sugar

You will need:

- a Bunsen burner
- a tripod, gauze and heat-proof mat
- a large beaker
- some glucose powder or food you wish to test for a reducing sugar
- Benedict's solution in a dropping bottle
- boiling tubes

Carry out the following:

1. Half fill the beaker with water and heat the water until it starts to boil.

2. Place a small sample of the food, or a small measure of glucose into the boiling tube. Add water to a depth of about 2 cm.

3. Add several drops of Benedict's solution so that the mixture turns blue.

4. Place the tube in the hot water and boil the mixture for a few minutes until there is no further colour change.

5. If a reducing sugar is present, the clear blue solution will gradually change colour, forming a cloudy orange or brick-red precipitate (Figure 10.7b).

Benedict's solution is an alkaline solution of copper(II) sulfate. Glucose is called a **reducing sugar** because it reduces the copper(II) ions in the Benedict's solution to copper(I) oxide, which forms the brick-red precipitate. All other monosaccharides, such as fructose, are reducing sugars, as well as some disaccharides, including maltose and the milk sugar lactose (try the Benedict's test on a sample of milk). Sucrose, the main sugar in ordinary table sugar is a **non-reducing sugar**. If you heat some pure sucrose with Benedict's solution it will stay a clear blue colour. However, if sucrose is first broken down into its component monosaccharides (glucose and fructose), they will then give a positive Benedict's test. This can be done by heating the sucrose with dilute acid.

C: Testing for a non-reducing sugar

You will need:

- a Bunsen burner

- a tripod, gauze and heat-proof mat

- a large beaker

- some sucrose or food you wish to test for a non-reducing sugar

- Benedict's solution in a dropping bottle

- boiling tubes

- some dilute hydrochloric acid in a dropping bottle

- some dilute sodium hydrogencarbonate in a dropping bottle

Carry out the following:

1. Half fill the beaker with water and heat the water until it starts to boil.

2. Place a small amount sucrose or food into a boiling tube. Add water to a depth of 2 cm.

3. Add a few drops of dilute hydrochloric acid and place the tube in the hot water. Boil the mixture for 2 minutes.

4. Remove the boiling tube and allow it to cool to room temperature.

5. Add a few drops of dilute sodium hydrogencarbonate to neutralise the acid (the mixture is neutral when the contents stop fizzing).

6. Carry out the reducing sugar test (steps 3–5 in Activity B, page 133).

D: Testing for a lipid

The simplest way to test for a lipid in a sample of food is to rub the food onto a piece of filter paper. If it contains enough lipid it will leave a translucent (see-through) spot on the paper. To tell if the spot is due to a lipid and not water, allow the paper to dry. A spot caused by water will disappear, but one made by a lipid will remain. However, a better test is the emulsion test. This test depends on the fact that lipids are insoluble in water, but soluble in ethanol.

Note that you must use pure sucrose for this test. Some table sugars may be contaminated with reducing sugar.

Note that the reducing sugar test should be carried out first. If you have a mixture of reducing and non-reducing sugars, the reducing sugar test will be positive, so it will not be possible to carry out the non-reducing sugar test.

Activity 1: Testing for some organic compounds (continued)

You will need:

- clean, dry test-tubes
- some ethanol
- the food you wish to test

Carry out the following:

1. Place a small sample of food in a test-tube (e.g. add a drop of cooking oil using a pipette).

2. Add some ethanol to a depth of 2 cm.

3. Shake the contents of the tube to dissolve the lipid in the ethanol.

4. Fill a second test-tube three-quarters full with water.

5. Carefully pour the contents of the first tube into the water.

6. If lipid is present, a white, cloudy layer forms on the top of the water (Figure 10.7d).

The white layer is caused by the ethanol dissolving in the water and leaving the fat behind as a suspension of tiny droplets, called an **emulsion**.

E: Testing for a protein

You will need:

- test-tubes
- dilute (5%) potassium (or sodium) hydroxide solution in a dropping bottle
- 1% copper sulfate solution in a dropping bottle

Carry out the following:

1. Add a small amount of the food (powdered egg albumin works well) to 2 cm depth of water in a test-tube. Shake to mix the powder with the water.

2. Add an equal volume of dilute (5%) potassium hydroxide solution and mix.

3. Add three drops of 1% copper sulfate solution. (Note that these two solutions may be added together as a mixture called Biuret reagent.)

4. A mauve (purple) colour develops (Figure 10.7c).

This test is called the Biuret test, which is the name of the mauve compound that is formed.

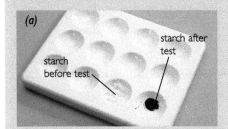

(a)

starch before test

starch after test

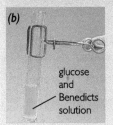
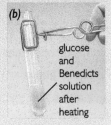

(b) glucose and Benedicts solution

(b) glucose and Benedicts solution after heating

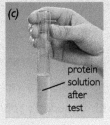

(c) protein solution before test

(c) protein solution after test

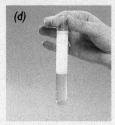

(d)

Figure 10.7 *Results of tests for a (a) starch, (b) reducing sugar, (c) protein and (d) lipid.*

---

9. Explain the difference between a reducing sugar and a non-reducing sugar.

10. When sucrose is boiled with dilute hydrochloric acid, what two monosaccharides are formed?

11. What is an emulsion?

Chapter summary

In this chapter you have learnt that:

- water acts as a solvent in cells and in transport systems. It is a reactant or product in many metabolic reactions
- organic molecules, which make up the material of living organisms, include carbohydrates, lipids and proteins
- simple sugars are called monosaccharides
- simple sugars can join with other monosaccharides by condensation reactions to form sugars called disaccharides
- carbohydrates made of many sugar units are called polysaccharides
- monosaccharides such as glucose, and disaccharides such as maltose, are small, soluble molecules
- all sugars are sweet tasting
- the polysaccharide starch is a polymer of glucose. It is a large, insoluble molecule and does not taste sweet
- other polysaccharides include glycogen, which is stored in some animal cells, and cellulose, which is used to make plant cell walls
- lipids are fats and oils. Fats are solid, and oils are liquid at room temperature
- lipids are composed of sub-units called fatty acids and glycerol, which join together through condensation reactions
- proteins are composed of long chains of amino acids, again linked as a result of condensation reactions
- there are 20 different amino acids, giving the possibility for almost infinite variation in protein structure and function
- the test for starch is to add iodine solution, which produces a blue-black colour
- the test for reducing sugars is to heat with Benedict's solution, which gives a brick-red precipitate
- non-reducing sugars will not give a positive Benedict's test until they have been broken down into reducing sugars by heating with dilute acid
- one test for lipids is that they leave a translucent spot when rubbed on filter paper. Another test is that they form a white emulsion when dissolved in ethanol and poured into water
- the test for proteins is to add Biuret reagent, which gives a mauve colour.

Questions

1. Maltose is a:

   A   monosaccharide reducing sugar
   B   disaccharide reducing sugar
   C   monosaccharide non-reducing sugar
   D   disaccharide non-reducing sugar

2. Which of the following would give a positive test when boiled with Benedict's solution?

   A   fructose
   B   lipid
   C   starch
   D   protein

3. Which of the following organic molecules contains carbon, hydrogen, oxygen, nitrogen and sulphur?

   A   cellulose
   B   saturated fats
   C   proteins
   D   glycogen

4. A student carried out four tests to investigate the composition of onion, biscuit, potato and banana. The results are shown in this table, a tick indicating a positive result and a cross a negative one.

Test	Onion	Biscuit	Potato	Banana
add iodine solution	✗	✔	✔	✔
boil with Benedict's solution	✔	✗	✗	✔
ethanol emulsion test	✗	✔	✗	✗
boil with acid then carry out Benedict's test	✔	✔	✔	✔

   Which food contained only a non-reducing sugar, starch and lipid?

   A   onion
   B   biscuit
   C   potato
   D   banana

5. Which substance makes up less than 1% of the body mass of a healthy person?

   A   water
   B   protein
   C   lipid
   D   carbohydrate

6. a) Copy and complete the following table.

Organic substance	Solubility in water	Sub-units	Elements present
glucose		(none)	
starch			C, H, O
fat			–
protein	some soluble, some insoluble	amino acids	

   (4)

b) Describe how you would test some powdered milk to show it contains protein. (4)

c) There are two tins containing different brands of powdered milk. One brand of milk contains 10% protein and the other 20%. The tins have lost their labels. Suggest how you could modify the test you have described in part (b) in order to find out which is which. How would you make sure it was a controlled experiment? (4)

7. Amylase is an enzyme that breaks down starch into maltose (see Chapter 12). A student set up an experiment as illustrated in the diagram below.

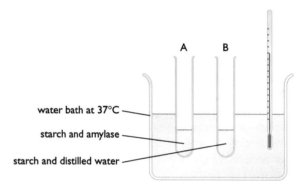

water bath at 37°C
starch and amylase
starch and distilled water

She took a suspension of starch in water and placed the same volume into each of two test-tubes, which she labelled A and B.

To tube A, she added a solution of amylase. To B she added the same volume of distilled water. She then placed each tube in a water bath at 37°C. After a period of 20 minutes, she took a small sample of the mixture from each tube, and tested the samples for starch, a reducing sugar and protein.

a) Suggest why the tubes were kept at 37°C. (2)

b) Describe the tests for:

   i)   starch

   ii) a reducing sugar. (4)

c) For each of the three food tests, say what you would expect the results to show in tubes A and B. Explain your answers. (6)

d) What is the name given to tube B? What is the point of this part of the experiment? (2)

8. a) Explain the meaning of each of the following terms:

   i)   inorganic substance

   ii) complex sugar

   iii) hydrolysis reaction. (3)

b) A lipid is formed by condensation reactions between glycerol and three fatty acid molecules. Construct a simple diagram to show how this happens. (3)

c) Explain the difference between a saturated fat and an unsaturated fat. (2)

Chapter 11: Enzymes

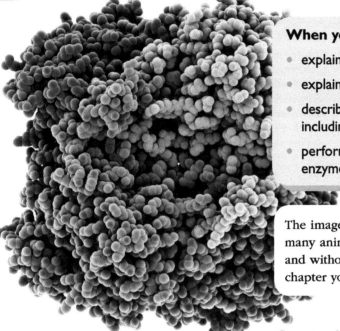

When you have completed this chapter, you will be able to:

- explain the role and importance of enzymes as biological catalysts
- explain how enzymes catalyse reactions
- describe the effects of different factors on the activity of enzymes, including: temperature, pH and concentration of enzyme and substrate
- perform experiments to show that these factors affect the rate of an enzyme-catalysed reaction.

The image on the left shows a model of catalase, an enzyme common in many animal and plant tissues. There are thousands of other enzymes, and without them reactions in cells would not be able to proceed. In this chapter you will look at the roles of enzymes, and their properties.

Cross-curricular links:
chemistry, rate of reaction

Controlling reactions in the cell

The chemical reactions that go on in a cell are controlled by a group of proteins called enzymes. Enzymes are *biological catalysts*. A catalyst is a chemical which speeds up a reaction without being used up. It takes part in the reaction, but afterwards is unchanged and free to catalyse more reactions. Cells contain hundreds of different enzymes, each catalysing a different reaction. This is how the activities of a cell are controlled – the nucleus contains the genes, which control the production of enzymes, which catalyse reactions in the cytoplasm:

genes → proteins (enzymes) → catalyse reactions

Everything a cell does depends on which enzymes it can make, which in turn depends on which genes in its nucleus are working.

What hasn't been mentioned is why enzymes are needed at all. This is because the temperatures inside organisms are low (e.g. the human body temperature is about 37°C), and without catalysts, most of the reactions that happen in cells would be far too slow to allow life to go on. Only when enzymes are present to speed them up do the reactions take place quickly enough.

You have probably heard of the enzymes involved in digestion of food. They are secreted by the intestine onto the food to break it down. They are called **extracellular** enzymes, which means 'outside cells'. However, most enzymes stay *inside* cells – they are **intracellular**. You will read about digestive enzymes in Chapter 12.

It is possible for there to be thousands of different sorts of enzymes because they are made of proteins, and protein molecules have an enormous range of structures and shapes. The molecule that an enzyme acts on is called its **substrate**. Each enzyme has a small area on its surface called the **active site**. The substrate attaches to the active site of the enzyme. The reaction then takes place and products are formed. When the substrate joins up with the active site, it lowers the energy needed for the reaction to happen, allowing the products to be formed more easily.

The substrate fits into the active site of the enzyme rather like a key fitting into a lock. That is why this is called the 'lock and key' model of enzyme action (Figure 11.1), page 139.

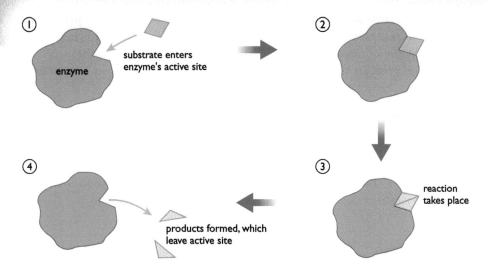

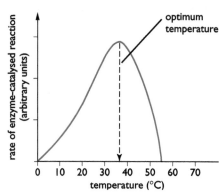

Figure 11.2 *Effect of temperature on the action of an enzyme.*

Figure 11.1 *Enzymes catalyse reactions at their active site. This acts like a 'lock' to the substrate 'key'. The substrate fits into the active site, and products are formed. This happens more easily than without the enzyme — so enzymes act as catalysts.*

Notice how, after it has catalysed the reaction once, the enzyme is free to act on more substrate molecules.

1. What is an enzyme?
2. Why do cells need enzymes?
3. Why do cells need so many different enzymes?
4. Explain the role of the active site in the action of an enzyme.

Factors affecting enzymes

Temperature affects the action of enzymes. This is easiest to see as a graph, in which we plot the rate of the reaction controlled by an enzyme against the temperature (Figure 11.2).

Enzymes in the human body have evolved to work best at about body temperature (37°C). The graph shows this, because the peak on the curve happens at about this temperature. In this case, 37°C is called the **optimum temperature** for the enzyme.

> Optimum temperature means the best temperature, in other words the temperature at which the reaction takes place most rapidly.

As the enzyme is heated up to the optimum temperature, the increasing temperature speeds up the rate of reaction. This is because higher temperatures give the molecules of the enzyme and substrate more energy, so they collide more often. More collisions mean that the reactions will take place more frequently. However, above the optimum temperature, another factor comes into play. Enzymes are made of protein, and proteins are broken down by heat. From 40°C upwards, the heat destroys the enzyme. We say that it is **denatured**. You can see the effect of denaturing when you boil an egg. The egg white is made of protein, and turns from a clear runny liquid into a white solid as the heat denatures the protein.

Temperature is not the only factor which affects the activity of enzymes. The rate of an enzyme-catalysed reaction also depends on the pH of the surroundings. The pH inside cells is around neutral (7) and, not surprisingly, most enzymes have evolved to work best at this pH. At extremes of pH either

> Not all enzymes have an optimum temperature near 37°C, just those of animals such as mammals and birds, which all have body temperatures close to this value. Enzymes have evolved to work best at the normal body temperature of the organism. Bacteria that always live at an average temperature of 10°C will probably have enzymes with an optimum temperature of 10°C.

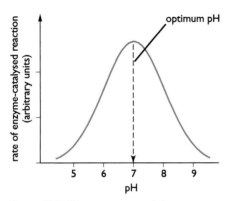

Figure 11.3 *Most enzymes work best at a neutral pH.*

Although most enzymes work best at a neutral pH, a few have an optimum below or above pH 7. The stomach produces hydrochloric acid, which makes its contents very acidic (see Chapter 12). Most enzymes stop working at a low pH like this, but the stomach makes an enzyme called pepsin, which has an optimum pH of about 2, so that it is adapted to work well in these unusually acidic surroundings.

Did you know?
The most abundant enzyme on the Earth is called ribulose bisphosphate carboxylase. It is an enzyme found in green plants, where it catalyses an important step in photosynthesis. The total mass of this enzyme on the planet is about 70 million tonnes.

side of neutral, their activity decreases, as shown by Figure 11.3. The pH at which the enzyme works best is called the **optimum pH** for that enzyme. Either side of the optimum, the pH affects the structure of the enzyme molecule, and changes the shape of its active site, so that the substrate will not fit as well into the active site.

The rate of most enzyme-controlled reactions depends on the concentration of the substrate. The rate increases as the substrate concentration increases (Figure 11.4), as more substrate molecules collide with the active sites of the enzyme molecules. Eventually though, the substrate concentration will become so high that all the active sites are working as quickly as possible. When this happens, only increasing the concentration of the enzyme will increase the rate.

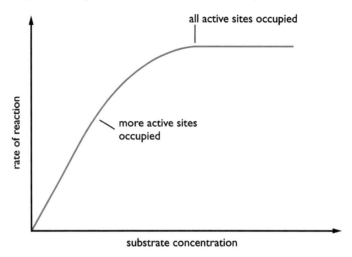

Figure 11.4 *The rate of an enzyme-controlled reaction increases as the substrate concentration increases, until all the active sites are occupied.*

5. Explain why an enzyme has an optimum temperature.
6. Why does pH affect enzyme activity?

Naming enzymes

There are thousands of different enzymes. Early scientists did not use a systematic way of naming enzymes, and many are still known by their old names, such as pepsin and catalase. These days, they are generally named after the substrate that they act upon. For example amylase breaks down amylose (a type of starch) and maltase digests the sugar maltose.

Digestive enzymes are classified into groups according to the type of food molecule that they break down (Table 11.1).

Digestive enzymes	Substrates
carbohydrases	carbohydrates
lipases	lipids
proteases	proteins

Table 11.1: *Digestive enzymes and their substrates.*

Activity 1: The effect of temperature on the activity of amylase

SBA skills

| ORR | MM | AI | PD | Dr |

The digestive enzyme amylase breaks down starch into the sugar maltose. If you record the speed at which the starch disappears, you have a measure of the activity of the amylase.

You will need:

* iodine solution
* a white spotting tile
* a suspension of starch
* amylase solution
* two 5-cm³ syringes or pipettes
* a stopwatch or clock
* a large beaker
* a Bunsen burner
* a tripod, gauze and heat-proof mat
* two boiling tubes
* a dropping pipette
* a thermometer

Carry out the following:

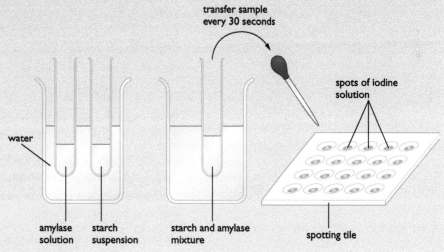

Figure 11.5 *Steps 1–6.*

1. Place 20 spots of iodine solution onto the depressions in a spotting tile.

2. Use a syringe or pipette to place 5 cm³ of starch suspension in one boiling tube.

3. Place 5 cm³ of amylase solution into another tube using a second syringe.

4. Half fill a beaker with water at room temperature. Stand both boiling tubes in the water for 5 minutes and record the temperature.

5. Pour the amylase into the starch suspension, leaving the tube containing the mixture in the water bath.

6. Immediately take a small sample of the mixture with the pipette and add it to the first drop of iodine on the spotting tile.

7. Record the colour of the iodine.

8. Take a sample of the mixture every 30 seconds for 10 minutes, testing the mixture for starch as in steps 6 and 7.

9. Repeat the experiment, maintaining the water bath at a different temperature. You might like to try a range of temperatures between 20°C and 60°C. You will find it easier if you record your results in a table like the one below.

| Time (min) | Colour of mixture at different temperatures | | | | |
	20°C	30°C	40°C	50°C	60°C
0.5					
1.0					
1.5					
2.0					
(etc)					
10.0					

10. Suppose the mixture took 2 minutes until the starch was all broken down by the enzyme, so that the iodine stopped changing colour. You can calculate the rate of the reaction by dividing the volume of the starch by the time (2 min):

time for 5 cm³ to be used up = 2 min

$$\text{rate} = \frac{5}{2} = 2.5 \text{ cm}^3 \text{ min}^{-1}$$

If you find the rate of the reaction at different temperatures, you can plot a graph of rate against temperature.

11. Use the figures from your results table to construct a graph to show the average rate of reaction at different temperatures. Explain the shape of your graph. Is the curve as you would have expected from Figure 11.2? If not, can you explain why?

12. How could you improve the experiment to get more reliable results?

Activity 2: The effect of pH on the activity of amylase

SBA skills

ORR	MM	AI	**PD**	Dr

Cross-curricular links: chemistry, acids and bases

You can modify the previous experiment (Activity 1) to investigate the effect of pH on the activity of amylase. You will need to keep the pH constant. This can be done by adding 5 cm³ of a buffer solution to the starch before adding the enzyme. Buffer solutions resist changes in pH, and can be obtained in a range of pH values.

Design an experiment to find out how pH values between pH 4 and pH 9 affect the activity of amylase.

Your experiment must be a properly controlled one, in other words you must changes *only* the pH. Explain:

• how you will change the pH and keep other conditions constant

• how you will measure the rate of the reaction

• how you will ensure that your results are reliable.

If your plan is a good one, your teacher may allow you to proceed with the experiment.

Activity 3: The effect of substrate concentration on the activity of the enzyme catalase

SBA skills

ORR	**MM**	**AI**	PD	Dr

Many cells produce a waste product called hydrogen peroxide (H_2O_2). This substance is poisonous and needs to be broken down. An enzyme called catalase breaks the hydrogen peroxide down into water and oxygen gas, both of which are harmless to the cell:

$$2H_2O_2 \rightarrow 2H_2O + O_2$$

hydrogen peroxide   water   oxygen

There is a very high concentration of catalase in a number of animal and plant tissues, including potato. If you add hydrogen peroxide to a piece of potato, 'fizzing' will occur, as oxygen gas is produced. The rate of production of the gas can be used as a way of measuring the activity of catalase. The experiment works well if the potato is homogenised by putting it in a blender with an equal volume of water (if a blender is not available, thin slices or small cubes of potato will work instead).

You will need:

• a potato

• a blender

• a large beaker

• a chopping board and knife

• two 5–cm³ syringes

• a boiling tube, bung and delivery tube (Figure 11.6, page 144)

• 5% hydrogen peroxide solution

• a stopwatch or clock

Carry out the following:

1. Cut the skin off a medium-sized potato. Chop the potato into small pieces and then homogenise the tissue with an equal volume of distilled water in a blender.

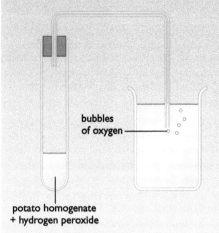

bubbles of oxygen

potato homogenate + hydrogen peroxide

Figure 11.6 *Measuring the rate of breakdown of hydrogen peroxide.*

2. Allow the solid potato debris to settle and remove the liquid layer above the debris. This liquid layer is the homogenate. It should contain catalase from the broken up potato cells.

3. Place 5 cm³ of the potato homogenate in a boiling tube with a delivery tube attached.

4. Remove the bung and delivery tube, and add 5 cm³ of 5% hydrogen peroxide solution to the potato homogenate. Quickly replace the bung and put the end of the delivery tube under water as shown in Figure 11.6.

5. Count the number of bubbles of oxygen gas produced in the first minute after adding the hydrogen peroxide.

6. Repeat the experiment using different concentrations of hydrogen peroxide of between 0.5 and 5%.

7. Plot a graph of the number of bubbles per minute against the concentration of hydrogen peroxide. Write up your experiment and explain your results.

Activity 4: Investigations involving catalase activity

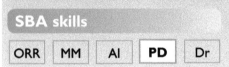

SBA skills

ORR MM AI **PD** Dr

You could use the method above (Activity 3) to answer further questions about catalase activity, for example:

- Do carrots contain as much catalase as potato?

- Do different potatoes contain different concentrations of catalase?

- Does an old, sprouting potato contain more catalase than a younger potato?

- Do different parts of a potato contain different levels of catalase?

Choose one of these questions and make a suggestion (hypothesis) that you could test. Design an experiment to test your hypothesis. Make sure that your experiment is properly controlled, and state how you will ensure that the results are reliable. You may be able to carry out the experiment if your teacher is happy with your design.

Activity 5: Finding out whether an enzyme is involved in the browning of yam

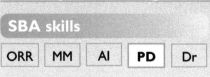

SBA skills

ORR MM AI PD Dr

A market lady noticed that freshly cut yam turned brown after exposure to air. However, when it was rubbed with lime or lemon juice, it did not change colour. Design an experiment to find out if an enzyme is involved in producing this colour change.

(Notice that this is a more difficult, open-ended problem. You might want to start by carrying out a preliminary experiment to find out whether the market lady's observations were correct. Then, if an enzyme *were* involved, think about which factors you could change that would affect its activity.)

Chapter summary

In this chapter you have learnt that:

- enzymes are biological catalysts. They increase the rate of reactions in cells. Without enzymes, the reactions would be too slow to sustain life

- the enzymes that a cell can make are determined by its genes

- enzymes are proteins. They catalyse specific reactions by having an active site where the substrate fits in a 'lock and key' method

- enzymes are affected by temperature. Up to an optimum temperature, raising the temperature increases the rate of reaction. Above the optimum, high temperatures decrease the rate of reaction due to denaturing of the enzyme

- enzymes also have an optimum pH at which they work best

- increasing substrate or enzyme concentrations can affect the rate of an enzyme-controlled reaction

- the rate of an enzyme-catalysed reaction can be measured by finding the rate at which a product is formed (e.g. oxygen by the action of catalase) or the rate at which a substrate is used up (e.g. starch by the action of amylase).

Questions

1. Which of the following statements about the characteristics of enzymes is/are true?

   I   they only work at body temperature
   II they are only made in the gut
   III they only act on specific substrates
   IV they only work over a certain range of pH

   A   I and II
   B   III and IV
   C   I and III
   D   I, II and IV

2. Which of the following enzymes digests starch?

   A   catalase
   B   lipase
   C   starchase
   D   amylase

3. The graph shows the activities (rates of reaction) of four enzymes at different pH values. Which enzyme is most likely to be pepsin?

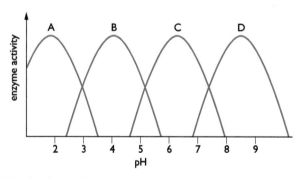

4. Which of the following is true of the 'lock and key' model of enzyme action?

   A   The enzyme is the 'key' to the substrate's 'lock'.
   B   The reaction can only take place in the active site.
   C   When the substrate joins with the active site, it reduces the energy needed to start the reaction.
   D   The enzyme is unable to carry out further reactions after the substrate leaves the active site.

5. Which of the following graphs best shows the effect of substrate concentration on the rate of an enzyme-controlled reaction?

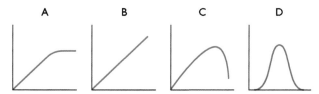

A   B   C   D

6. The graph shows the effect of temperature on an enzyme. The enzyme was extracted from a micro-organism that lives in hot mineral springs near a volcano.

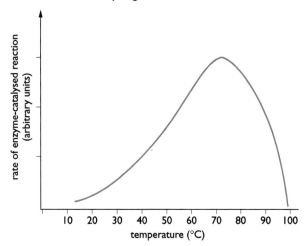

a) What is the optimum temperature of this enzyme? *(1)*

b) Explain why the activity of the enzyme is greater at 60°C than at 30°C. *(2)*

c) The optimum temperature of enzymes in the human body is about 37°C. Explain why this enzyme is different. *(2)*

d) What happens to the enzyme at 90°C? *(2)*

7. Catalase is an enzyme found in many plant and animal cells. It catalyses the breakdown of hydrogen peroxide into water and oxygen.

$$\text{hydrogen peroxide} \xrightarrow{\text{catalase}} \text{water} + \text{oxygen}$$

a) In an investigation into the action of catalase in potato, 20 g potato tissue was put into a small beaker containing hydrogen peroxide weighing 80 g in total. The temperature was maintained at 20°C throughout the investigation. As soon as the potato was added, the mass of the beaker and its contents was recorded until there was no further change in mass. The results are shown in the graph.

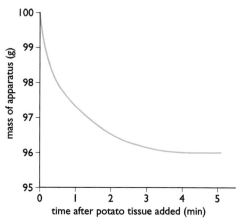

i) How much oxygen was formed in this investigation? Explain your answer. *(2)*

ii) Estimate the time by which half this mass of oxygen had been formed. *(2)*

iii) Explain, in terms of collisions between enzyme and substrate molecules, why the rate of reaction changes during the course of the investigation. *(2)*

b) The students repeated the investigation at 30°C. What difference, if any, would you expect in:

i) the mass of oxygen formed

ii) the time taken to form this mass of oxygen?

Explain your answers. *(4)*

8. The results in the table were obtained by incubating samples of starch suspension with amylase at different temperatures.

Temperature (°C)	20	30	39	53	62	68
Amount of maltose (arbitrary units)	0.8	2.3	3.5	3.3	1.6	0.3

a) Using graph paper, plot a graph of the results shown in the table. *(5)*

b) Suggest the hypothesis that this experiment was designed to test. *(2)*

c) Explain the changes observed in your graph. *(4)*

d) Estimate the optimum temperature for amylase activity from your graph. *(1)*

e) State *one* factor (other than temperature) that could affect the amount of maltose produced. *(1)*

f) If the enzyme pepsin were used instead of amylase, would you expect results similar to those in the table? Explain your answer. *(2)*

9. a) What is an enzyme? *(3)*

b) By means of a labelled and annotated diagram, describe the 'lock and key' model of enzyme action. *(4)*

c) Describe, with the aid of simple sketch graphs, how i) temperature and ii) pH affect the activity of a typical enzyme. *(4)*

d) The enzyme **urease** breaks down urea into carbon dioxide gas and ammonia. Ammonia is an alkali that will change the colour of a pH indicator such as universal indicator solution. You are provided with a solution of urea, a solution of urease enzyme and some universal indicator solution. Suggest how you could carry out an experiment to find out the effect of temperature on the activity of urease. *(6)*

When you have completed this chapter, you will be able to:

- discuss the meaning of a balanced diet and how malnutrition can lead to deficiency diseases, obesity and other problems
- recall the role of minerals and vitamins in a balanced diet
- discuss the advantages and disadvantages of a vegetarian diet in humans
- discuss the effects of age, sex and occupation on dietary needs, including energy requirements
- recall the method of measuring energy content of a sample of food using a calorimeter
- investigate the energy content of foods by a simple calorimetric method
- describe the structure and function of the different regions of the human alimentary canal, including:
  - mastication and the role of the teeth
  - the internal structure of a tooth
  - peristalsis
  - the roles and importance of digestive enzymes
  - the processes of digestion, absorption and egestion
- describe what happens to the products of digestion after their absorption, including transport to the liver, assimilation and the fate of products that are in excess of body requirements
- construct a Visking tubing model of the ileum.

In many ways it is true to say that 'we are what we eat'. This statement means that the human body can only grow and maintain itself in a healthy condition if it is supplied with the right nutrients. This chapter looks at the components of a balanced diet, what happens to food in the body, and some of the health issues that occur when the body doesn't receive the right food.

Humans are heterotrophs

The difference between autotrophic and heterotrophic nutrition was dealt with in Chapter 9. All animals, including humans, carry out heterotrophic nutrition. This means that they feed on organic materials that have been made by other organisms. We need food for three main reasons:

- to supply us with a 'fuel' for energy
- to provide materials for growth and repair of tissues
- to help fight disease and keep our bodies healthy.

A balanced diet

The food that we eat is called our **diet**. No matter what you like to eat, if your body is to work properly and stay healthy, your diet must include five groups of food substance. They are **carbohydrates**, **lipids**, **proteins**, **minerals** and **vitamins**, as well as **water** and **fibre**. Food should provide you with all of these substances, but they must also be present in the right amounts. A diet that provides enough of these substances and in the correct proportions to keep you healthy is called a **balanced diet** (Figure 12.1).

Figure 12.1 *A balanced diet contains all the types of food the body needs, in just the right amounts.*

Malnutrition

Malnutrition is often taken to mean the same as under-nourishment. A person who does not receive enough food does show malnutrition, but so does a person who eats too much, or eats the wrong proportions of the different kinds of food. Malnutrition just means 'bad diet'.

People whose weight is more than 20% greater than is average for their age, height and sex are said to suffer from **obesity**. Obesity is caused by a person eating more food than is needed for their level of activity (see Table 12.4, page 154) and is often made worse through lack of exercise. Obesity is quite common in the Caribbean. The excess food that is eaten is stored as fat around many internal organs, and under the skin. Obesity is a problem in itself, the extra weight putting a strain on the heart. It is also linked to a number of other serious health problems, such as **coronary heart disease** (see Chapter 23) and **diabetes**.

Protein energy malnutrition (PEM) is a condition which affects millions of children throughout the world, particularly in poorer countries of Africa and Asia. It is most often seen in small children after they stop breast-feeding, and results in stunted growth, tiredness and poor development. It is caused by a general lack of nutrients, resulting in muscle proteins having to be used as a source of energy, rather than for growth and repair.

Deficiency diseases

Lack of a particular nutrient from the diet can result in a **deficiency disease**. We do not need much protein in our diet to stay healthy. Doctors recommend a maximum daily intake of about 70 grams. In developed countries, people often eat far more protein than they need, whereas in many poorer countries, a protein-deficiency disease called **kwashiorkor** is common (Figure 12.2).

> 1. What are the components of a balanced diet?
> 2. What is a deficiency disease?

Two other components of food: minerals and vitamins

Chapter 10 dealt with the major components of the bodies of organisms and consequently the food of animals: carbohydrates, lipids, and proteins, as well as water and fibre. There are two other classes of food substance that form a small part of our diet, but are extremely important to the normal functioning and health of the body. These are minerals and vitamins.

Diabetes is a disease where a person does not make enough of the hormone insulin. Insulin controls blood glucose levels (see Chapter 23). More serious forms of diabetes can only be treated by injections of insulin, without which the patient would die. However, some forms of the disease can be controlled by careful monitoring of the patient's diet and reducing their carbohydrate and lipid intake.

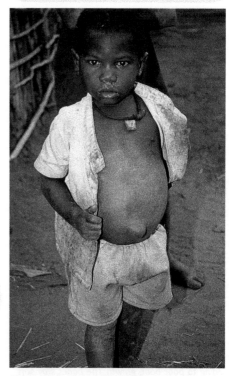

Figure 12.2 *This child is suffering from a lack of protein in his diet, a disease called kwashiorkor. His swollen belly is not due to a full stomach, but is caused by fluid collecting in the tissues. Other symptoms include loss of weight, poor muscle growth, general weakness and flaky skin.*

Minerals

Carbohydrates, lipids and proteins are made from just five chemical elements: carbon, hydrogen, oxygen, nitrogen and sulphur. Our bodies contain many other elements which we get from our food. Some are present in large amounts in the body, for example calcium, which is used for making teeth and bones. Others are present in much smaller amounts, but still have essential jobs to do. For instance, our bodies contain about 3 grams of iron, but without it our blood would not be able to carry oxygen. Table 12.1 shows just a few of these elements and the reasons they are needed. They are called **minerals** or **mineral elements**.

Most people add salt to food during cooking. However, what is not widely recognised is how much additional salt is present in processed and preserved foods such as canned meats and vegetables. The sodium ions from salt have been shown to be a contributory factor to high blood pressure (hypertension), which can be fatal. Sufferers from this condition are advised to go on a salt-free diet. In fact, we should all eat a salt-free diet, since there is enough salt already present in unsalted foods to supply our needs.

Mineral	Approximate mass in an adult body	Location or role in body	Examples of foods rich in mineral
calcium	1 000 g	making teeth and bones	dairy products, fish, bread, vegetables
phosphorus	650 g	making teeth and bones; part of many chemicals, e.g. DNA	most foods
sodium	100 g	in body fluids, e.g. blood	common salt, most foods
chlorine	100 g	in body fluids, e.g. blood	common salt, most foods
magnesium	30 g	making bones; found inside cells	green vegetables
iron	3 g	part of haemoglobin in red blood cells, helps carry oxygen	red meat, liver, eggs, some vegetables, e.g. spinach

Table 12.1: *Some examples of minerals needed by the body.*

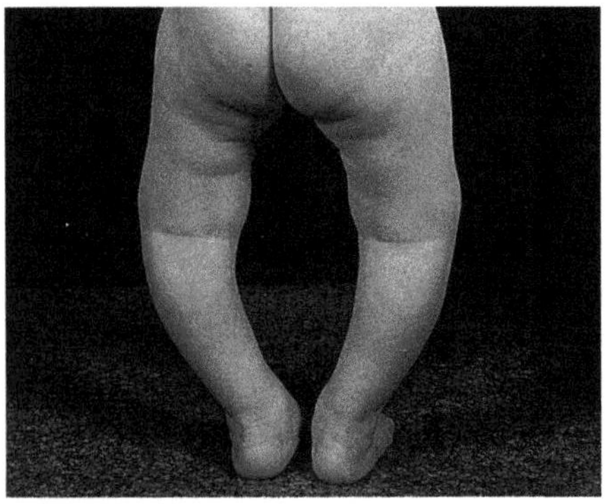

If a person doesn't get enough of a mineral from his or her diet, that person will show the symptoms of a **mineral deficiency disease**. For example, a one-year-old child needs to consume about 0.6 g (600 mg) of calcium every day, to make the bones grow properly and harden. Anything less than this over a prolonged period could result in poor bone development. The bones will become deformed, resulting in a disease called **rickets** (Figure 12.3). Rickets can also be caused by lack of vitamin D in the diet (see below).

Similarly, 16-year-olds need about 12 mg of iron in their daily food intake. If they don't get this amount, they can't make enough haemoglobin for their red blood cells (see Chapter 14). This causes a condition called **anaemia**. People who are anaemic become tired and lack energy, because their blood doesn't carry enough oxygen.

Figure 12.3 *The legs of this child show the symptoms of rickets. This is due to lack of calcium or a lack of vitamin D in the diet, leading to poor bone growth. The bones stay soft and can't support the weight of the body, so they become deformed.*

Vitamins

During the early part of the 20th century, experiments were carried out that identified another class of food substances. When young laboratory rats were fed a diet of pure carbohydrate, fat and protein, they all became ill and died. If they were fed on the same pure foods with a little added milk, they grew normally. The milk contained chemicals that the rats needed in small amounts to stay healthy. These chemicals are called **vitamins**. The results of one of these experiments are shown in Figure 12.4.

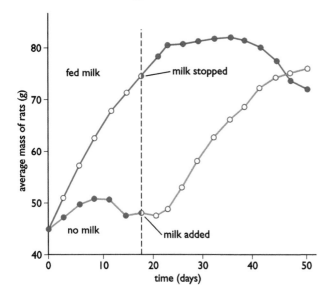

Figure 12.4 *Rats were fed on a diet of pure carbohydrate, fat and protein, with and without added milk. Vitamins in the milk had a dramatic effect on their growth.*

At first, the chemical nature of vitamins was not known, and they were given letters to distinguish between them, such as vitamin A, vitamin B, and so on. Each was identified by the effect a lack of the vitamin, or **vitamin deficiency**, had on the body. For example, **vitamin D** is needed for growing bones to take up calcium salts. A deficiency of this vitamin can result in rickets (Figure 12.3), just as a lack of calcium can.

We now know the chemical structure of the vitamins and the exact ways in which they work in the body. As with vitamin D, each has a particular function. **Vitamin A** is needed to make a light-sensitive chemical in the retina of the eye (see Chapter 18). A lack of this vitamin causes **night blindness**, where a person finds it difficult to see in dim light. **Vitamin C** is needed to make fibres of a material called connective tissue. This acts as a 'glue', bonding cells together in a tissue. It is found in the walls of blood vessels, and in the skin and lining surfaces of the body. Vitamin C deficiency leads to a disease called **scurvy**, where wounds fail to heal, and bleeding occurs in various places in the body. This is especially noticeable in the gums.

Vitamin B is not a single substance, but a collection of many different substances called the vitamin B group. It includes **vitamins B1 (thiamine)**, **B2 (riboflavin)** and **B3 (niacin)**. These compounds have roles in helping with the process of cell respiration. A different deficiency disease is produced if any of them is lacking from the diet. For example, a lack of vitamin B1 results in weakening of the muscles and paralysis, a disease called **beri-beri**.

The main vitamins, their role in the body and some foods which are good sources of each are summarised in Table 12.2, page 152.

Notice that the amounts of vitamins that we need are very small, but we cannot stay healthy without them.

Did you know?
The cure for scurvy was discovered as long ago as 1753. Sailors on long voyages often got scurvy because they ate very little fresh fruit and vegetables (the main source of vitamin C). A ship's doctor called James Lind wrote an account of how the disease could quickly be cured by eating fresh oranges and lemons. The famous explorer Captain Cook, on his world voyages in 1772 and 1775, kept his sailors healthy by making sure that they ate fresh fruit.

Vitamin	Recommended daily amount in diet[1]	Use in the body	Effect of deficiency	Some foods that are a good source of the vitamin
A	0.8 mg	making a chemical in the retina; also protects the surface of the eye	night blindness, damaged cornea of eye	fish liver oils, liver, butter, margarine, carrots
B1	1.1 mg	helps with cell respiration	beri-beri	yeast extract, cereals
B2	1.0 mg	helps with cell respiration	poor growth, dry skin	green vegetables, eggs, fish
B3	16 mg	helps with cell respiration	pellagra (dry red skin, poor growth, and digestive disorders)	liver, meat, fish.
C	45 mg	sticks together cells lining surfaces such as the mouth	scurvy	fresh fruit and vegetables
D	5 µg	helps bones absorb calcium and phosphate	rickets, poor teeth	fish liver oils; also made in skin in sunlight

[1]Figures are the World Health Organization's recommended daily intake for an adolescent girl aged 10–18 years (2004). The recommended values differ slightly with the age and sex of the individual. 'mg' stands for milligram (a thousandth of a gram) and 'µg' for microgram (a millionth of a gram).

Table 12.2: Summary of the main vitamins.

A vegetarian diet

Vegetarians are people who restrict their diet to plant materials, and do not eat meat or fish. Some vegetarians also eat dairy products (milk, butter, cheese) and sometimes eggs, while others do not eat any products of animals, and are called **vegans**.

People have many reasons for being vegetarian. A vegetarian diet, especially one which does not include dairy products or eggs, is very healthy. It is low in saturated fats, cholesterol and salt, and high in dietary fibre. These factors reduce the threat of coronary heart disease, high blood pressure and diseases of the intestine. Other people have moral or religious grounds for vegetarianism, believing that it is wrong to kill animals for food.

Whatever their reasons, vegetarians can achieve a balanced diet from plant products. They must eat a good variety of vegetables, pulses (peas and beans), nuts and cereals. Nuts and pulses are high in protein, replacing the protein that would be supplied by meat or fish, and they substitute the more 'healthy' unsaturated plant oils for saturated animal fats (see Chapter 10). One problem with a vegetarian diet is the lack of vitamin B12 in plant materials. Vegetarians may have to take vitamin supplements or eat yeast extract to gain enough of this vital vitamin to stay healthy. Lack of B12 causes anaemia (low blood cell count) and can result in serious damage to the nervous system.

3. What are the functions of iron and calcium in the diet?
4. What are the functions of vitamin A and vitamin C in the diet?
5. Give three reasons why some people prefer to be vegetarians.
6. How do vegetarians ensure that they eat a balanced diet?

Energy from food

Some foods contain more energy than others. It depends on the proportions of carbohydrate, lipid and protein that they contain. Their energy content is measured in **kilojoules (kJ)**. If one gram of carbohydrate is fully oxidised, it produces about 17 kJ, whereas one gram of lipid yields over twice as much as this (39 kJ). Protein can produce about 18 kJ. If you look on a food label, it usually shows the energy content of the food, along with the amounts of different nutrients that it contains (Figure 12.5).

Baked Beans in tomato sauce

Nutritional information

Typical values	Amount per 100 g	Amount per serving (207 g)
Energy	313 kJ / 76 kcal	646 kJ / 155 kcal
Protein	4.7 g	9.7 g
Carbohydrate (of which sugars)	13.6 g (6.0 g)	28.2 g (12.4 g)
Fat (of which saturates)	0.2 g (Trace)	0.4 g (0.1 g)
Fibre	3.7 g	7.7 g
Sodium	0.5 g	1.0 g

HIGH IN FIBRE

210g e

Figure 12.5 *Food packaging is labelled with the proportions of different food types that it contains, along with its energy content. The energy in units called kilocalories (kcal) is also shown, but scientists no longer use this old-fashioned unit.*

Foods with a high percentage of lipid, such as butter or nuts, contain a large amount of energy. Others, like fruits and vegetables, which are mainly composed of water, have a much lower energy content (Table 12.3).

Food	kJ per 100 g	Food	kJ per 100 g
margarine	3 200	fried beefburger	1 100
butter	3 120	white bread	1 060
peanuts	2 400	chips	990
samosa	2 400	grilled beef steak	930
chocolate	2 300	fried cod	850
Cheddar cheese	1 700	roast chicken	770
grilled bacon	1 670	boiled potatoes	340
table sugar	1 650	milk	270
grilled pork sausages	1 550	baked beans	270
cornflakes	1 530	yoghurt	200
rice	1 500	boiled cabbage	60
spaghetti	1 450	lettuce	40

Table 12.3: *Energy content of some common foods.*

Even while you are asleep, you need a supply of energy for keeping warm, for your heartbeat, to allow messages to be sent through your nerves, and for other body functions. However, the energy you need at other times depends

Food scientists measure the amount of energy in a sample of food by burning it in a calorimeter (Figure 12.6). The calorimeter is filled with oxygen to make sure that the food will burn easily. A heating filament carrying an electrical current ignites the food. The energy given out by the burning food is measured by using it to heat up water flowing through a coil in the calorimeter.

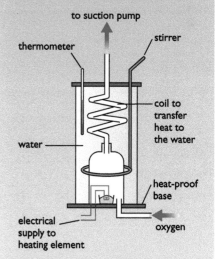

Figure 12.6 *A food calorimeter.*

If you have samples of food that will easily burn in air, you can measure the energy in them by a similar method, using the heat from the burning food to warm up water in a test-tube (see Activities 1 and 2 on pages 154 and 155).

on the physical work that you do. The total amount of energy that people need to keep healthy depends on their age and body size, and also on the amount of activity they do. Table 12.4 shows some examples of how much energy is needed each day by people of different ages, sex and occupations.

Age/sex/occupation of person	Energy needed per day/kJ
newborn baby	2 000
child aged 2	5 000
child aged 6	7 500
girl aged 12–14	9 000
boy aged 12–14	11 000
girl aged 15–17	9 000
boy aged 15–17	9 000
female office worker	9 500
male office worker	10 500
heavy manual worker	15 000
pregnant woman	10 000
breast-feeding woman	11 300

Table 12.4: *The daily energy needs of different types of people.*

Remember that these are approximate figures, and they are averages. Generally, the greater a person's weight, the more energy that person needs. This is why men, with a greater average body mass, need more energy than women. The energy needs of a pregnant woman are increased, mainly because of the extra weight that she has to carry. A heavy manual worker such as a labourer needs extra energy for increased muscle activity.

It is not only the recommended energy requirements that vary with age, sex and pregnancy, but also the *content* of the diet. For instance, during pregnancy a woman may need extra iron or calcium in her diet for the growth of the fetus. In younger women, the blood loss during menstruation (periods) can result in anaemia, producing a need for extra iron in the diet.

Activity 1: Experiment to find the energy content of a food

SBA skills

ORR	MM	AI	PD	Dr

If a sample of food will burn well in air, you can measure its energy content using a simplified version of the food calorimeter (Figure 12.7, page 155).

You will need:

- a boiling tube
- some peanuts
- a Bunsen burner and heat-proof mat
- a small measuring cylinder to measure 20 cm³ of water
- a thermometer
- a mounted needle
- a balance accurate to at least 0.1 g
- a stand, clamp and boss

Activity 1: Experiment to find the energy content of a food (continued)

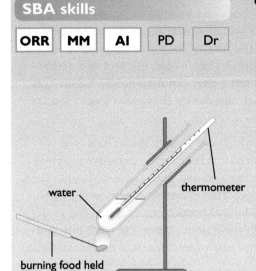

water

thermometer

burning food held
on mounted needle

Figure 12.7 *Measuring the energy produced by burning a sample of food.*

Carry out the following:

1. Find the mass of a peanut.

2. Place 20 cm³ of water into a boiling tube.

3. Support the boiling tube in a clamp on a stand as shown in Figure 12.7.

4. Measure the temperature of the water.

5. Spear a peanut on the end of a mounted needle.

6. Light the Bunsen burner and hold the peanut in the flame until it catches fire (this may take 30 seconds or so).

7. When the peanut is alight, hold it underneath the boiling tube of water so that the flame heats up the water.

8. If the nut stops burning, relight it in the Bunsen flame.

9. Continue until the peanut will no longer burn.

10. Measure the final temperature of the water (use the thermometer to stir the water gently to make sure the heat is evenly distributed).

11. Two facts are needed to allow you to calculate the energy content of the peanut:

    • 4.2 J of energy raises the temperature of 1 g of water by 1°C.

    • 1 cm³ of water has a mass of 1 g.

    So, if you multiply the rise in temperature of the water by the mass of the water and then by 4.2, you have the number of joules of heat that were transferred to the water. Dividing this by the mass of the peanut gives the energy/gram of nut:

    energy (joules per gram) =
    $$\frac{(\text{final temperature} - \text{temperature at start})}{\text{mass of peanut (g)}} \times 20 \times 4.2 \quad \text{joules}$$
    Now calculate the energy content of the peanut in this way.

12. How does your value compare with the value given in Table 12.3?

13. Can you think of any reasons why the values are different?

Activity 2: Comparing of the energy content of different foods

You could use the method above (Activity 1) to compare the energy in different foods. Any solid foods that will burn easily in air are suitable, such as corn curls, pasta or biscuits. Suggest a hypothesis that you could test about the energy content of the foods and design an experiment to test your hypothesis. Make sure that your experiment is properly controlled, and state how you will ensure that the results are reliable. You may be able to carry out the experiment if your teacher is happy with your design.

Chapter 12: How Humans Obtain Nutrition

Digestion

Food, such as a piece of bread, contains carbohydrates, lipids and proteins, but they are not the same carbohydrates, lipids and proteins as in our tissues. The components of the bread must first be broken down into their 'building blocks' before they can be absorbed through the wall of the gut. This process is called **digestion**. The digested molecules – sugars, fatty acids, glycerol and amino acids – along with minerals, vitamins and water, can then be carried around the body in the blood. When they reach the tissues they are reassembled into the molecules that make up our cells.

Digestion is speeded up by enzymes, which are biological catalysts (see Chapter 11). Although most enzymes stay inside cells, the digestive enzymes are made by the tissues and glands in the gut, and pass out of cells onto the gut contents, where they act on the food. This **chemical** digestion is helped by **mechanical** digestion. This is the physical breakdown of food. The most obvious place where this happens is in the mouth, where the teeth bite and chew the food, cutting it into smaller pieces, which increases the surface area.

At the same time the food is mixed with saliva, which lubricates it, allowing it to be swallowed more easily. Saliva also contains the enzyme amylase, which starts the breakdown of starch (see later in this chapter).

Teeth

In mammals, only the lower jaw is movable, while the upper jaw is fused to the skull. we chew using our cheek muscles, which move the lower jaw up and down, and allow some side-to-side movement.

The arrangement of an animal's teeth is called its **dentition**.

As with most mammals, humans have four types of teeth called incisors, canines, premolars and molars (Figure 12.8).

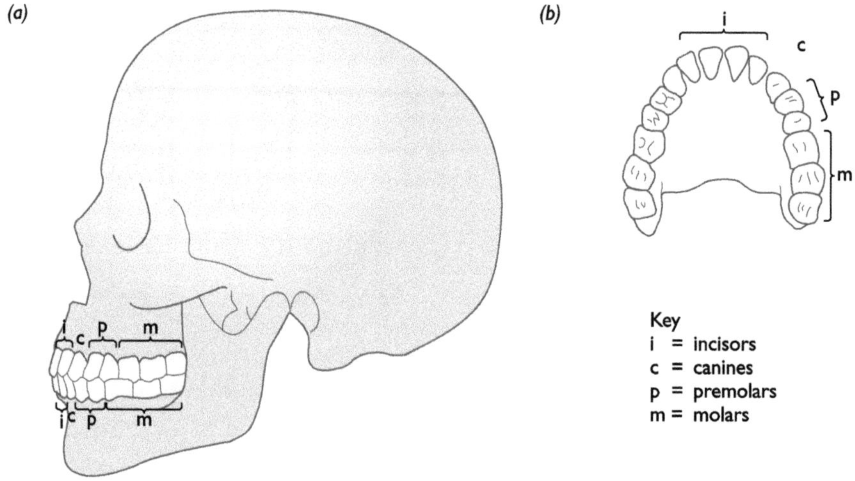

Key
i = incisors
c = canines
p = premolars
m = molars

Figure 12.8 (a) Side view of human skull and teeth, and (b) Upper teeth shown from below.

At the front of the mouth are the **incisors**. They are relatively sharp, chisel-shaped teeth, used for biting off pieces of food. Behind these in each jaw is a pair of **canines** (one on each side of the mouth). In humans the canines are similar in shape to the incisors, and have the same biting function. Behind the canines are the **cheek teeth** (**molars** and **premolars**). The cheek teeth have a flatter top surface or **crown**, and the top and bottom sets of cheek teeth meet crown to crown, and are used for chewing or crushing food. Human teeth are adapted for dealing with a wide range of food types, from meat and fish to plant roots, stems and leaves.

The crown of a tooth is covered with a non-living material called **enamel** (Figure 12.9), which is the hardest substance in the body. Underneath the enamel is a softer material (but still about as hard as bone) called **dentine**. The middle of the tooth is called the **pulp cavity**. It contains blood vessels and nerves. There are fine channels running through the dentine, that are filled with cytoplasm. These cytoplasmic strands are kept alive by nutrients and oxygen from the blood vessels in the pulp cavity. The root of the tooth is covered with **cement**, containing **fibres**. This material anchors the tooth in the jawbone but allows a slight degree of movement when the person is chewing.

During their lifetime, mammals have two sets of teeth. The first set is called the **milk** or **deciduous teeth**. In humans, these start to grow through the gum when a child is a few months old. By the age of about 2½, he or she will have 20 teeth, mainly milk teeth. From around the age of 7, the milk teeth are pushed out by permanent teeth growing underneath them. The molars at the back are only present as permanent teeth. Eventually a full set of 32 adult teeth is formed (Figure 12.8, page 156).

Tooth decay

Tooth decay (dental **caries**) is one of the most common diseases in the world. It is caused by bacteria in the mouth feeding on sugar. The bacteria break down the sugar, forming acids which dissolve the tooth enamel. Once the enamel is penetrated, the acid breaks down the softer dentine underneath. Eventually, a cavity is formed in the tooth (Figure 12.10). Bacteria can then enter this cavity and enlarge it until the decay reaches the nerves in the pulp cavity. Then you feel the pain! These bacteria are also the cause of **periodontal disease**, where the gums become inflamed and so sensitive that they bleed when the teeth are brushed. Periodontal disease can also lead to loss of teeth.

> 7. Explain the difference between chemical and mechanical digestion.
> 8. What are molars and what are they used for?
> 9. Explain how bacteria cause tooth decay.

The role of muscles in the gut wall

Other parts of the gut also help with mechanical digestion. For example, muscles in the wall of the stomach contract to churn up the food while it is being chemically digested.

Muscles are also responsible for moving the food along the gut. The walls of the intestine contain two layers of muscles. One layer has fibres running in rings around the gut. This is the **circular** muscle layer. The other has fibres running down the length of the gut, and is called the **longitudinal** muscle layer. Together, these two layers act to push the food along. When the circular muscles contract and the longitudinal muscles relax, the gut is made narrower. When the opposite happens, i.e. the longitudinal muscles contract and the

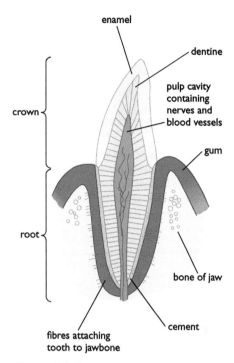

Figure 12.9 *The internal structure of an incisor tooth.*

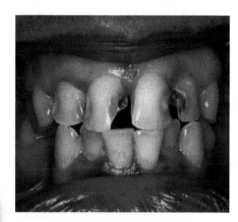

Figure 12.10 *A bad case of tooth decay. One of the causes was too much sugar in the person's diet.*

> This is a good definition of digestion, useful for exam answers:
> 'Digestion is the chemical and mechanical breakdown of food. It converts large insoluble molecules into small soluble molecules, which can be absorbed into the blood.'

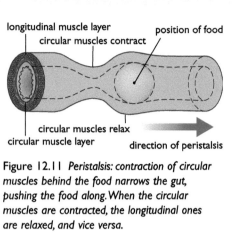

longitudinal muscle layer
circular muscles contract
position of food
circular muscles relax
circular muscle layer
direction of peristalsis

Figure 12.11 *Peristalsis: contraction of circular muscles behind the food narrows the gut, pushing the food along. When the circular muscles are contracted, the longitudinal ones are relaxed, and vice versa.*

circular muscles relax, the gut becomes wider. Waves of muscle contraction like this pass along the gut, pushing the food along, rather like squeezing toothpaste from a tube (Figure 12.11). This is called **peristalsis**. It means that movement of food in the gut doesn't depend on gravity – we can still eat standing on our heads!

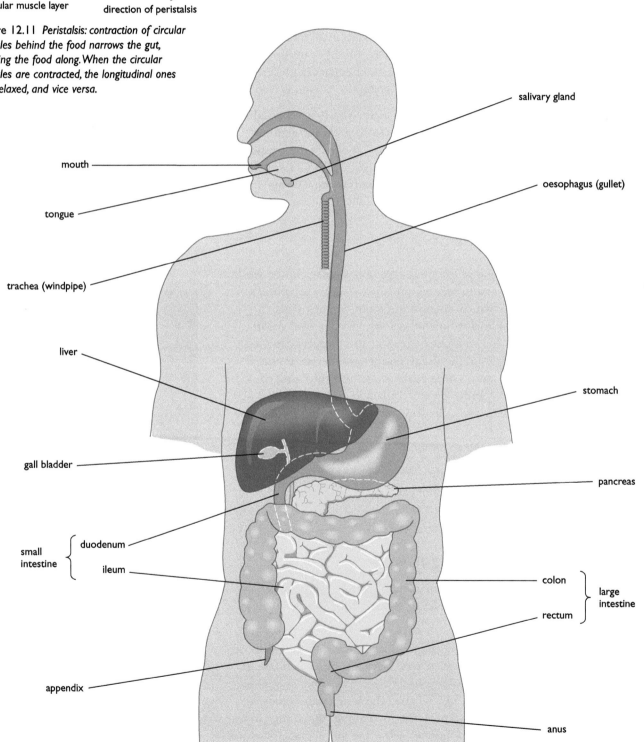

salivary gland
mouth
oesophagus (gullet)
tongue
trachea (windpipe)
liver
stomach
gall bladder
pancreas
small intestine { duodenum
ileum
colon
large intestine
rectum
appendix
anus

Figure 12.12 *The human digestive system, or alimentary canal.*

Figure 12.12 shows a diagram of the human digestive system. The gut is coiled up so that it fills the whole space of the abdomen. Overall, its length in an adult is about 8 m. This gives plenty of time for the food to be broken down and absorbed as it passes through.

The mouth, stomach and the first part of the small intestine (called the **duodenum**) all break down the food using enzymes, either made in the gut wall itself, or by glands such as the **pancreas**. Digestion continues in the last part of the small intestine (the **ileum**) and it is here that the digested food is absorbed. The last part of the gut, the large intestine, is mainly concerned with absorbing water out of the remains, and storing the waste products (**faeces**) before they are removed from the body.

The three main classes of food are broken down by three classes of enzymes. Carbohydrates are digested by enzymes called **carbohydrases**. Proteins are acted upon by **proteases**, and enzymes called **lipases** break down lipids (fats and oils). Some of the places in the gut where these enzymes are made are shown in Table 12.5.

Class of enzyme	Examples	Digestive action	Source of enzyme	Where it acts in the gut
carbohydrases	amylase	starch → maltose[1]	salivary glands	mouth
		starch → maltose	pancreas	small intestine
	maltase	maltose → **glucose**	wall of small intestine	small intestine
proteases	pepsin	proteins → peptides[2]	stomach wall	stomach
	trypsin	proteins → peptides	pancreas	small intestine
	peptidases	peptides → **amino acids**	wall of small intestine	small intestine
lipases	lipase	lipids → **glycerol** and **fatty acids**	pancreas	small intestine

[1]Maltose is a disaccharide made of two glucose molecules joined together.
[2]Peptides are short chains of amino acids.

Table 12.5: *Some of the enzymes that digest food in the human gut. The substances shown in bold are the end products of digestion that can be absorbed from the gut into the blood.*

---

10. Explain how peristalsis works.
11. What are the three main categories of digestive enzymes?

Digestion begins in the mouth. **Saliva** helps moisten the food and contains the enzyme **amylase**, which starts the breakdown of starch. The chewed lump of food, mixed with saliva, then passes along the **oesophagus** (gullet) to the stomach.

The food is held in the stomach for several hours, while initial digestion of protein takes place. The stomach wall secretes **hydrochloric acid**, so the stomach contents are strongly acidic. This has a very important function. It kills bacteria that are taken into the gut along with the food, helping to protect us from food poisoning. The protease enzyme that is made in the stomach, called **pepsin**, has to be able to work in these acidic conditions, and has an optimum pH value of about 2. This is unusually low – most enzymes work best at near neutral conditions (see Chapter 11).

Amylase digests starch into maltose. In this reaction, we say that starch is the **substrate** and maltose is the **product**.

The semi-digested food is held back in the stomach by a ring of muscle at the outlet of the stomach, called a **sphincter** muscle. When this relaxes, it releases the food into the first part of the small intestine, called the **duodenum** (Figure 12.13).

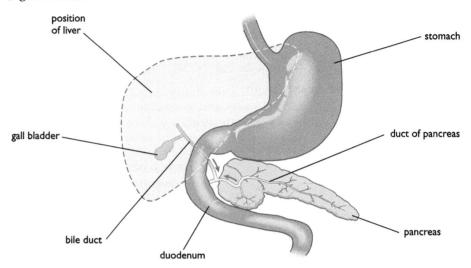

Figure 12.13 *The first part of the small intestine, the duodenum, receives digestive juices from the liver and pancreas through tubes called ducts.*

Several digestive enzymes are added to the food in the duodenum. These are made by the **pancreas**, and digest starch, proteins and fats (Table 12.5). As well as this, the **liver** makes a digestive juice called **bile**. Bile is a green liquid that is stored in the **gall bladder** and passes down the **bile duct** onto the food. Bile does not contain enzymes, but has another important function. It turns any large lipid globules in the food into an emulsion of tiny droplets (Figure 12.14). This increases the surface area of the lipid, so that **lipase** enzymes can break it down more easily.

Bile and pancreatic juice have another function. They are both alkaline. The mixture of semi-digested food and enzymes coming from the stomach is acidic, and needs to be neutralised by addition of alkali before it continues on its way through the gut.

As the food continues along the intestine, more enzymes are added, until the parts of the food that can be digested have been fully broken down into soluble end products, which can be absorbed. This is the role of the last part of the small intestine, the **ileum**.

Figure 12.14 *Bile turns lipids into an emulsion of tiny droplets for easier digestion.*

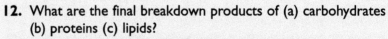

12. What are the final breakdown products of (a) carbohydrates (b) proteins (c) lipids?
13. Why does the stomach make hydrochloric acid?
14. What is a sphincter muscle?
15. What is the function of bile?
16. How is stomach acid neutralised in the duodenum?

Absorption in the ileum

The ileum is highly adapted to absorb the digested food. The lining of the ileum has a very large surface area, which means that it can quickly and efficiently absorb the soluble products of digestion into the blood. The length of the intestine helps to provide a large surface area, and this is aided by folds in its lining, but the greatest increase in area is due to tiny projections from the lining, called **villi** (Figure 12.15). The singular of villi is villus. Each villus is only about 1–2 mm long, but there are millions of them, so that the total area of the lining is thought to be about 300 m². This provides a massive area in contact with the digested food. As well as this, high-powered microscopy has revealed that the surface cells of each villus themselves have hundreds of minute projections, called **microvilli**, which increase the surface area for absorption even more (Figure 12.16).

Each villus contains a network of blood capillaries. Most of the digested food enters these blood vessels, but the products of lipid digestion, as well as tiny lipid droplets, enter a tube in the middle of the villus, called a **lacteal**. The lacteals form part of the body's **lymphatic** system, which transports a liquid called lymph. This **lymph** eventually drains into the blood system too.

The blood vessels from the ileum join up to form a large blood vessel called the **hepatic portal vein**, which leads to the liver (see Chapter 13). The liver acts rather like a food processing works, breaking some molecules down, and building up and storing others. For example, glucose from carbohydrate digestion is converted into **glycogen** and stored in the liver. Later, the glycogen can be converted back into glucose when the body needs it.

The digested food molecules are distributed around the body by the blood system (see Chapter 14). The soluble food molecules are absorbed from the blood into cells of tissues, and are used to build new parts of cells. This is called **assimilation**.

(a)

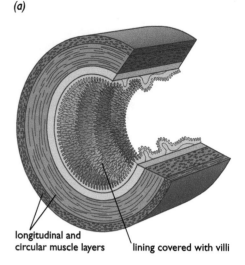

longitudinal and
circular muscle layers   lining covered with villi

(b)

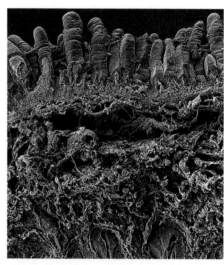

Figure 12.15 *(a) The inside lining of the ileum is adapted to absorb digested food by the presence of millions of tiny villi. (b) A section through the lining, showing the villi.*

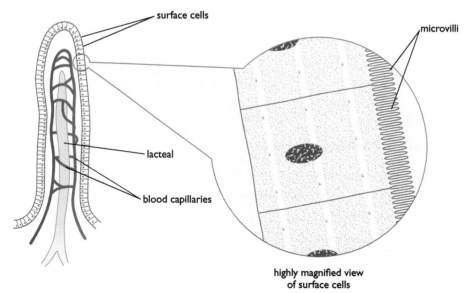

surface cells

microvilli

lacteal

blood capillaries

highly magnified view
of surface cells

Figure 12.16 *Each villus contains blood vessels and a lacteal, which absorb the products of digestion. The surface cells of the villus are covered with microvilli, which further increase the surface area for absorption.*

The large intestine – elimination of waste

By the time that the contents of the gut have reached the end of the small intestine, most of the digested food, as well as most of the water, has been absorbed. The waste material consists mainly of cellulose (fibre) and other indigestible remains, water, dead and living bacteria, and cells lost from the lining of the gut. The function of the first part of the large intestine, called the **colon**, is to absorb most of the remaining water from the contents, leaving a semi-solid waste material called **faeces**. This is stored in the **rectum**, until expelled out of the body through the **anus**.

> **17.** List three ways in which the internal surface area of the ileum is increased.
>
> **18.** What is a lacteal?
>
> **19.** What is the role of the hepatic portal vein?

Activity 3: Making a Visking tubing model of the gut

SBA skills

ORR	MM	AI	PD	Dr

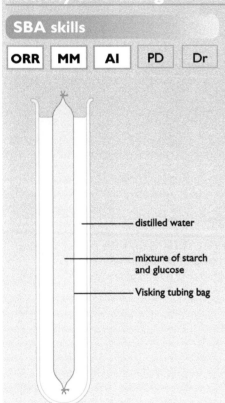

Figure 12.17 *A Visking tubing model of the ileum.*

Labels: distilled water; mixture of starch and glucose; Visking tubing bag

Visking tubing is an artificial membrane that is used in kidney dialysis machines to filter the blood of patients who have kidney failure. It has microscopic holes in it that smaller molecules can pass through. In this experiment you will be using Visking tubing to model the action of the intestine wall.

You will need:

- a piece of Visking tubing about 20 cm long
- some thread
- a boiling tube
- a 20-cm³ syringe
- a spotting tile
- iodine solution in a dropping bottle
- test-tubes
- a large beaker, half-full with water
- a Bunsen burner
- a tripod, gauze and heat-proof mat
- several dropping pipettes
- Benedict's solution
- a mixture of starch suspension and glucose solution

Carry out the following:

1. Fill a boiling tube about two-thirds full with water.

2. Open up the Visking tubing by soaking it in water and rubbing the end between your fingers.

3. Carefully tie a knot in one end of the tubing.

4. Use the syringe to fill the tubing with a mixture of starch suspension and glucose solution.

5. Tie a piece of thread around the top of the tubing, leaving a length with which you can handle it.

6. Wash the outside of the tubing under a tap to ensure there are no drops of the mixture adhering to it.

7. Place the tubing in the boiling tube of water as shown in Figure 12.17.

8. Immediately take a sample from the water surrounding the tubing. Test the water for (a) starch, using iodine solution, and (b) glucose, by heating with Benedict's solution (see page 114).

9. Take samples of the water at 5-minute intervals and repeat the tests for starch and glucose.

   The mixture of starch and glucose represents food inside the Visking tubing 'gut'. The water around the tubing represents the blood. The boiling tube is maintained at about 40°C to keep it close to the temperature of the human body.

10. Write up your experiment and explain your findings. You will need to look again at the difference in structure between starch and glucose (Chapter 10) and the reasons why digestion of food is necessary (in this chapter).

Chapter summary

In this chapter you have learnt that:

- a balanced diet is one that contains the right proportions of carbohydrates, lipids, proteins, minerals, vitamins, water and fibre, to keep the body healthy

- malnutrition means having a poor diet. It can result from the lack of a major nutrient, such as protein, or of a nutrient needed in much smaller quantities, such as a mineral or vitamin. This can cause specific deficiency symptoms

- some foods contain more energy than others, for example. Lipid-rich foods have a particularly high energy content

- a person's energy requirements and dietary content depend upon their age, sex and occupation

- digestion means the breakdown of large insoluble food materials into small soluble molecules which can be absorbed into the blood

- digestion is both chemical, as a result of the action of enzymes, and mechanical, by the action of teeth and gut muscles

- peristalsis is caused by the rhythmic contraction of the muscles in the wall of the gut; a wave of contraction pushes along the contents of the alimentary canal

- carbohydrase, protease and lipase enzymes are produced by different regions of the gut. They digest carbohydrates, proteins and lipids respectively, producing monosaccharides, amino acids, fatty acids and glycerol

- the liver makes bile, a liquid emulsifies lipid. This makes it easier for the lipid to be digested by enzymes

- the ileum is the site of absorption of most of the products of digestion. To facilitate this, the internal surface of the ileum has a large surface area, due to its length, folds, villi and microvilli

- each villus contains a blood capillary network that absorbs digested monosaccharides and amino acids. The products of lipid digestion enter a lacteal in the villus, part of the lymphatic system

- products of digestion are taken to the liver via the hepatic portal vein, where the liver processes the products, for example storing carbohydrates as glycogen

- the colon absorbs water from the remaining gut contents, leaving the waste faeces, which are eliminated via the anus.

Questions

1. Which of the following structures represents the bile duct?

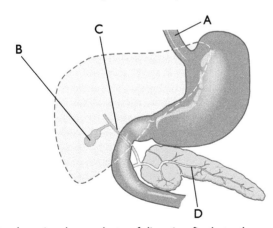

2. Blood carries the products of digestion firstly to the:

  A brain
  B liver
  C heart
  D anus

3. Which of the following organs makes a juice which helps to neutralise stomach acids?

  A liver
  B small intestine
  C gall bladder
  D stomach

4. What is the normal number of teeth in an adult human?

  A 20
  B 24
  C 30
  D 32

5. Which of the following regions of the alimentary canal produces **no** digestive enzymes?

  A mouth
  B oesophagus
  C stomach
  D pancreas

6. Which of the following statements about digestion is **not** correct?

   A   digestion releases energy from food
   B   digestion converts proteins into amino acids
   C   digestion makes insoluble materials into soluble molecules
   D   digestion changes starch into glucose

7. A student carried out an experiment to find the best conditions for the enzyme pepsin to digest protein. For the protein, she used egg white powder, which forms a cloudy white suspension in water. The table below shows how the four tubes were set up.

Tube	Contents
A	5 cm³ egg white suspension, 2 cm³ pepsin, 3 drops of dilute acid. Tube kept at 37 °C
B	5 cm³ egg white suspension, 2 cm³ distilled water, 3 drops of dilute acid. Tube kept at 37 °C
C	5 cm³ egg white suspension, 2 cm³ pepsin, 3 drops of dilute acid. Tube kept at 20 °C
D	5 cm³ egg white suspension, 2 cm³ pepsin, 3 drops of dilute alkali. Tube kept at 37 °C

The tubes were left for 2 hours and the results were then observed. Tubes B, C and D were still cloudy. Tube A had gone clear.

a) Three tubes were kept at 37 °C. Why was this temperature chosen? (2)

b) Explain what had happened to the protein in tube A. (2)

c) Why did tube D stay cloudy? (2)

d) Tube B is called a **control**. Explain what this means. (2)

e) Tube C was left for another 3 hours. Gradually it started to clear. Explain why digestion of the protein happened more slowly in this tube. (3)

f) The lining of the stomach secretes hydrochloric acid. Explain the function of this. (1)

g) When the stomach contents pass into the duodenum, they are still acidic. How are they neutralised? (2)

8. Copy and complete the following table of digestive enzymes.

Enzyme	Food on which it acts	Products
amylase		
trypsin		
		fatty acids and glycerol

(6)

9. The diagram shows a method that can be used to measure the energy content of some types of food. A student placed 20 cm³ of water in a boiling tube and measured its temperature. He weighed a small piece of pasta, and then held it in a Bunsen burner flame until it caught alight. He then used the burning pasta to heat the boiling tube of water, until the pasta had finished burning. Finally, he measured the temperature of the water at the end of the experiment.

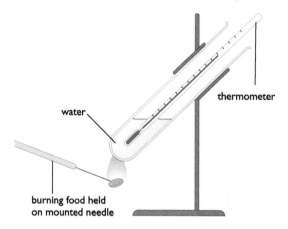

To answer the questions that follow, use the following information:

- The density of water is 1 g/cm³.
- The pasta weighed 0.22 g.
- The water temperature at the start was 21 °C and at the end was 39 °C.
- The heat energy supplied to the water can be found from the formula:

   energy (in joules)
   = mass of water × temperature change × 4.2

a) Calculate the energy supplied to the water in the boiling tube in joules (J). Convert this to kilojoules (kJ) by dividing by 1 000. (2)

b) Calculate the energy released from the pasta as kilojoules per gram of pasta (kJ/g). (1)

c) The correct figure for the energy content of pasta is 14.5 kJ/g. The student's result is an underestimate. Write down *three* reasons why he may have got a lower than expected result. (Think about how the design of the apparatus might introduce errors.) (3)

d) Suggest *one* way the apparatus could be modified to reduce these errors. (2)

e) The energy in a peanut was measured using the method described above. The peanut was found to contain about twice as much energy per gram as the pasta. Explain why this is the case. (2)

10. The diagram shows an experiment that was set up as a model to show why food needs to be digested.

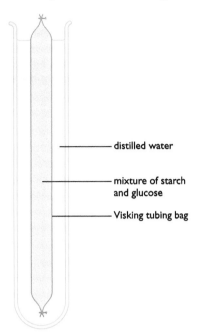

distilled water

mixture of starch and glucose

Visking tubing bag

The Visking tubing acts as a model of the small intestine because it has tiny holes in it that some molecules can pass through. The tubing was left in the boiling tube for an hour, then the water in the tube was tested for starch and glucose.

a) Describe how you would test the water for starch, and for glucose. What would the results be for a 'positive' test in each case? (4)

b) The tests showed that glucose was present in the water, but starch was not. Explain why. (2)

c) If the tubing takes the place of the intestine, what part of the body does the water in the boiling tube represent? (1)

d) What does 'digested' mean? (2)

11. Bread is made mainly of starch, protein and lipid. Imagine a piece of bread about to start its journey through the alimentary canal.

a) Describe what happens to the bread (or its breakdown products) as it passes through the mouth, stomach and duodenum. (6)

b) Explain the ways that the ileum is adapted for absorbing the breakdown products from the digestion of the bread. (4)

c) Explain, with the help of a labelled diagram, how food is moved through the gut by the process of peristalsis. (3)

d) The killer disease cholera is caused by an infectious bacterium that lives in the human gut. (loss of water in the faeces), which leads to dehydration and death, particularly amongst children. Suggest why cholera causes diarrhoea. (2)

12. a) Explain the meaning of the following terms:
i) digestion, ii) assimilation, and iii) egestion. (6)

b) One function of the liver is the formation of bile. What is the role of bile in digestion? (2)

c) Name the blood vessel that carries blood from the ileum to the liver. (1)

d) After a meal rich in carbohydrates, what happens to glucose in the liver? (2)

e) Describe how you could test for glucose (reducing sugar) in a sample of food. (4)

f) If reducing sugar was not present in the sample in (e), explain how you could modify the test to find out if the food contained a non-reducing sugar. (3)

Chapter 13: Breathing and Respiration

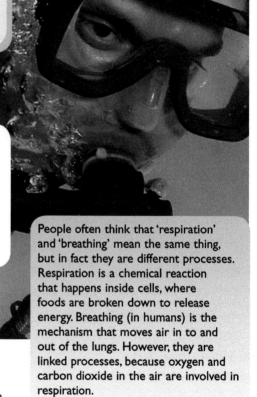

When you have completed this chapter, you will be able to:

- understand the meaning of cell respiration and the difference between respiration and breathing
- describe the process of aerobic respiration
- understand the function of ATP as an energy 'currency' in the cell
- distinguish between aerobic and anaerobic respiration
- carry out controlled experiments to demonstrate the products of aerobic and anaerobic respiration
- describe the structure of the respiratory system
- describe the mechanism of ventilation of the lungs
- identify characteristics common to gas exchange surfaces
- discuss the effects of cigarette smoking on the body.

When we breathe, air is drawn in to and out of our lungs so that gas exchange can take place between the air and the blood. Oxygen is taken to the body cells for use in a chemical reaction called respiration and carbon dioxide is removed. This chapter looks at the processes of respiration and breathing, as well as gas exchange in some other organisms. It also deals with some ways in which smoking can damage the lungs and stop these vital organs from working properly.

People often think that 'respiration' and 'breathing' mean the same thing, but in fact they are different processes. Respiration is a chemical reaction that happens inside cells, where foods are broken down to release energy. Breathing (in humans) is the mechanism that moves air in to and out of the lungs. However, they are linked processes, because oxygen and carbon dioxide in the air are involved in respiration.

Cross-curricular links: chemistry, endothermic and exothermic reactions

Respiration: how the cell gets its energy

To be able to carry out all the processes needed for life, a cell needs a source of energy. It gets this energy by breaking down food molecules to release the stored chemical energy that they contain. This process is called **cell respiration**, or just **respiration**. Many people think that respiration means breathing, but although there are links between the two processes, the biological meaning of respiration is very different.

The process of respiration happens in all the cells of our body. Oxygen is used to oxidise food, and carbon dioxide (and water) are released as waste products. The main food oxidised is the sugar glucose. Glucose contains stored chemical energy that can be converted into other forms that the cell can use. It's rather like burning a fuel to get the energy out of it, except that burning releases all its energy as heat, whereas respiration releases some as heat, but most is trapped as energy in other chemicals. This chemical energy can be used for a variety of purposes, such as:

- contraction of muscle cells, producing movement
- active transport of molecules and ions
- building large molecules, such as proteins
- cell division.

The overall reaction for respiration is:

glucose + oxygen ⟶ carbon dioxide + water (+ energy)

$$C_6H_{12}O_6 \ + \ 6O_2 \ \longrightarrow \ 6CO_2 \ + \ 6H_2O \quad (+ \text{ energy})$$

This is called **aerobic** respiration, because it uses oxygen. The waste products are carbon dioxide gas and water. Respiration is not just carried out by human cells, but by all animals and plants and many other organisms. The glucose is available from digestion of food in animals, and is the first product of photosynthesis in plants.

It is important to realise that the equation above is just a summary of the process. In reality, it takes place gradually, as a sequence of small steps which release the energy of the glucose in small amounts. Each step in the process is catalysed by a different enzyme. The later steps in the process are the aerobic ones, and these release the most energy. They happen in the mitochondria of the cells (see Chapter 7).

> 1. Explain the difference between respiration and breathing.
> 2. What is meant by *aerobic* respiration?
> 3. Write a balanced symbol equation for aerobic respiration.
> 4. List three uses of the energy from respiration.

ATP: the energy 'currency' of the cell

We have seen that respiration gives out energy, while other processes such as protein synthesis and active transport use it up. Cells must have a way of passing the energy from respiration across to these other processes that need it. The way that they do this is through a substance called **adenosine triphosphate**, or **ATP**, which is present in all cells.

ATP is made up from an organic molecule (adenosine) attached to three inorganic phosphate groups (hence triphosphate). ATP can be broken down in the cell, losing a phosphate and producing a similar molecule called adenosine diphosphate, or ADP (Figure 13.1a, page 169).

When this reaction happens, energy is released and is available for the processes that demand energy.

More ATP is made during the reactions of respiration, using the energy from the oxidised glucose to add a phosphate back onto ADP.

Because of its role, ATP is described as the energy currency of a cell. It exchanges chemical energy between the process that produces the energy (respiration) and the processes that use it up.

Cross-curricular links: physics, forms of energy

Cross-curricular links: chemistry, water, carbon dioxide

Cross-curricular links: physics, forms of energy

Water is needed to break down ATP to ADP and phosphate, so this is an example of a hydrolysis reaction (see Chapter 10, page 132).

(a) When energy is needed ATP is broken down into ADP and phosphate (P):

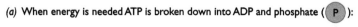

$$\text{adenosine} - P - P - P + H_2O \longrightarrow \text{adenosine} - P - P + P$$

(b) During respiration ATP is made from ADP and phosphate:

$$\text{adenosine} - P - P + P \longrightarrow \text{adenosine} - P - P - P + H_2O$$

Figure 13.1 *ATP is the energy currency of the cell.*

The reactions of respiration are not 100% efficient, and some energy is not used to make ATP, but instead is lost as heat. Animals such as mammals and birds use this heat to keep their bodies warm, maintaining a constant body temperature.

---

5. Why do you think mitochondria are sometimes called the 'power houses' of the cell?

6. Explain why ATP is an energy currency.

Activity 1: Measuring the rate of oxygen uptake during respiration

SBA skills

| ORR | MM | AI | PD | Dr |

You can measure the rate of oxygen uptake (and therefore the rate of respiration) of small organisms, such as maggots, woodlice or germinating seeds, using a simple piece of apparatus called a **respirometer**. The organisms are placed in a tube attached to a capillary tube containing an oil drop (Figure 13.2). As the organisms use up the oxygen in the tube, the oil moves along the capillary. This will only work if there is a chemical in the tube which absorbs the carbon dioxide that is produced from the respiration of the organisms. A suitable chemical for this is called soda lime. Sodium or potassium hydroxide solution can also be used, but they are very caustic and you must take care that they do not come into contact with the organisms in the respirometer.

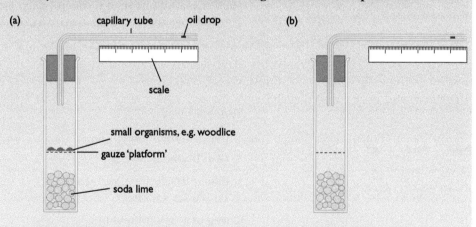

Figure 13.2 *Using a simple respirometer to measure the rate of uptake of oxygen during respiration.*

You will need:

- two respirometer tubes
- two bungs with holes for the capillary tubes

- two L-shaped capillary tubes
- some soda lime
- a ruler
- suitable small organisms, such as maggots, woodlice or germinating seeds (before their leaves appear, otherwise they will photosynthesise as well as respire, which will upset the results)
- a stopwatch or clock

Carry out the following:

1. Set up both respirometers as shown in Figure 13.2 (page 169). Make sure that all connections are airtight. You can use vaseline to ensure a good seal around the bung and where the capillary tube passes through the bung.

2. Introduce an oil drop into the end of each capillary tube by dipping the end into some oil.

3. Leave each respirometer set up for a few minutes to allow the contents to equilibrate to the temperature of the surroundings, and for any carbon dioxide in the air in tube B to be absorbed by the soda lime.

4. After the period of equilibration, mark the position of the oil drop in each capillary tube.

5. Record the position of the oil drop in each capillary tube at regular time intervals.

6. Plot a graph of the distance travelled by the oil drop against time. You can plot both sets of results on the same axes.

7. Write up your experiment and explain your findings.

8. What is the function of the soda lime?

9. Respirometer B contains no organisms. What is this part of the experiment called, and what is its purpose?

10. What would happen to the oil drop if the respirometer contained seedlings that were photosynthesising?

Activity 2: Demonstrating that heat is produced by respiration

SBA skills

ORR	MM	AI	PD	Dr

You will need:

- two vacuum flasks
- two thermometers
- dilute disinfectant, such as 1% bleach solution
- some cotton wool
- some peas

Carry out the following:

1. Soak some peas in water for 24 hours so that they start to germinate.

2. Boil a second batch of peas to kill them.

3. Wash each set of peas in the disinfectant to surface-sterilise them, killing any bacteria that may be present.

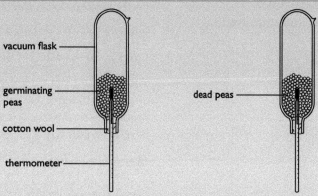

Figure 13.3 *Experiment to show that heat is produced during respiration in germinating peas.*

4. Place each batch of peas in a vacuum flask as shown in Figure 13.3. Leave some air in each flask.

5. Support each flask upside down, as shown. The seeds will produce carbon dioxide gas, which is denser than air. The inverted flasks and porous cotton wool allows this to escape. It might otherwise kill the peas.

6. Record the temperature inside each flask at the start.

7. Leave the apparatus set up for a couple of days and record the temperature inside the flask at the end of the experiment.

8. Write up the experiment and explain your results.

9. The purpose of using vacuum flasks is to insulate the contents, so that any temperature change can be measured. Which flask showed a higher temperature at the end of the experiment? Why?

10. Why is it necessary to kill any micro-organisms on the surface of the peas?

11. Explain the importance of the flask containing dead peas.

Anaerobic respiration

There are some situations in which cells can respire without using oxygen. This is called **anaerobic** respiration. In anaerobic respiration, glucose is not completely broken down, and less energy is released than from aerobic respiration, in other words fewer ATP molecules are formed. However, the advantage of anaerobic respiration is that it can go on in situations where oxygen is in short supply. Two important examples of this are in yeast cells and muscle cells.

Anaerobic respiration in yeast

Yeasts are single-celled fungi (Figure 13.4). They are used in commercial processes such as making wine and beer, and baking bread.

Did you know?
Aerobic respiration produces about 16 times as much ATP as anaerobic respiration.

Figure 13.4 *Yeast cells, highly magnified. Some can be seen reproducing by forming 'buds'. Each bud will eventually break off to form a new yeast cell.*

Cross-curricular links:
chemistry, reactions involved in
baking dough

When yeast cells are prevented from getting enough oxygen, they stop respiring aerobically, and start to respire anaerobically instead. The glucose is partly broken down into ethanol (alcohol) and carbon dioxide:

glucose $\longrightarrow$ ethanol + carbon dioxide (+ some energy)

$C_6H_{12}O_6 \longrightarrow 2C_2H_5OH + 2CO_2$

The ethanol from this respiration is the alcohol in wine and beer. The carbon dioxide is the gas that makes bread rise when it is baked. Think about the properties of ethanol – it makes a good fuel and will burn to produce a lot of heat, so it still has a lot of stored chemical energy in it.

Aerobic respiration	Anaerobic respiration in yeast
Breaks down glucose	Breaks down glucose
Uses oxygen	Does not use oxygen
Produces carbon dioxide and water	Produces carbon dioxide and ethanol
Releases a large amount of energy and much ATP	Produces a small amount of energy and little ATP

Table 13.1: *A comparison of aerobic respiration and anaerobic respiration in yeast.*

Anaerobic respiration in muscle cells

Muscle cells can also respire anaerobically. This also happens when they are short of oxygen. If muscles are overworked, the blood cannot reach them fast enough to deliver enough oxygen for aerobic respiration. This happens when a person does a 'burst' activity, such as a sprint, or quickly lifting a heavy weight. This time, the glucose is broken down into a substance called **lactic acid**:

glucose $\longrightarrow$ lactic acid (+ some energy)

$C_6H_{12}O_6 \longrightarrow 2C_3H_6O_3$

Anaerobic respiration has two main disadvantages over aerobic respiration. It converts much less of the energy stored in food into a form of chemical energy that cells can use. It also produces toxic waste products, such as lactic acid or ethanol.

Anaerobic respiration provides enough energy to keep the overworked muscles going for a short period, but continuing the 'burst' of activity means that too much lactic acid builds up in the bloodstream. When the person stops the activity the lactic acid is broken down (oxidised). This process uses oxygen. The volume of oxygen needed to completely oxidise the lactic acid that builds up in the body during anaerobic respiration is called the **oxygen debt**.

Aerobic respiration	Anaerobic respiration in muscle cells
Breaks down glucose	Breaks down glucose
Uses oxygen	Does not use oxygen
Produces carbon dioxide and water	Produces lactic acid
Releases a large amount of energy and much ATP	Produces a small amount of energy and little ATP

Table 13.2: *A comparison of aerobic respiration and anaerobic respiration in muscle.*

7. Name two useful products of anaerobic respiration in yeast.
8. What are the similarities and differences between anaerobic respiration in yeast and in muscle?
9. A 100 m sprinter gains almost 100% of his energy from anaerobic respiration, whereas a marathon runner gains almost 100% of his energy aerobically. Explain why this is the case.

Activity 3: Showing the products of anaerobic respiration in yeast

SBA skills

ORR	MM	AI	PD	Dr

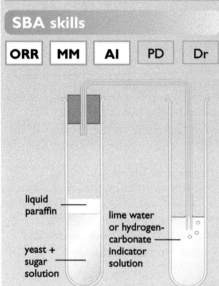

Figure 13.5 *Apparatus to show the products of anaerobic respiration in yeast.*

You will need:

- two boiling tubes
- delivery tubes and bungs
- some sugar (sucrose)
- some yeast
- a Bunsen burner and heat-proof mat
- some lime water or hydrogencarbonate indicator solution
- a pipette
- a glass stirring rod
- some liquid paraffin
- a test-tube holder

Note: Addition of carbon dioxide gas to lime water turns the lime water from clear to cloudy (milky). Addition of the same gas to hydrogencarbonate indicator solution changes the colour of the solution from orange to yellow.

Carry out the following:

1. Carefully boil some water in a boiling tube to drive off any air that is dissolved in the water.

2. Dissolve a small amount of sugar in the boiled water and allow it to cool.

3. Add a little yeast and stir.

4. Set up the apparatus as shown in Figure 13.5. Carefully add a thin layer of liquid paraffin to the surface of the yeast/sugar mixture using the pipette.

5. Set up a control apparatus exactly as shown in Figure 13.5, but using boiled (killed) yeast instead of living yeast.

6. Leave the apparatus in a warm place for an hour or two. Observe any changes in the indicator solution.

7. Take the bung out of the tube containing the yeast and use a pipette to remove the layer of liquid paraffin. Gently sniff the contents of the tube. Can you smell alcohol?

8. Write up your experiment and explain the results.

9. Why is the yeast added to boiled water?

10. What is the function of the liquid paraffin?

SBA skills

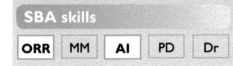

ORR	MM	AI	PD	Dr

You will need:

- nothing, except your arms!

Carry out the following:

1. Raise one arm above your head and lower the other one by your side.

2. Clench and unclench both hands as hard and fast as you can, at the same rate, for as long as you can bear it.

3. Which arm tires first?

4. Rest both arms on the table after this 'burst' exercise. What can you feel?

5. Write up the experiment and explain your observations.

6. The blood flow to the raised arm is reduced by the blood having to flow 'uphill'. How will this affect the build-up of lactic acid in the muscles of each arm?

7. Why do you experience muscle fatigue in one arm before the other?

The tingling feeling in your arms after the exercise is the blood flowing through your arm muscles. This 'flushes out' the lactic acid from the muscle cells. Eventually, the lactic acid will be broken down aerobically.

Using anaerobic respiration to make biogas

Caribbean countries obtain most of their energy from imported oil products such as petrol, diesel and natural gas, but there is a drive to increase the use of alternative energy supplies, such as sunlight and wind power. One alternative energy source is **biogas**. Biogas is produced by the anaerobic respiration of bacteria feeding on organic waste material. Biogas production takes place in a large tank called a **digester**. Biogas can be produced on a small scale, for a village or individual household, or for large-scale industrial purposes. Digesters are particularly successful in disposing of large quantities of waste produced by farming, such as slurry from pig farms.

Biogas is a clean and renewable fuel. It consists of approximately 60% methane and 40% carbon dioxide, and can be burnt as a fuel or used to drive electricity generators. The material used in the digester can come from many different sources, such as animal dung, waste products from growing crops, sewage, or domestic organic rubbish.

There is a variety of different types of digester, but they all have certain features in common:

- a digester tank, where the bacteria break down the waste

- an inlet pipe, for adding waste to the tank, which often has a mixing pit for mixing the waste before it is added to the tank

- a gas holder above the tank to collect the gas (in simple digesters the gas simply collects at the top of the tank)

- an outlet pipe for the biogas
- an outlet pipe leading to an overflow tank.

One example of a small-scale digester is shown in Figure 13.6.

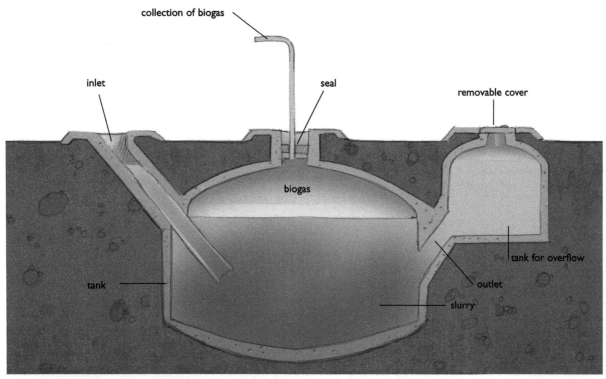

Figure 13.6 A small-scale anaerobic digester for producing biogas.

Breathing in humans

Breathing consists of **ventilation**, the mechanism that moves air in and out of the lungs, and enables **gas exchange** to take place between the air in the lungs and the gases dissolved in the blood. It would be more accurate to call the lungs and associated organs the breathing system. However, they are usually called the respiratory system, which can be confusing!

The structure of the respiratory system

The lungs are enclosed in the chest or **thorax** by the ribcage and a muscular sheet of tissue called the **diaphragm** (see Figure 13.8, page 176). As you will see, the actions of these two structures bring about the movements of air into and out of the lungs. Joining each rib to the next one are two sets of muscles called **intercostal muscles** ('costals' are rib bones). If you eat meat, you will have seen intercostal muscles attached to the long bones of 'spare ribs'. The diaphragm separates the contents of the thorax from the abdomen. It is not flat, but a shallow dome shape, with a fibrous middle part forming the 'roof' of the dome, and muscular edges forming the walls.

The air passages of the lungs form a highly branching network (Figure 13.7). This is why it is sometimes called the **bronchial tree**.

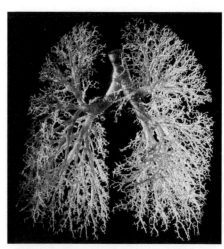

Figure 13.7 This cast of the human lungs was made by injecting a pair of lungs with a liquid plastic. The plastic was allowed to set, then the lung tissue was dissolved away with acid.

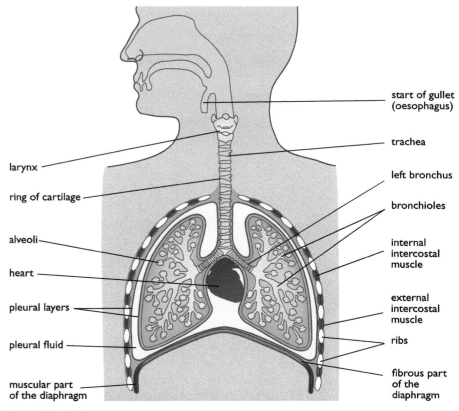

Figure 13.8 *The human respiratory system.*

In the bronchi, the cartilage forms complete, circular rings. In the trachea, the rings are incomplete, and shaped like a letter 'C'. The open part of the ring is at the back of the trachea next to the oesophagus (gullet), which passes through the thorax, where it lies behind. When food passes along the oesophagus by peristalsis (see Chapter 12), the gaps in the rings allow the lumps of food to pass through more easily, without the peristaltic wave 'catching' on the rings (Figure 13.9).

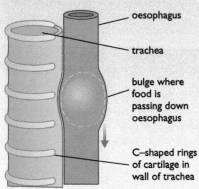

Figure 13.9 *C-shaped cartilage rings in the trachea.*

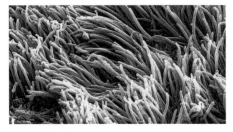

Figure 13.10 *This electron microscope picture shows cilia from the lining of the trachea.*

When we breathe in, air enters our nose or mouth and passes down the windpipe or **trachea**. The trachea splits into two tubes called the **bronchi**, one leading to each lung. Each **bronchus** divides into smaller and smaller tubes called **bronchioles**, eventually ending at microscopic air sacs, called **alveoli**. It is here that gas exchange with the blood takes place.

The walls of trachea and bronchi contain rings of gristle or **cartilage**. These support the airways and keep them open when we breathe in. They are rather like the rings in a vacuum cleaner hose – without them the hose would squash flat when the cleaner sucks air in.

The inside of the thorax is separated from the lungs by two thin, moist membranes called the **pleural layers**. They make up a continuous envelope around the lungs, forming an airtight seal. Between the two layers is a space called the **pleural cavity**, filled with a thin layer of liquid called **pleural fluid**. This acts as lubrication, so that the surfaces of the lungs don't stick to the inside of the chest wall when we breathe.

Keeping the airways clean

The trachea and larger airways are lined with a layer of cells that have an important role in keeping the airways clean. Some cells in this lining secrete a sticky liquid called **mucus**, which traps particles of dirt or bacteria that are breathed in. Other cells are covered with tiny hair-like structures called **cilia** (Figure 13.10). The cilia beat backwards and forwards, sweeping the mucus and trapped particles out towards the mouth. In this way, dirt and bacteria are prevented from entering the lungs, where they might cause an infection. As you will see, one of the effects of smoking is that it destroys the cilia and stops this protection mechanism from working properly.

Ventilation of the lungs

Ventilation means moving air in to and out of the lungs. This requires a difference in air pressure – the air moves from a place where the pressure is high to one where it is low. Ventilation depends on the fact that the thorax is an airtight cavity. When we breathe, we change the volume of our thorax, which alters the pressure inside it. This causes air to move in to or out of the lungs.

There are two movements that bring about ventilation, those of the ribs and the diaphragm. If you put your hands on your chest and breathe in deeply, you can feel your ribs move upwards and outwards. They are moved by the intercostal muscles (Figure 13.11). The outer (external) intercostals contract, pulling the ribs up. At the same time the muscles of the diaphragm contract, pulling the diaphragm down into a more flattened shape (Figure 13.12a, page 178). Both these movements increase the volume of the chest and cause a slight drop in pressure inside the thorax compared with the air outside. Air then enters the lungs.

The opposite happens when you breathe out deeply. The external intercostals relax, and the internal intercostals contract, pulling the ribs down and in. At the same time, the diaphragm muscles relax and the diaphragm goes back to its normal dome shape. The volume of the thorax decreases, and the pressure in the thorax is raised slightly above atmospheric pressure. This time, the difference in pressure forces air out of the lungs (Figure 13.12b, page 178). Exhalation is helped by the fact that the lungs are elastic, so that they tend to empty like a balloon.

During normal (shallow) breathing, the elasticity of the lungs and the weight of the ribs acting downwards is enough to cause exhalation. The internal intercostals are only really used for deep (forced) breathing out, for instance when we are exercising.

It is important that you remember the changes in volume and pressure during ventilation. If you have trouble understanding these, think of what happens when you use a bicycle pump. If you push the pump handle, the air in the pump is squashed, its pressure rises and it is forced out of the pump. If you pull on the handle, the air pressure inside the pump falls a little, and air is drawn in from outside. This is similar to what happens in the lungs. In exams, students sometimes talk about the lungs *forcing* the air in and out – they don't!

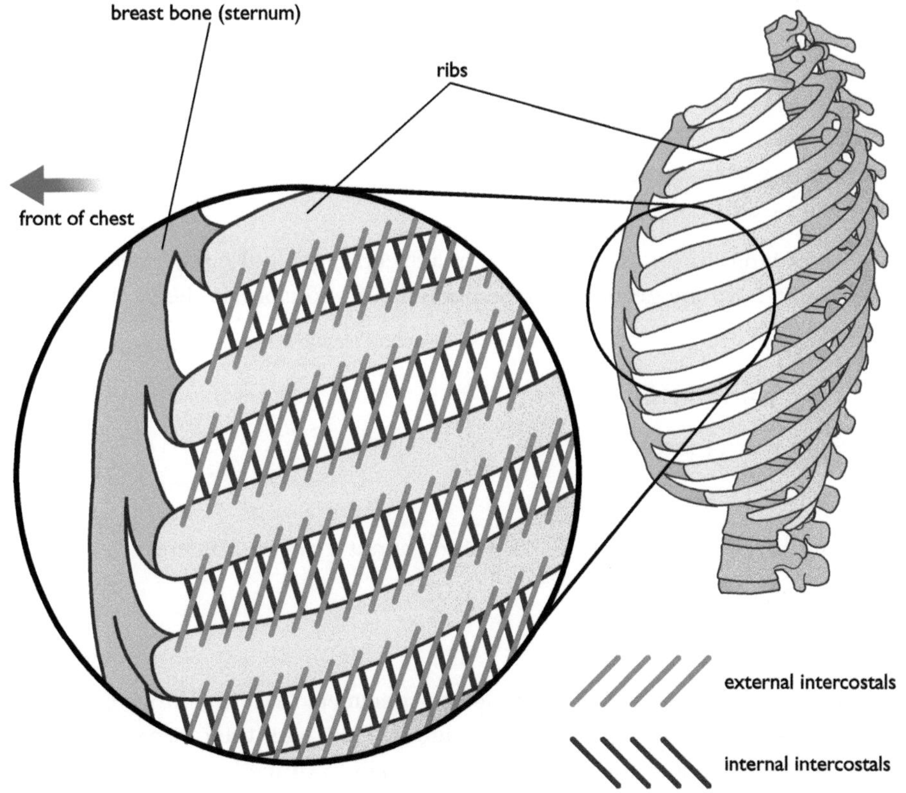

Figure 13.11 *X-ray of side view of the chest wall, showing the ribs. The diagram shows how the two sets of intercostal muscles run between the ribs. When the external intercostals contract, they move the ribs upwards. When the internal intercostals contract, the ribs are moved downwards.*

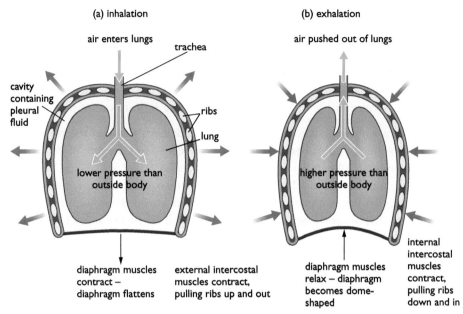

Figure 13.12 *Changes in the position of the ribs and diaphragm during breathing.*
(a) Breathing in (inhalation) and (b) breathing out (exhalation).

10. Where does gas exchange take place in the lungs?

11. Why are there C-shaped rings of cartilage in the trachea?

12. How is the trachea kept clean of dust and bacteria?

13. What happens to your external intercostal muscles and your diaphragm when you breathe in?

Activity 5: Examining the lungs of a mammal

SBA skills

ORR MM AI PD Dr

You will need:

- a set of lungs, bronchi and trachea of a sheep or goat (the specimen will probably have the oesophagus, diaphragm and possibly the heart attached)
- a dissecting board
- a sharp pair of scissors
- a blunt pair of forceps
- a scalpel
- a long piece of rubber or plastic tubing about 1 cm in diameter
- surgical gloves

Carry out the following:

1. Use surgical gloves when handling the specimen. Place the lungs on the dissecting board and arrange them side by side, with their curved surfaces facing upwards and the trachea towards the top of the board.
   You are now looking at the dorsal surface of the lungs (i.e. from the back of the animal). The heart may still be in place between the lungs.

2. Identify the following:
   - the right and left lungs
   - trachea
   - bronchi
   - larynx (voice box)
   - diaphragm
   - oesophagus
   - heart (if present).

3. Make a drawing of the lungs and other organs. Label the drawing, and add a title and a scale.

4. Examine the trachea and note how it divides to form the bronchi.

5. Examine the outside of the trachea and note the rings of cartilage in the trachea wall. Using a pair of scissors, cut open the trachea near the top and examine the inside of the trachea. Note the difference between the trachea and oesophagus.

6. Your teacher may be able to demonstrate the lungs inflating by passing a long tube down into the trachea and into one bronchus, and then blowing into the tube. Observe what happens. Note the elastic nature of the lungs – they immediately collapse when the blowing stops.

7. Cut through the lung tissue with a scalpel and examine its appearance. Note the smaller airways (bronchioles) and blood vessels in the lungs.

Gas exchange in the alveoli

You can tell what is happening during gas exchange if you compare the amounts of different gases in atmospheric air with the air breathed out (Table 13.3).

Gas	Atmospheric air	Exhaled air
nitrogen	78	79
oxygen	21	16
carbon dioxide	0.04	4
other gases (mainly argon)	1	1

Table 13.3: *Approximate percentage volume of gases in atmospheric (inhaled) and exhaled air.*

Exhaled air is also warmer than atmospheric air, and is saturated with water vapour. The amount of water vapour in the atmosphere varies, depending on weather conditions.

Clearly, the lungs are absorbing oxygen into the blood and removing carbon dioxide from it. This happens in the alveoli. To do this efficiently, the alveoli must have a structure which brings the air and blood very close together, over a very large surface area. There are enormous numbers of alveoli. It has been

Be careful when interpreting percentages! The *percentage* of a gas in a mixture can vary, even if the actual *amount* of the gas stays the same. This is easiest to understand from an example. Imagine you have a bottle containing a mixture of 20% oxygen and 80% nitrogen. If you used a chemical to absorb all the oxygen in the bottle, the nitrogen left would now be 100% of the gas in the bottle, despite the fact that the *amount* of nitrogen would still be the same. That is why the percentage of nitrogen in inhaled and exhaled air is slightly different.

calculated that the two lungs contain about 700 000 000 of these tiny air sacs, giving a total surface area of 60 m². That's bigger than the floor area of an average classroom! Viewed through a high-powered microscope, the alveoli look rather like bunches of grapes, and they are covered with tiny blood capillaries (Figure 13.13).

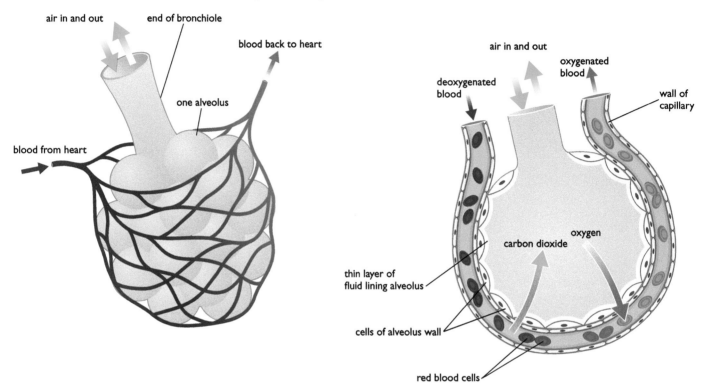

Figure 13.13 *An alveolus and its surrounding capillary network. Diffusion of oxygen and carbon dioxide takes place between the air in the alveolus and the blood in the capillaries.*

Cross-curricular links:
physics, diffusion

The thin layer of fluid lining the inside of the alveoli comes from the blood. The capillaries and cells of the alveolar wall are 'leaky' and the blood pressure pushes fluid out from the blood plasma into the alveolus. Oxygen dissolves in this moist surface before it passes through the alveolar wall into the blood.

Blood is pumped from the heart to the lungs and passes through the capillaries surrounding the alveoli. The blood has come from the respiring tissues of the body, where it has given up some of its oxygen to the cells, and gained carbon dioxide. Around the lungs, the blood is separated from the air inside each alveolus by only two cell layers; the cells making up the wall of the alveolus, and the capillary wall itself. This is a distance of less than one thousandth of a millimetre.

Because the air in the alveolus has a higher concentration of oxygen than the blood entering the capillary network, oxygen diffuses from the air, across the wall of the alveolus and into the blood. At the same time there is more carbon dioxide in the blood than there is in the air in the lungs. This means that there is a diffusion gradient for carbon dioxide in the other direction, so carbon dioxide diffuses the other way, out of the blood and into the alveolus. The result is that the blood which leaves the capillaries and flows back to the heart has gained oxygen and lost carbon dioxide. The heart then pumps the blood around the body again, to supply the respiring cells.

14. What is the total internal surface area of the alveoli?
15. How many cell layers does oxygen pass through when it diffuses from the air in the alveolus into the blood?

Activity 6: Comparing the carbon dioxide content of inhaled and exhaled air

SBA skills

ORR	MM	AI	PD	Dr

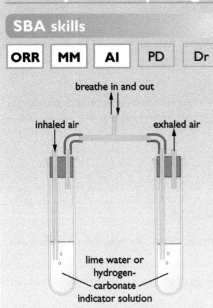

Figure 13.14 *Apparatus to compare the amount of carbon dioxide in inhaled and exhaled air.*

You will need:

* two boiling tubes
* delivery tubes and bungs as shown in Figure 13.14
* some lime water

Carry out the following:

1. Set up the apparatus as shown in Figure 13.14.
2. Gently breathe through the rubber tube. Air should be drawn in through one boiling tube and pass out through the other.
3. Observe any changes in the lime water after a few minutes.
4. Write up the investigation and explain the results.
5. In which tube did the lime water change first, the one receiving inhaled or exhaled air? What does this tell you about the concentration of carbon dioxide in inhaled and exhaled air?

Activity 7: Do other organisms produce carbon dioxide?

SBA skills

ORR	MM	AI	PD	Dr

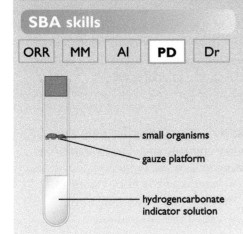

Figure 13.15 *Apparatus for measuring carbon dioxide production by small animals.*

Design your own experiment to find out if other living organisms (such as maggots, germinating seeds or woodlice) produce carbon dioxide. You do not need to use apparatus as complicated as that shown in Figure 13.14. You could use a boiling tube with a platform to support the organisms (Figure 13.15), with lime water or hydrogencarbonate indicator solution underneath.

From your observations of these three organisms, you could predict which of them will respire the quickest, and show the highest rate of production of carbon dioxide. When you have developed your hypothesis, you can plan how to test it. You will need to ensure that your comparison between the three organisms is 'fair'. Don't forget to include a description of a control that you would set up. After you have shown your plan to your teacher, you may be able to carry out the investigation.

ORR	MM	AI	PD	Dr

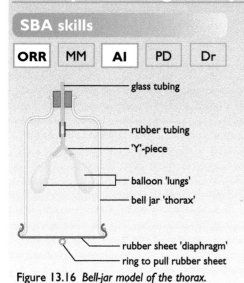

- glass tubing
- rubber tubing
- 'Y'-piece
- balloon 'lungs'
- bell jar 'thorax'
- rubber sheet 'diaphragm'
- ring to pull rubber sheet

Figure 13.16 *Bell-jar model of the thorax.*

You will need:

- the apparatus shown in Figure 13.16

Carry out the following:

1. Grip the centre of the rubber sheet and pull gently downwards. Observe what happens to the balloon 'lungs'.

2. Push gently upwards on the rubber sheet and again watch what happens to the 'lungs'.

3. Write up the experiment, explaining your observations in terms of volume and pressure changes.

4. This model of the thorax mimics the action of the diaphragm and lungs, but which part of the ventilation mechanism is not shown by the model?

Gas exchange in other organisms

The human alveoli show several features which allow them to be an efficient gas exchange surface. These are:

- they present a very large surface area for gas diffusion

- their walls are thin and porous for rapid diffusion of gases

- they have an large intimate blood system to deliver carbon dioxide and remove oxygen from the gas exchange surface, maintaining concentration gradients of these gases

- they are ventilated (by the breathing movements), which also maintains concentration gradients of the respiratory gases.

Many other animals have evolved specialised gas exchange surfaces, such as lungs, gills and other structures. These too are usually thin, ventilated surfaces with a good blood supply. A common feature of all of them is their large surface area. A good example is seen in the gills of a fish.

Gas exchange in fish

Fish extract oxygen that is dissolved in water, using their **gills**. Carbon dioxide is excreted out into the water in the opposite direction. Volume for volume, water contains only about one-twentieth of the oxygen that air does, so fish need to pass very large volumes of water over the gills. Figure 13.17 shows the location of the gills of a fish. They are in a cavity behind the mouth, covered by a flap of skin and bone called the **operculum**.

Water flows in through the fish's mouth and over the gills, and leaves via the gap between the operculum and the body of the fish. To keep water flowing over the gills, some fish swim with their mouths open. Other species ventilate their gills by breathing movements of the mouth cavity. Whichever method

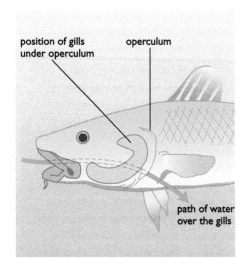

position of gills under operculum

operculum

path of water over the gills

Figure 13.17 *The location of the gills of a fish.*

they use, enough oxygen is only extracted from the water because of the very large surface area of the gills. Each gill is supported by a bony piece of tissue called a **gill bar**, which has many thin **gill filaments** attached to it (Figure 13.18). Each bar has two rows of filaments, and each filament has secondary filaments or lamellae attached to it. These lamellae are the gas exchange surface of the fish. Each **lamella** is well supplied with blood capillaries, and the wall of each lamella is only one cell thick, so that oxygen can easily diffuse through into the blood. Carbon dioxide diffuses out the other way, into the water. Ventilation of the gills ensures that diffusion gradients are maintained.

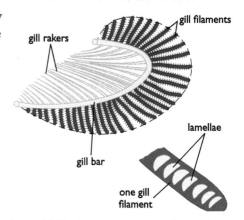

Figure 13.18 *The structure of the gills of a fish.*

Activity 9: Dissection of a fish to show the gills

SBA skills

| ORR | MM | AI | PD | Dr |

You will need:

- a fresh, dead fish
- a dissecting board
- a sharp pair of scissors
- a hand lens
- a seeker
- a pair of forceps
- a small dish (e.g. a Petri dish)

Carry out the following:

1. Find the operculum of the fish and lift it gently with a seeker. Observe the gills lying underneath. (Why are they this colour?)

2. Push the seeker into the fish's mouth so that it passes out under the operculum. This is the pathway that water takes when the fish is breathing.

3. Use scissors and forceps to cut away the operculum, exposing the gills. Cut away one gill, removing as much as possible. Place it in a Petri dish of water.

4. Examine the gill filaments and lamellae with a hand lens. Draw the gill, adding labels and annotations. Don't forget to add a scale to your drawing.

5. What happens to the gill filaments and lamellae when the gill is out of water? Suggest one reason why a fish is unable to breathe out of water.

Gas exchange in plants

The main gas exchange surface of green plants is the spongy mesophyll tissue of the leaves (Figure 13.19, page 184). Plants carry out both photosynthesis and respiration, and both of these processes require exchange of gases with the air. In daylight, the rate of photosynthesis exceeds the rate of respiration, so that carbon dioxide is taken up by the leaves and oxygen is lost to the

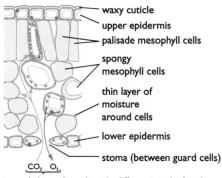

waxy cuticle
upper epidermis
palisade mesophyll cells
spongy mesophyll cells
thin layer of moisture around cells
lower epidermis
stoma (between guard cells)

CO_2 O_2

in sunlight carbon dioxide diffuses into leaf and oxygen diffuses out

Figure 13.19 *Section through a leaf showing gas exchange at the spongy mesophyll.*

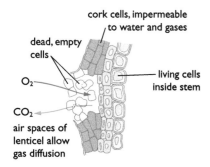

cork cells, impermeable to water and gases
dead, empty cells
living cells inside stem
O_2
CO_2
air spaces of lenticel allow gas diffusion

Figure 13.20 *Structure of a lenticel in a woody stem.*

air around the leaves (see Chapter 9). Carbon dioxide diffuses into the leaf through the stomata, entering the air spaces in the spongy mesophyll. The carbon dioxide then dissolves in the thin layer of moisture on the surface of the mesophyll cells, and enters the cells to be used in the process of photosynthesis:

$$\text{carbon dioxide} \quad + \quad \text{water} \quad \xrightarrow[\text{chlorophyll}]{\text{light}} \quad \text{glucose} \quad + \quad \text{oxygen}$$

The oxygen diffuses out in the other direction, and exits from the leaf through the stomata (Figure 13.19). In dim light, respiration exceeds photosynthesis and the direction of diffusion of these two gases is reversed. You can observe the structure of the spongy mesophyll and stomata in Chapter 9, Activity 3, page 118.

The shape and number of leaves already provides a large surface area in comparison to their volume. This means that the large internal surface area of the spongy mesophyll is adequate for gas exchange to take place by diffusion alone, so that ventilation of the leaves is not necessary.

Other parts of a flowering plant that are not green, such as roots and woody stems, do not photosynthesise, but they still need a supply of oxygen for cell respiration. Roots get their oxygen by diffusion from air spaces in the soil. Woody stems are covered with a layer of cork cells making up the bark. These dead cells are not permeable to gases, so there are regions where the cork cells are loosely packed, allowing oxygen to diffuse through. These regions are called **lenticels** (Figure 13.20).

The harmful effects of smoking

If the lungs are to be able to exchange gases properly, the air passages need to be clear, the alveoli to be free from dirt particles and bacteria, and they must have as big a surface area as possible in contact with the blood. There is one habit that can upset all of these conditions – smoking.

Links between smoking and diseases of the lungs are now a proven fact. Smoking is associated with the diseases lung cancer, bronchitis and emphysema. It is also a major contributing factor to other problems, such as coronary heart disease and ulcers of the stomach and duodenum (part of the intestine). Pregnant women who smoke are more likely to give birth to underweight babies. We need to deal with some of these effects in more detail.

Effects of smoke on the lining of the air passages

You saw above how the lungs are kept free of particles of dirt and bacteria by the action of mucus and cilia (Figure 13.10, page 176). In the trachea and bronchi of a smoker, the cilia are destroyed by the chemicals in cigarette smoke. The reduced numbers of cilia mean that the mucus is not swept away from the lungs, but remains to clog the air passages. This is made worse by the fact that the smoke irritates the lining of the airways, stimulating the cells to secrete more mucus. This clogging mucus is the source of the 'smoker's cough'. Irritation of the bronchial tree, along with infections from bacteria in the mucus can cause the lung disease **bronchitis**. Bronchitis blocks the normal flow of air, so the sufferer has difficulty breathing properly.

Emphysema is another fatal lung disease caused by smoking. Smoke damages the walls of the alveoli, which break down and fuse together again, forming enlarged, irregular air spaces (Figure 13.21). This greatly reduces the surface area for gas exchange, which then becomes very inefficient. The blood of a person with emphysema carries less oxygen. In serious cases, this leads to the sufferer being unable to carry out even mild exercise, such as walking. Emphysema patients often have to have a supply of oxygen nearby at all times. There is no cure for emphysema, and usually the sufferer dies after a long and distressing illness.

Lung cancer

Cigarette smoke contains a strongly addictive drug – **nicotine**. It also contains at least 17 chemicals that are known to cause cancer, and several more that are suspected of doing so. These chemicals are called **carcinogens**, and are contained in the **tar** that collects in a smoker's lungs. Cancer happens when cells mutate and start to divide uncontrollably, forming a **tumour** (Figure 13.22). If a lung cancer patient is lucky, they may have the tumour removed by an operation before the cancer cells spread to other tissues of the body. Unfortunately, tumours in the lungs usually cause no pain, so they are not discovered until it is too late: it may be inoperable, or tumours may have developed elsewhere.

If you smoke, you are not bound to get lung cancer, but the risks that you will get it are much greater. In fact, the more cigarettes you smoke, the greater is the risk that you will get the disease (Figure 13.23).

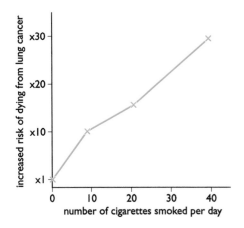

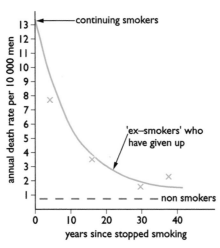

Figure 13.23 The more cigarettes a person smokes, the more likely it is they will die of lung cancer. For example, smoking 20 cigarettes a day increases the risk by about 15 times.

Figure 13.24 Death rates from lung cancer for smokers, non-smokers and ex-smokers.

The obvious thing to do is not to start smoking. However, if you are a smoker, giving up the habit soon improves your chance of survival (Figure 13.24). After a few years, the likelihood of your dying from a smoking-related disease is almost back to the level of a non-smoker.

Carbon monoxide in smoke

One of the harmful chemicals in cigarette smoke is the poisonous gas **carbon monoxide**. When this gas is breathed in with the smoke, it enters the bloodstream and interferes with the ability of the blood to carry oxygen.

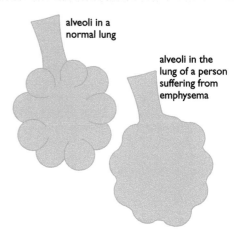

Figure 13.21 The alveoli of a person suffering from emphysema have a greatly reduced surface area and inefficient gas exchange.

People often talk about 'yellow nicotine stains'. In fact it is the tar that stains a smoker's fingers and teeth. Nicotine is a colourless, odourless chemical.

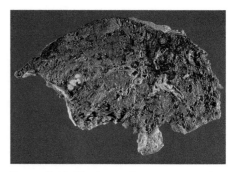

Figure 13.22 This lung is from a patient with lung cancer.

Studies have shown that the type of cigarette smoked makes very little difference to the smoker's risk of getting lung cancer. Filtered and 'low tar' cigarettes only reduce the risk slightly.

Oxygen is carried around in the blood in the red blood cells, attached to a chemical called **haemoglobin** (see Chapter 14). Carbon monoxide can combine with the haemoglobin much more tightly than oxygen can, forming a compound called **carboxyhaemoglobin**. The haemoglobin will combine with carbon monoxide in preference to oxygen. When this happens, the blood carries much less oxygen around the body. Carbon monoxide from smoking is also a major cause of heart disease.

If a pregnant woman smokes, she will be depriving her unborn fetus of oxygen. This has an effect on its growth and development, and leads to the mass of the baby at birth being lower, on average, than the mass of babies born to non-smokers.

Figure 13.25 shows apparatus that demonstrates some of the harmful products of cigarette smoke. Smoke from the cigarette is drawn through the U-tube by a filter pump. Cotton wool in the U-tube traps smoke particles and tar, which shows up as a grey-brown stain. The Universal Indicator solution changes colour from green to yellow, showing that smoke contains acidic gases. After the demonstration, if you remove the cotton wool with forceps, its smell is disgusting! This is the material that collects in a smoker's lungs.

> Take care not to get the tar from the cotton wool on your skin. It is very carcinogenic.

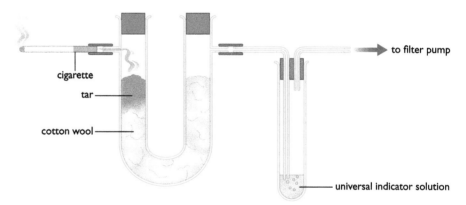

Figure 13.25 A 'smoking machine'.

> **16.** Nicotine is addictive. What does this mean?
> **17.** Name three lung diseases caused by smoking.
> **18.** What does the term carcinogenic mean?

Caribbean smoking statistics

Despite all the evidence of the harm caused by smoking, and the many action plans that have been introduced to reduce tobacco consumption, people in the Caribbean continue to smoke. The figures vary from country to country. Taking Jamaica as an example, a World Health Organisation (WHO) report in 2010 showed that in adults (aged 15 and over) 22.9% of men smoked tobacco (mostly cigarettes), and 7.5% of women.

The situation in children is particularly worrying. The WHO Global Youth Tobacco Survey (2010) found that in Jamaica, 40% of 13–15-year-olds had tried smoking cigarettes, 20% before the age of 10. Of the boys, 21.5% continued to smoke, as did 14.3% of the girls.

Another WHO report (2011) showed that deaths from lung cancer in Jamaica are relatively high (ranked 78th in the world), at about 24 deaths per 100 000.

Activity 10: Interpreting smoking data

SBA skills

ORR MM AI PD Dr

You will need:

* access to the Internet

Carry out the following:

1. Carry out an Internet search to find data on smoking statistics in your country, and how they compare with other countries throughout the world. Try to find the most up-to-date data for cigarette use and for deaths from lung cancer and other smoking-related diseases.

2. Write a report of your findings. Illustrate your report using graphs and tables.

Marijuana

Marijuana (or **cannabis**) is a preparation made from leaves and flowers of the cannabis plant. It is normally mixed with tobacco and smoked. It is a **psychoactive** drug, affecting the nervous system (see Chapter 18). It produces effects such as relaxation and heightened sociability, although in some people it can induce feelings of anxiety and paranoia.

It has been estimated that about 4% of the world's population use marijuana regularly and about 0.6% use it daily, making it one of the most widely used illegal drugs. There is much evidence that marijuana users become dependent on or **addicted** to the drug. The cultivation, sale and use of marijuana are illegal in most Caribbean countries, but this is often overlooked and marijuana may be sold openly.

Many people think that smoking marijuana is safer than smoking tobacco, but this is not true. Long-term use carries the same health risks as smoking tobacco, and the tar contains similar cancer-causing chemicals.

Figure 13.26 *An illegal cannabis plantation.*

Chapter summary

In this chapter you have learnt that:

- aerobic respiration is a chemical reaction that takes place in living cells. Foods such as glucose are oxidised, using oxygen gas, to give energy. The waste products are carbon dioxide and water. Heat energy is also produced

- breathing does not mean the same thing as respiration. Breathing is the way that gases are exchanged with the atmosphere

- respiration takes place in a series of small steps, catalysed by enzymes. During these reactions, the energy released from the glucose is used to make a compound called ATP

- ATP is a universal energy currency in cells, and can be 'spent' on other processes that require energy

- respiration can take place without oxygen. This is called anaerobic respiration. In muscles anaerobic respiration of glucose produces lactic acid. In yeast, it produces ethanol and carbon dioxide

- anaerobic respiration in certain bacteria can be used to make biogas, which is a renewable fuel

- the respiratory system or bronchial tree consists of the trachea, bronchi, bronchioles and alveoli. Air is drawn into the air passages leading to the alveoli as a result of changes in volume of the chest cavity

- ventilation of the lungs takes place due to the action of the ribs and intercostal muscles, and the diaphragm

- the alveoli are the human gas exchange surface. They have a very large surface area, which allows for rapid and efficient exchange of the respiratory gases, as well as thin walls and intimate blood capillary networks

- other gas exchange surfaces are similarly adapted by having a large surface area, such as the gill filaments and lamellae of a fish and the spongy mesophyll in the leaves of plants

- smoking causes nicotine addiction. The tar in smoke damages the lungs, causing a number of respiratory diseases, including bronchitis, emphysema and lung cancer

- carbon monoxide in smoke reduces the oxygen carrying capacity of the blood

- marijuana is an addictive drug, and causes the same health problems as smoking tobacco.

Questions

1. Which of the following statements about aerobic respiration is/are true? Aerobic respiration produces:

   I   oxygen
   II   carbon dioxide
   III energy
   IV   excretory products

   A   I, II, III and IV
   B   I, II and III only
   C   II, III and IV only
   D   I, II and IV only

2. The apparatus below was set up to find out if germinating seeds produce carbon dioxide.

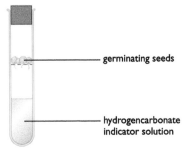

   — germinating seeds

   — hydrogencarbonate indicator solution

   Which of the following would be the best control for this experiment?

   A   a tube containing dead seeds instead of germinating ones
   B   a tube containing no seeds
   C   a tube containing small living animals
   D   a tube containing glass beads instead of germinating seeds

3. Which of the following changes take place during inhalation (breathing in)?

   A   The diaphragm contracts, the internal intercostals contract, the pressure in the chest cavity decreases.
   B   The diaphragm relaxes, the external intercostals contract, the pressure in the chest cavity decreases.
   C   The diaphragm relaxes, the internal intercostals relax, the pressure in the chest cavity increases.
   D   The diaphragm contracts, the external intercostals contract, the pressure in the chest cavity decreases.

4. Respiratory enzymes are mostly located in the:
   A   nucleus
   B   alveolus
   C   chloroplasts
   D   mitochondria

5. After vigorous exercise, which of the following will accumulate in muscle cells?

   A   lactic acid
   B   ethanol
   C   glucose
   D   haemoglobin

6. Which of the following is **not** a feature of respiratory surfaces?

   A   large surface area
   B   close proximity to blood capillaries
   C   thick walls
   D   moist lining

7. Copy and complete the table, which shows what happens in the thorax during ventilation of the lungs. Two boxes have been completed for you. *(12)*

	Action during inhalation	Action during exhalation
external intercostal muscles	contract	
internal intercostal muscles		
ribs		move down and in
diaphragm		
volume of thorax		
pressure in thorax		
volume of air in lungs		

8. Some people injured in an accident such as a car crash suffer from a *pneumothorax*. This is an injury where the chest wall is punctured, allowing air to enter the pleural cavity (see Figure 13.7, page 176). A patient was brought to the casualty department of a hospital, suffering from a pneumothorax on the left side of his chest. His left lung had collapsed, but he was able to breathe normally with his right lung.

   a)   Explain why a pneumothorax caused the left lung to collapse. *(3)*

   b)   Explain why the right lung was not affected. *(2)*

   c)   If a patient's lung is injured or infected, a surgeon can sometimes 'rest' it by performing an operation called an *artificial pneumothorax*. What do you think might be involved in this operation? *(2)*

9. Briefly explain the importance of the following:

   a)   The trachea wall contains C-shaped rings of cartilage. *(4)*

   b)   The distance between the air in an alveolus and the blood in an alveolar capillary is less than 1/1000th of a millimetre. *(2)*

   c)   The lining of the trachea contains mucus-secreting cells and cells with cilia. *(4)*

d) Smokers have a lower concentration of oxygen in their blood than non-smokers. *(3)*

e) The alveoli of the lungs have a surface area of about 60 m² and a good blood supply. *(3)*

10. A long-term investigation was carried out into the link between smoking and lung cancer. The smoking habits of male doctors aged 35 or over were determined while they were still alive, then the number and causes of deaths among them were monitored over a number of years. (Note that this survey was carried out in the 1950s – very few doctors smoke these days!). The results are shown in the graph.

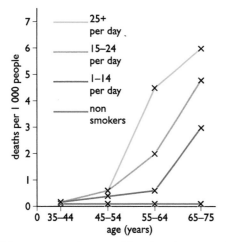

a) Write a paragraph to explain what the researchers found out from the investigation. *(6)*

b) How many deaths from lung cancer would be expected for men aged 55 who smoked 25 cigarettes a day up until their death? How many deaths from lung cancer would be expected for men in the same age group smoking 10 cigarettes a day? *(2)*

c) Explain the difference between the lung diseases bronchitis and emphysema. *(5)*

11. The table shows the concentration of gases in inhaled and exhaled air.

Gas	Inhaled air	Exhaled air
nitrogen	78	79
oxygen		
carbon dioxide		
other gases (mainly argon)	1	1

a) Copy the table and fill in the gaps by choosing from the following numbers:

21   4   0.04   16 *(2)*

b) Explain why the concentration of carbon dioxide is so different. *(2)*

c) Explain why exhaling is a form of excretion. *(2)*

d) The following features can be seen in the lungs:

i) thin membranes between the alveoli and the blood supply

ii) a good blood supply

iii) a large surface area.

In each case explain how the feature helps gas exchange to happen quickly. *(6)*

12. A respirometer is used to measure the rate of respiration. The diagram shows a simple respirometer. The sodium hydroxide solution in the apparatus absorbs carbon dioxide. Some results from the investigation are also shown.

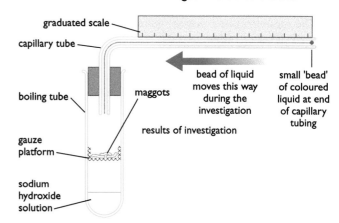

Experiment	Distance moved by bead /mm
1	20
2	3
3	18

Assume that the maggots in the apparatus respire aerobically.

i) Write the symbol equation for aerobic respiration. *(4)*

ii) From the equation, what can you assume about the amount of oxygen taken in and carbon dioxide given off by the maggots? Explain your answer. *(3)*

iii) Result 2 is significantly different from the other two results. Suggest a reason for this. *(2)*

iv) How would the results be different if the organisms under investigation respired anaerobically? *(2)*

13. a) Explain fully what is meant by the term respiration. *(4)*

b) Adenosine triphosphate is sometimes described as the energy currency of the cell. Explain what this means. *(3)*

c) Give *three* reasons why animals and plants need ATP. *(3)*

d) Describe the involvement of the ribs, intercostal muscles and diaphragm in the ventilation of the lungs. *(8)*

14. **a)** What is meant by the term anaerobic respiration? (3)

   b) Write a balanced symbol equation for the process of anaerobic respiration in yeast. (3)

   c) Describe a simple method you could use to demonstrate anaerobic respiration in yeast. Include a diagram of the apparatus. (6)

   d) What is the product of anaerobic respiration in muscle? Suggest why muscles do not respire anaerobically all the time. (4)

   e) A runner in a 100-m race obtains 95% of his energy from anaerobic respiration, whereas a runner in a 10 000-m race only gets 10% of his energy this way. Explain why there is this difference. (4)

15. **a)** Compare the characteristics of the structures involved in gas exchange in humans and fish. (5)

   b) Suggest *two* ways that cigarette smoking can reduce the efficiency of gas exchange in the lungs. Explain your answers fully. (6)

Chapter 14: Transport in Mammals

When you have completed this chapter, you will be able to:

- understand why small organisms do not need a transport system, but large organisms do
- identify the materials that need to be transported in mammals
- describe the structure and functions of the human circulatory system
- describe the structure and function of the heart
- explain how the structures of the three types of blood vessel are suited to their function
- list the components of blood and identify the main types of blood cells
- describe the functions of the main components of blood
- explain how immunisation is used to control disease.

Large, multicellular animals like mammals need a circulatory system to transport substances to and from the cells of the body. This chapter looks at the structure and function of the circulatory system and the composition and functions of the blood.

The need for circulatory systems

Figure 14.1 shows the human circulatory system.

Blood is pumped around a closed circuit made up of the heart and blood vessels. As it travels around, it collects materials from some places and unloads them in others. In mammals, blood transports:

- oxygen from the lungs to all other parts of the body
- waste carbon dioxide from all parts of the body to the lungs
- nutrients from the gut to all parts of the body
- the nitrogenous waste product urea from the liver to the kidneys.

Hormones, antibodies, alcohol, drugs and many other substances are also transported by the blood. Blood also distributes heat around the body.

Single-celled organisms, like the ones shown in Figure 14.2, do not have circulatory systems.

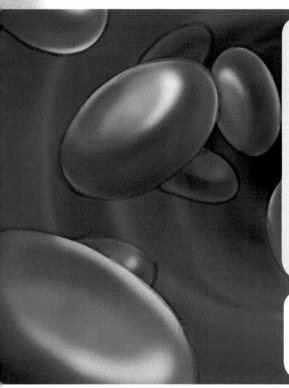

pulmonary artery
pulmonary vein
heart
vena cava
aorta

← oxygenated blood
→ deoxygenated blood

Figure 14.1 *The human circulatory system.*

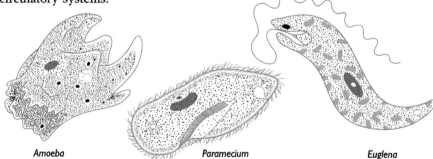

Amoeba   *Paramecium*   *Euglena*

Figure 14.2 *Unicellular organisms do not have circulatory systems.*

Materials can easily move around the cell without a special system. There is no need for lungs or gills to obtain oxygen from the environment either. One-celled organisms obtain oxygen through the cell surface membrane. The rest of the cell then uses the oxygen. The area of the cell surface determines how much oxygen the organism can get (the supply rate), and the volume of the cell determines how much oxygen the organism uses (the demand rate).

The ratio of supply to demand can be written as: $\dfrac{\text{surface area}}{\text{volume}}$.

This is called the **surface area to volume ratio** (s.a.:vol.) and it is affected by the size of an organism. Single-celled organisms have a high surface area to volume ratio. Their cell surface membrane has a large enough area to supply all the oxygen that their volume demands. In larger animals, the surface area to volume ratio is lower.

Large animals cannot get all the oxygen they need through their surface (even if the body surface would allow it to pass through) – there just isn't enough surface to supply all that volume. To overcome this problem, large organisms have evolved special gas exchange organs and circulatory systems. The gills of fish and the lungs of mammals are linked to a circulatory system that carries oxygen to all parts of the body. The same idea applies to obtaining nutrients – the gut obtains nutrients from food and the circulatory system distributes the nutrients around the body.

The circulatory systems of different animals

One of the main functions of a circulatory system in animals is to transport oxygen. Blood is pumped to a gas exchange organ to load up on oxygen. It is then pumped to other parts of the body where it unloads the oxygen. There are two main types of circulatory systems in animals:

- In **single circulatory systems**, the blood is pumped from the heart to the gas exchange organ and then directly to the rest of the body.

- In **double circulatory systems**, the blood is pumped from the heart to the gas exchange organ, back to the heart and then to the rest of the body.

Figure 14.4 shows the difference between these systems.

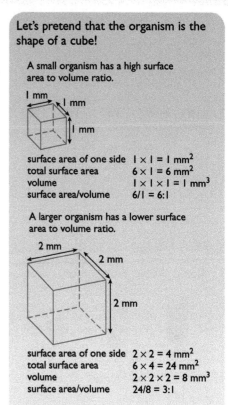

Let's pretend that the organism is the shape of a cube!

A small organism has a high surface area to volume ratio.

surface area of one side $1 \times 1 = 1$ mm^2
total surface area $6 \times 1 = 6$ mm^2
volume $1 \times 1 \times 1 = 1$ mm^3
surface area/volume $6/1 = 6{:}1$

A larger organism has a lower surface area to volume ratio.

surface area of one side $2 \times 2 = 4$ mm^2
total surface area $6 \times 4 = 24$ mm^2
volume $2 \times 2 \times 2 = 8$ mm^3
surface area/volume $24/8 = 3{:}1$

Figure 14.3 *An illustration of surface area to volume ratio.*

The bigger cube has a smaller surface area to volume ratio. It would be less able than the smaller cubes to obtain all the oxygen it needs through its surface.

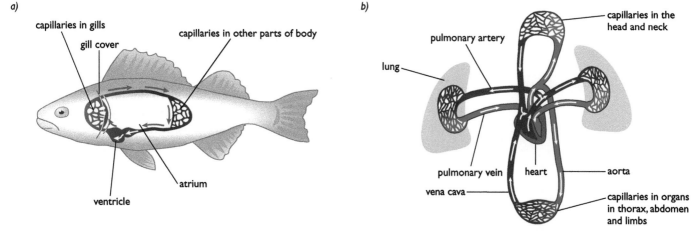

Figure 14.4 *(a) The single circulatory system of a fish. The blood passes through the heart once only in a complete circuit of the body. (b) The double circulatory system of a human (and other mammals). The blood passes through the heart twice in one complete circuit of the body.*

Activity 1: Does diffusion take longer in structures with a smaller surface area to volume ratio?

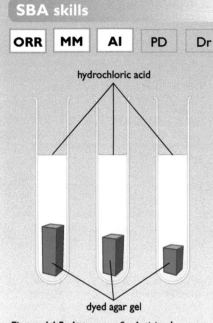

Figure 14.5 *Apparatus for Activity 1.*

The hydrochloric acid molecules diffuse into the gel cubes. They move from a high concentration outside the cube of gel to a lower concentration inside the cube. The amount of hydrochloric acid *needed* to decolourise the cube depends on the *volume*. The surface area determines the rate of supply.

You will need:

- some agar gel which has been dyed using potassium permanganate (this may be already cut into different sized cubes for you)

- dilute hydrochloric acid

- test-tubes or 100 cm³ beakers

- a clock

Carry out the following:

1. Cut the agar gel into four different shapes:

    - 1 cm × 1 cm × 1 cm (s.a.:vol. = 6:1)

    - 2 cm × 1 cm × 0.5 cm (s.a.:vol. = 7:1)

    - 4 cm × 0.5 cm × 0.5 cm (s.a.:vol: = 8.5:1)

2. Place equal amounts of hydrochloric acid into each test-tube or beaker – one for each piece of gel.

3. Place the pieces of agar gel into the test-tubes/beakers *at the same time.* Note the time.

4. Observe until each piece of gel turns completely colourless. Note the times for each piece to change.

5. Record how long it took for each piece of gel to turn colourless.

6. Write up your experiment and try to explain the differences in the times taken for the cubes to turn colourless.

7. How is this similar to the problem of obtaining oxygen in different sized animals?

Cross-curricular links:
physics, diffusion; chemistry, particulate nature of matter

Pulmonary means concerning the lungs.

There are two distinct parts to a double circulation:

- the **pulmonary** circulation, in which blood is circulated through the lungs where it is oxygenated

- the **systemic** circulation, in which blood is circulated through all other parts of the body where it unloads its oxygen.

A double circulatory system is more efficient than a single circulatory system. The heart pumps the blood twice, so higher pressures can be maintained. The blood travels more quickly to organs. In the single circulatory system of a fish, blood loses pressure as it passes through the gills. It then travels relatively slowly to the other organs.

The human circulatory system comprises:

- the **heart** – this is a pump

- **blood vessels** – these carry the blood around the body; **arteries** carry blood away from the heart and towards other organs, **veins** carry blood towards the heart and away from other organs, **capillaries** carry blood through organs

- **blood** – the transport medium.

Figure 14.6 shows the main blood vessels in the human circulatory system.

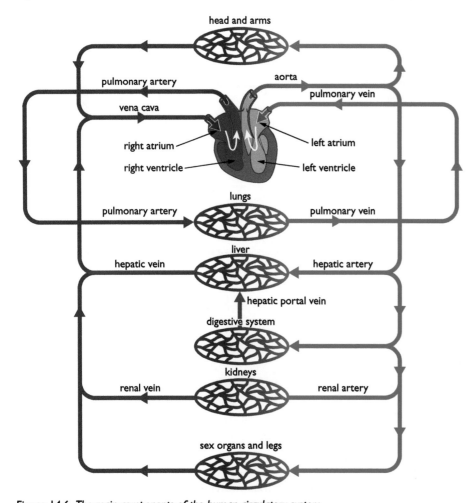

Figure 14.6 *The main components of the human circulatory system.*

1. Explain why small organisms do not need transport systems.
2. What is the main advantage of a double circulatory system over a single circulatory system?
3. List the main components of the human circulatory system and give the functions of each.
4. What are the pulmonary and systemic circulations?
5. Name the organs to which the following arteries carry blood: renal artery, hepatic artery, pulmonary artery.

Cardiac means 'o do with the heart.

The structure and function of the human heart

The human heart is a pump. It pumps blood around the body at different speeds and at different pressures according to the body's needs. It can do this because the wall of the heart is made from **cardiac muscle**. Cardiac muscle is unlike any other muscle in our bodies. It never gets fatigued (tired) like skeletal muscle. On average, cardiac muscle fibres contract and then relax again about 70 times a minute. In a lifetime of 70 years, this special muscle will contract over 2 billion times – and never take a rest!

The bicuspid (mitral) and tricuspid valves are both sometimes called **atrio-ventricular** valves, as each controls the passage of blood from an atrium to a ventricle.

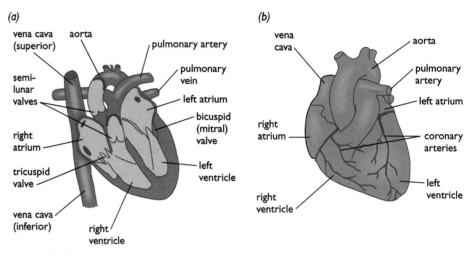

(a)

vena cava (superior)   aorta   pulmonary artery
semi-lunar valves   pulmonary vein
right atrium   left atrium
bicuspid (mitral) valve
tricuspid valve   left ventricle
vena cava (inferior)   right ventricle

(b)

vena cava   aorta
pulmonary artery
right atrium   left atrium
coronary arteries
right ventricle   left ventricle

Figure 14.7 *The human heart (a) vertical section, and (b) external view.*

Activity 2: Dissecting a sheep's heart

SBA skills

ORR | MM | AI | PD | Dr

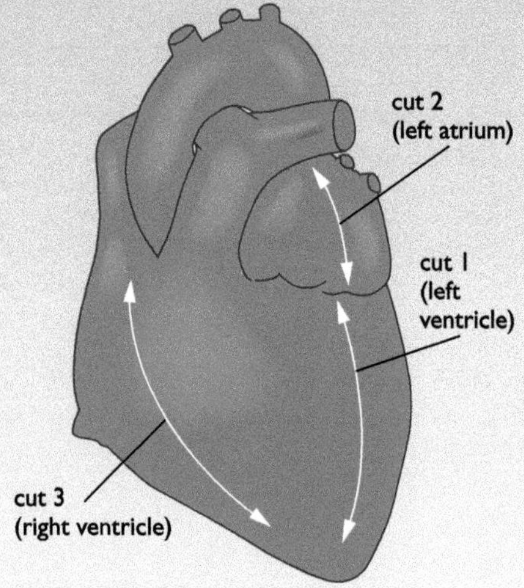

cut 2 (left atrium)

cut 1 (left ventricle)

cut 3 (right ventricle)

Figure 14.8 *Guide to dissection.*

You will need:

- a board on which to dissect the heart
- a sheep's heart
- dissecting equipment

Carry out the following:

1. Lay the heart on the board so that you can see the structures shown in Figure 14.8.

2. Make a cut through the wall of the heart, shown in the diagram as cut 1.

3. Pull the cut edges of the heart wall apart. Try to find the following:
   a) the valve between the atrium and ventricle
   b) the valve at the base of the aorta
   c) the thick ventricle wall.

4. Make another cut, shown in the diagram as cut 2.

5. Open the cut edges and notice the thin wall of the left atrium.

6. Make a final cut (cut 3) and open the right ventricle wall. Notice that it is thinner than the wall of the left ventricle.

Blood is moved through the heart by a series of contractions and relaxations of cardiac muscle in the walls of the four chambers. These events form the **cardiac cycle**. The main stages are illustrated in Figure 14.9.

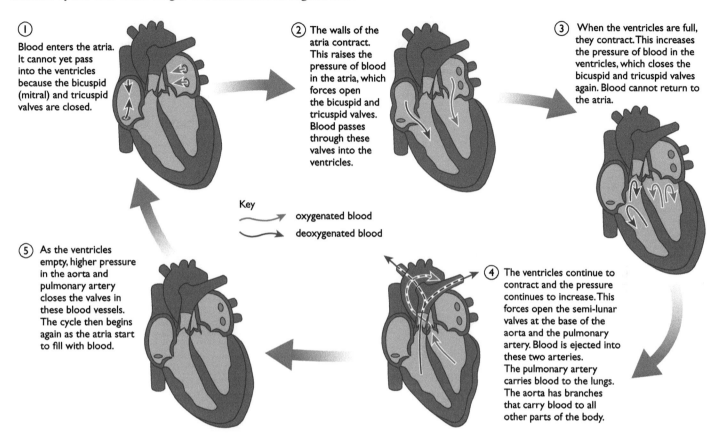

① Blood enters the atria. It cannot yet pass into the ventricles because the bicuspid (mitral) and tricuspid valves are closed.

② The walls of the atria contract. This raises the pressure of blood in the atria, which forces open the bicuspid and tricuspid valves. Blood passes through these valves into the ventricles.

③ When the ventricles are full, they contract. This increases the pressure of blood in the ventricles, which closes the bicuspid and tricuspid valves again. Blood cannot return to the atria.

Key
oxygenated blood
deoxygenated blood

⑤ As the ventricles empty, higher pressure in the aorta and pulmonary artery closes the valves in these blood vessels. The cycle then begins again as the atria start to fill with blood.

④ The ventricles continue to contract and the pressure continues to increase. This forces open the semi-lunar valves at the base of the aorta and the pulmonary artery. Blood is ejected into these two arteries. The pulmonary artery carries blood to the lungs. The aorta has branches that carry blood to all other parts of the body.

Figure 14.9 *The cardiac cycle.*

When a chamber of the heart is contracting, we say it is in **systole**. When it is relaxing, we say it is in **diastole**.

The structure of the heart is adapted to its function in several ways:

- It is divided into a left side and a right side by the **septum**. The right ventricle pumps blood only to the lungs while the left ventricle pumps blood to all other parts of the body. This requires much more pressure, which is why the wall of the left ventricle is much thicker than that of the right ventricle.

- Valves ensure that blood can flow only in one direction through the heart.

- The walls of the atria are thin. They can be stretched to receive blood as it returns to the heart, but can also contract with enough force to push blood through the bicuspid and tricuspid valves into the ventricles.

- The walls of the heart are made of cardiac muscle, which can contract and then relax continuously, without becoming fatigued.

- The cardiac muscle has its own blood supply – the **coronary circulation**. Blood reaches the muscle via **coronary arteries**. These carry blood to capillaries that supply the heart muscle with oxygen and nutrients. Blood is returned to the right atrium via **coronary veins**.

Cross-curricular links:
physics, pressure

Atria is the plural of **atrium**.

You will need:

- a timer that can time to an accuracy of 1 second
- permission to carry out the exercise that your teacher will suggest

Carry out the following:

1. Construct a table for your results like this one.

Time		Beats in 15 seconds	Pulse rate (beats per min)
before exercise	1		
	2		
	3		
	mean		
after exercise/s	0		
	30		
	60		
	90		
	60		
	90		
	120		
	150		
	180		
	210		
	240		

2. Find your pulse and record the number of beats in 15 seconds. Do this three times and find the mean (average) number of beats.

3. Carry out the exercise described by your teacher.

4. As soon as you have completed the exercise, find your pulse and record the number of beats in 15 seconds.

5. Do you notice any other difference in your pulse besides the change in rate?

6. Repeat step 4 every 30 seconds until your heart rate has returned to its pre-exercise level.

7. Write up your experiment.

> Each pulse that you feel is caused by the left ventricle of your heart contracting (beating) once.

Arteries, veins and capillaries

Arteries carry blood from the heart to the organs of the body. This blood (**arterial blood**) has been pumped out by the ventricles and puts a lot of pressure on the walls of the arteries. They must be able to 'give' under the pressure and allow their walls to stretch. They must also have the ability to recoil (pull back into shape) and help to push the blood along.

Veins carry blood from organs back towards the heart. The pressure of this blood (**venous blood**) is much lower than that in the arteries. It puts very little pressure on the walls of the veins. Veins must be able to allow the blood to pass through easily and prevent it from flowing in the wrong direction. Figure 14.10 shows the structure of a typical artery and a typical vein with the same diameter.

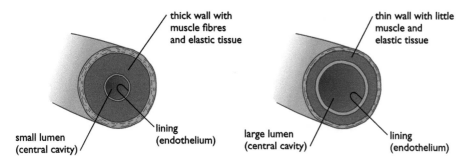

Figure 14.10 *The structure of (a) an artery and (b) a vein as seen in cross section.*

Veins also have valves called 'watch-pocket valves', which prevent the backflow of blood. The action of these valves is explained in Figure 14.11.

vein in longitudinal section

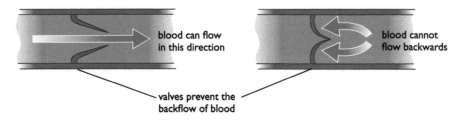

Figure 14.11 *The action of watch-pocket valves in veins.*

Capillaries carry blood through organs, bringing the blood close to every cell in the organ. Substances are transferred between the blood in the capillary and the cells. To do this, capillaries must be small enough to fit between cells, and allow materials to pass through their walls easily. Figure 14.13 shows the structure of a capillary and how exchange of substances takes place between the capillary and nearby cells.

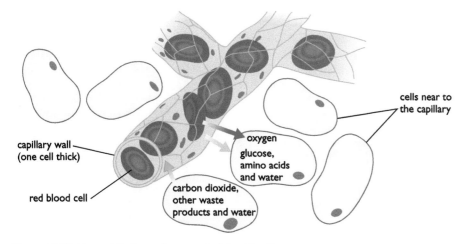

Figure 14.13 *How capillaries exchange materials with cells.*

Arterioles are small arteries. They carry blood into organs from arteries. Their structure is similar to the larger arteries, but they have a larger proportion of muscle fibres in their walls. They are also **innervated** (have nerve endings in their walls) and so can be made to dilate (become wider) or constrict (become narrower) to allow more or less blood into the organ.

If *all* the arterioles constrict, it is harder for blood to pass through them – there is more resistance. This increases blood pressure. Prolonged stress can cause arterioles to constrict and so increase blood pressure.

All arteries carry **oxygenated blood** (blood containing a lot of oxygen), except the pulmonary artery and the umbilical artery of an unborn baby. All veins carry **deoxygenated blood** (blood containing less oxygen) except the pulmonary vein and umbilical vein.

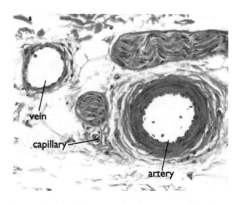

Figure 14.12 *Photograph of a section through an artery, vein and capillary.*

Activity 4: Examining a slide of blood vessels

SBA skills

ORR | MM | AI | PD | Dr

You will need:

- a microscope
- a prepared slide of a section through an artery, a vein and a capillary

Carry out the following:

1. Examine the slide of sections through the blood vessels. Use Figure 14.12, page 199, to help you identify the artery, vein and capillary.

2. Use Figure 14.10, page 199, to help you locate the different tissues in the wall of the artery and vein.

3. Make a drawing of the three blood vessels. Label the drawing and give it a title.

4. Construct a table to show the differences in appearance between the three blood vessels.

5. Explain how the structures of the blood vessels are adapted for their functions.

6. Name the four chambers of the heart and state the function of each one.
7. Describe five ways in which the heart is suited to its function.
8. State three differences between arteries and veins.
9. Explain how and why our heart rate changes when we exercise and when we sleep.
10. Describe two ways in which capillaries are adapted to their function of exchanging substances between blood and surrounding cells.

The composition of blood

Blood is a lot more than just a red liquid flowing through your arteries and veins. In fact, blood is a complex tissue. Figure 14.14 illustrates the main types of cells found in blood.

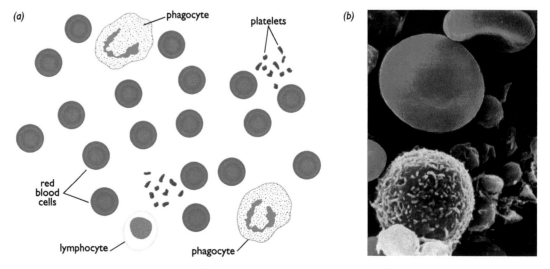

Figure 14.14 *The different types of blood cells (a) drawings of the different cells and (b) as seen in a photomicrograph.*

The different parts of blood have different functions. These are described in Table 14.1.

Component of blood	Description of component	Function of component
plasma	liquid part of blood: mainly water	carries the blood cells around the body; carries dissolved nutrients, hormones, carbon dioxide and urea; also distributes heat around the body
red blood cells	biconcave, disc-like cells with no nucleus; millions in each mm³ of blood	transport oxygen – contain mainly haemoglobin, which loads oxygen in the lungs and unloads it in other regions of the body
white blood cells:		
lymphocytes	about the same size as red blood cells with a large spherical nucleus	produce antibodies to destroy micro-organisms – some lymphocytes persist in our blood after infection and give us immunity to specific diseases
phagocytes	much larger cells with a large spherical or lobed nucleus	engulf bacteria and other micro-organisms that have infected our bodies
platelets	the smallest cells – are really fragments of other cells	release chemicals to make blood clot when we cut ourselves

Table 14.1: *Functions of the different components of blood.*

Red blood cells

Red blood cells are highly specialised cells made in the bone marrow. They have a limited life span of about 100 days, after which time they are destroyed in the spleen. They have only one function – to transport oxygen. Several features enable them to carry out this function very efficiently.

Red blood cells contain **haemoglobin**. This is an iron-containing protein that associates (combines) with oxygen to form **oxyhaemoglobin** when there is a high concentration of oxygen in the surroundings. We say that the red blood cell is loading oxygen. When the concentration of oxygen is low, oxyhaemoglobin turns back into haemoglobin and the red blood cell unloads its oxygen.

$$\text{haemoglobin} + \text{oxygen} \underset{\text{low oxygen concentration (in tissues)}}{\overset{\text{high oxygen concentration (in lungs)}}{\rightleftharpoons}} \text{oxyhaemoglobin}$$

As red blood cells pass through the lungs, they load oxygen. As they pass through active tissues, they unload oxygen.

Red blood cells do not contain a nucleus. It is lost during their development in the bone marrow. This means that more haemoglobin can be packed into each red blood cell and so more oxygen can be transported. Their biconcave shape allows efficient exchange of oxygen in and out of the cell. Each red blood cell has a high surface area to volume ratio, giving a large area for diffusion. The thinness of the cell gives a short diffusion distance to the centre of the cell. In addition, red blood cells have very thin cell surface membranes which allow oxygen to diffuse through them easily.

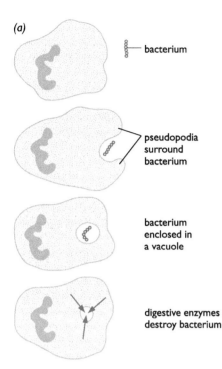

(a)

bacterium

pseudopodia surround bacterium

bacterium enclosed in a vacuole

digestive enzymes destroy bacterium

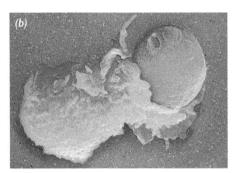

(b)

Figure 14.15 *(a) Phagocytosis by a white blood cell. (b) A phagocyte ingesting a yeast cell.*

White blood cells

There are several types of white blood cell. Their main role is to protect the body against invasion by disease-causing micro-organisms (pathogens), such as bacteria and viruses. They do this in two main ways: **phagocytosis** and **antibody production**.

About 70% of white blood cells can ingest (take in) micro-organisms such as bacteria. This is called phagocytosis, and the cells are called **phagocytes**. They do this by changing their shape, producing extensions of their cytoplasm, which are called **pseudopodia**. The pseudopodia surround and enclose the micro-organism in a vacuole. Once it is inside, the phagocyte secretes enzymes into the vacuole to break the micro-organism down (Figure 14.15). Phagocytosis means 'cell eating'; you can see in the diagram why it is called this.

Approximately 25% of white blood cells are **lymphocytes**. Their function is to make chemicals called **antibodies**. Antibodies are soluble proteins that pass into the plasma. Pathogens such as bacteria and viruses have telltale chemical 'markers' on their surfaces, which the antibodies recognise. These markers are called **antigens**. The antibodies stick to the surface antigens and destroy the pathogen. They do this in a number of ways, for example by:

- causing bacteria to stick together so that phagocytes can ingest them more easily
- acting as a 'label' on the pathogen so that it is more easily recognised by a phagocyte
- causing bacterial cells to burst open
- neutralising poisons (toxins) produced by pathogens.

Some lymphocytes do not get involved in killing micro-organisms immediately. Instead, they develop into **memory cells**. Memory cells make us **immune** to a disease. These cells remain in the blood for many years, sometimes a lifetime. If the same micro-organism re-infects the body, the memory lymphocytes start to reproduce and produce antibodies so that the pathogen can be quickly dealt with.

This secondary immune response is much faster and more effective than the first (primary) response. The number of antibodies in the blood quickly rises to a high level, killing the micro-organisms before they have a chance to multiply to a point where they would cause disease. This is shown in Figure 14.16.

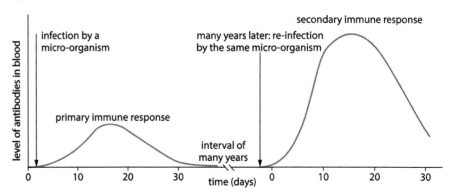

Figure 14.16 *The primary and secondary immune responses.*

A person can be given artificial immunity to a disease without ever actually contracting (having) the disease itself. This is done by **vaccination**. A person is injected with an 'agent' that carries the same antigens as a specific disease-causing pathogen. Lymphocytes recognise the antigens and multiply exactly as if that micro-organism had entered the bloodstream. They produce memory cells and make the person immune to the disease. Some agents used as vaccines are:

- a weakened strain of the actual micro-organism, e.g. vaccines against polio, tuberculosis (TB) and measles
- dead micro-organisms, e.g. typhoid and whooping cough vaccines
- modified toxins of the bacteria, e.g. tetanus and diphtheria vaccines
- just the antigens themselves, e.g. influenza vaccine
- harmless bacteria, genetically engineered to carry the antigens of a different disease-causing micro-organism, e.g. the vaccine against hepatitis B.

Platelets

Platelets are not whole cells, but fragments of large cells made in the bone marrow. If the skin is cut, exposure to the air stimulates the platelets and damaged tissue to produce a chemical. This chemical causes the soluble plasma protein **fibrinogen** to change into insoluble fibres of another protein, **fibrin**. The fibrin forms a network across the wound, in which red blood cells become trapped. This forms a **clot**, which prevents further loss of blood and entry of pathogens. The clot develops into a scab, which protects the damaged tissue while new skin grows.

11. Give two different functions of blood plasma.
12. Describe three ways in which red blood cells are adapted to their function of transporting oxygen around the body.
13. Explain how red blood cells load and unload oxygen.
14. What is the main function of platelets?
15. What is the main function of phagocytes?
16. What are antibodies?
17. Explain how antibodies destroy pathogens such as bacteria.
18. Explain how vaccination helps to protect against a disease.

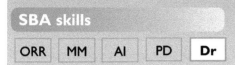

SBA skills

| ORR | MM | AI | PD | **Dr** |

You will need:

- a microscope
- a prepared slide of human blood cells

Carry out the following:

1. Examine the prepared slide of human blood using a microscope. The blood cells are very small, so you will need to use the high-power lens of the microscope to see them.

2. Most of the cells visible will be red blood cells. Look out for the larger white blood cells. These are usually stained purple in prepared slides.

3. Make a drawing of a few blood cells, including both red and white cells. Label your drawing and give it a title.

Chapter summary

In this chapter you have learnt that:

- as organisms increase in size, their surface area to volume ratio decreases. As a result of this, they are unable to exchange materials efficiently through their body surface. They have specialised exchange organs and a transport system

- the human circulatory system is a double circulatory system. It is made up of heart, blood, arteries, veins and capillaries

- the heart pumps blood to all organs in the body. The right ventricle pumps deoxygenated blood to the lungs. The left ventricle pumps oxygenated blood to all other organs

- arteries carry blood away from the heart. They are thick walled to withstand the pressure

- veins carry blood towards the heart and are thin walled with large lumens to allow easy flow of blood. Valves prevent backflow

- capillaries are thin-walled microscopic vessels that carry blood close to every cell in an organ

- red blood cells carry oxygen around the body

- white blood cells combat disease-causing organisms by phagocytosis and production of antibodies

- memory cells provide immunity to diseases

- vaccination is used to provide artificial immunity to a disease

- platelets stimulate blood clotting when a blood vessel is damaged

- blood plasma distributes cells, heat and dissolved substances around the body.

Questions

1. A very small organism does not need a circulatory system because:

   A it has a large surface area to volume ratio
   B it has a small surface area to volume ratio
   C the surface area and volume are equal
   D the volume is greater than the surface area

2. The main parts of the human circulatory system are:

   A the heart, veins and capillaries
   B the heart, arteries and capillaries
   C the heart, arteries and veins
   D the heart, arteries, veins and capillaries

3. Which most correctly describes the functions of the three main types of blood vessels?

	Artery	Vein	Capillary
A	carries blood away from heart to other organs	carries blood through an organ	carries blood away from an organ towards heart
B	carries blood away from heart to other organs	carries blood away from an organ towards heart	carries blood through an organ
C	carries blood through an organ	carries blood away from an organ towards heart	carries blood away from heart to other organs
D	carries blood away from an organ towards heart	carries blood away from heart to other organs	carries blood through an organ

4. What are the structures labelled A, B, C and D on the diagram?

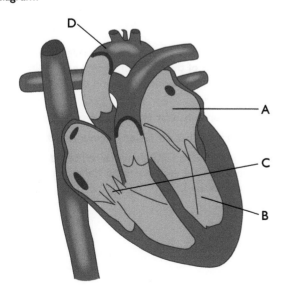

   A A – left atrium, B – left ventricle,
      C – tricuspid valve, D – aorta
   B A – right atrium, B – left ventricle,
      C – tricuspid valve, D – pulmonary artery
   C A – left atrium, B – bicuspid valve,
      C – tricuspid valve, D – aorta
   D A – left ventricle, B – left atrium,
      C – bicuspid valve, D – vena cava

5. Blood consists of:

   A plasma, red blood cells and white blood cells
   B platelets, red blood cells, water and white blood cells
   C plasma, red blood cells and platelets
   D plasma, red blood cells, platelets and white blood cells

6. a) The diagram shows a section through a human heart.

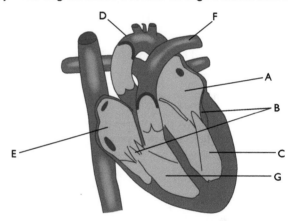

   i) Name the structures labelled A, B, C, D and E. (5)
   ii) What is the importance of the structure labelled B? (2)
   iii) Which letters represent the chambers of the heart to which blood returns:
       1 from the lungs
       2 from all the other organs of the body? (2)

   b) The diagram shows three types of cells found in human blood.

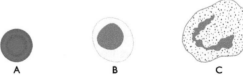

   Giving a reason for each answer, identify the blood cell which:

   i)   transports oxygen around the body
   ii) produces antibodies to destroy bacteria
   iii) engulfs and digests bacteria. (6)

   c) Name *one* other component of blood found in the plasma and state its function. (2)

7. a) The graph shows the changes in a person's heart rate over a period of time.

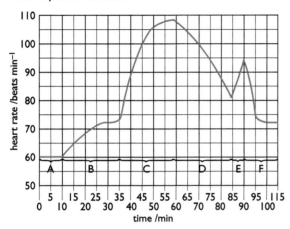

Giving reasons for your answers, give the letter of the time period when the person was:

i) running

ii) frightened by a sudden loud noise

iii) sleeping

iv) waking. *(8)*

b) The graph shows the changes that take place in the heart rate before, during and after a period of exercise.

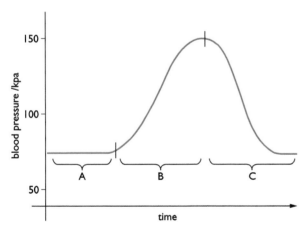

Describe and explain the blood pressures found:

i) at rest, before exercise (period A)

ii) as the person commences the exercise (period B)

iii) as the person recovers from the exercise (period C). *(6)*

c) How can the recovery period (period C) be used to assess a person's fitness? *(3)*

8. a) The diagram shows a single-celled organism.

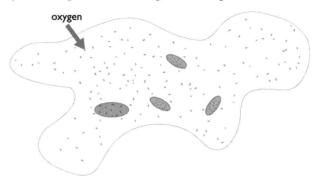

i) Name the process by which oxygen enters the organism. *(1)*

ii) Explain which features of a single-celled organism determine the amount of oxygen it needs and the amount of oxygen it can obtain. *(4)*

iii) Explain why this organism does not need lungs or gills to obtain the oxygen, or a transport system to distribute the oxygen. *(3)*

b) The diagram shows plans of two different transport systems.

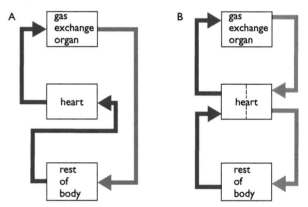

i) Name *one* animal which has transport system type A and *one* which has type B. *(2)*

ii) Which type is most efficient in transporting oxygen around the body? Explain why. *(3)*

9. a) The following observations were made about the composition of the blood of several people. In each case, describe one likely effect on the functioning of their bodies.

i) Person A had far fewer platelets than usual. *(2)*

ii) Person B had far fewer lymphocytes than usual. *(2)*

iii) Person C had slightly more red blood cells than normal, but each was spherical rather than disc-shaped. *(2)*

b) The graph shows the percentage of haemoglobin that is carrying oxygen under different conditions.

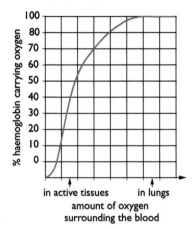

i) What percentage of haemoglobin is carrying oxygen in the lungs? *(1)*

ii) What percentage of haemoglobin is carrying oxygen in the active tissues? *(1)*

1 dm³ (1 000 cm³) of blood in the lungs carries 200 cm³ of oxygen.

iii) How much oxygen is carried by 1 dm³ of blood in the active tissues? *(3)*

iv) How much oxygen passes from the haemoglobin to the active tissues from each dm³ of blood? *(2)*

10. The diagram shows the levels of antibodies produced by a person during an initial infection, and later, when re-infected by the same micro-organism.

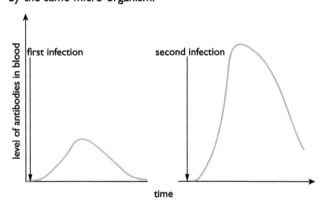

a) Give the names of the initial and subsequent responses by the body to infection. *(2)*

b) Describe the differences between the two responses in relation to level, speed and duration. *(4)*

c) Explain how the initial response is produced. *(4)*

11. The circulation system carries nutrients, oxygen and carbon dioxide around the body.

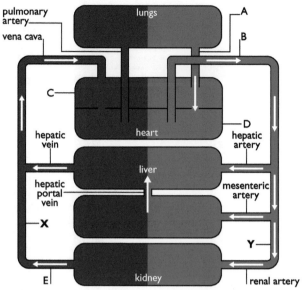

direction of blood flow ■ oxygenated blood ■ deoxygenated blood

a) Write out the correct labels for A to E. *(5)*

b) Give *two* differences between the blood vessels at point X and point Y. *(2)*

c) The diagram shows a double circulatory system. Some animals only have a single circulatory system.

i) Explain the difference between single and double circulatory systems. *(3)*

ii) Name *one* type of animal with a single circulatory system and *one* with a double circulatory system. *(2)*

d) List, in order, the blood vessels and heart chambers that a red blood cell passes through on its way from the liver to the kidney. *(4)*

Chapter 15: Transport in Plants

When you have completed this chapter, you will be able to:

- explain how the structure of xylem vessels is suited to their function
- discuss the role of the process of transpiration in plants
- describe and explain the effects of external factors on transpiration
- explain how the structure of phloem is suited to its function
- discuss adaptations of plants to conserve water.

To carry out photosynthesis plants need a supply of water and minerals from the roots, and a way of transporting sugars and other products of photosynthesis away from the leaves. This chapter describes the plant transport systems involved in these processes.

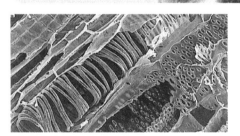

Figure 15.1 *Xylem vessels in a stem.*

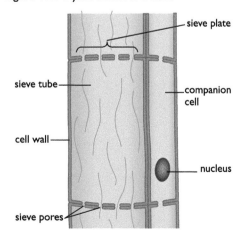

Figure 15.2 *Phloem is living tissue, responsible for carrying the products of photosynthesis around the plant.*

How plants transport substances

Plants don't have a true circulatory system with a pump like humans do. However, they do transport materials around. In fact, they have two different transport systems.

- The **xylem** transports water and dissolved mineral ions upwards from the roots. Most of the water evaporates from the leaves. Xylem cells are dead, empty cells with very thick walls that contain a woody substance called **lignin**. This makes them very strong and they act as a kind of skeleton, supporting the plant.

Xylem cells are well adapted to transporting water through a plant because:

 – they are tubular cells

 – they are hollow so there are no cell contents to impede the flow

 – the end walls between cells break down so the tubes are continuous through the plant

 – they have thick, strong walls to withstand the tension as water is pulled upwards through them.

- The **phloem** transports sugars, amino acids and other nutrients from leaves to other parts of the plant. These can move upwards to flowers and the tips of shoots as well as downwards to the roots. Phloem cells are also tubular, but they are living, active cells.

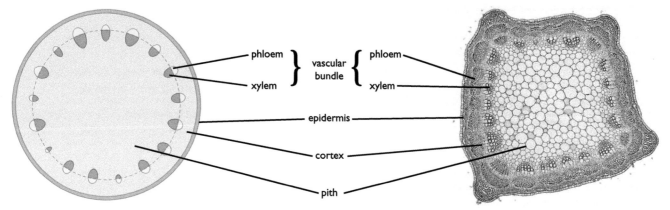

Figure 15.3 *This cross-section of a stem shows the arrangement of xylem and phloem tissue in vascular bundles.*

In a young stem, xylem and phloem are grouped together in areas called **vascular bundles**. Unlike in the root, where the vascular tissue is in the central core, the vascular bundles are arranged in a circle around the outer part of the stem (Figure 15.3).

Figure 15.4 shows how xylem and phloem transport substances through a plant.

> Vascular means made of vessels. A vascular bundle is a group of vessels or tubes (xylem and phloem).

Cross-curricular links:
physics, change of state

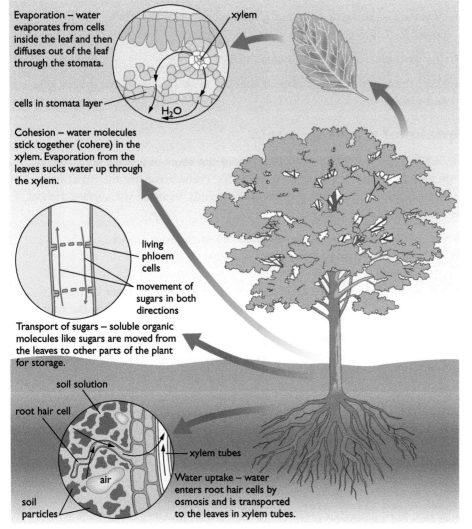

Evaporation – water evaporates from cells inside the leaf and then diffuses out of the leaf through the stomata.

cells in stomata layer

Cohesion – water molecules stick together (cohere) in the xylem. Evaporation from the leaves sucks water up through the xylem.

living phloem cells

movement of sugars in both directions

Transport of sugars – soluble organic molecules like sugars are moved from the leaves to other parts of the plant for storage.

soil solution

root hair cell

air

xylem tubes

Water uptake – water enters root hair cells by osmosis and is transported to the leaves in xylem tubes.

soil particles

Figure 15.4 *An overview of transport in plants.*

> Osmosis is the movement of water from a solution of high water potential to a solution of lower water potential across a partially permeable membrane. See Chapter 8 for a full explanation.

SBA skills

ORR	MM	AI	PD	Dr

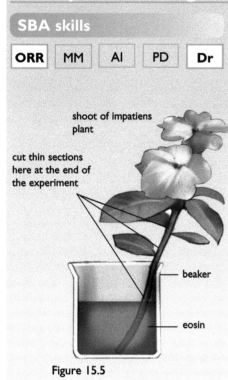

shoot of impatiens plant

cut thin sections here at the end of the experiment

beaker

eosin

Figure 15.5

You will need:

- a leafy shoot of a non-woody (herbaceous) plant – species belonging the balsam family (*Impatiens*) have soft stems that are easy to cut
- a beaker of water containing a red dye such as eosin
- a sharp scalpel
- a white tile
- a hand lens or microscope
- slides and coverslips

Carry out the following:

1. Place a freshly cut shoot of the plant in a beaker of eosin.
2. Leave the shoot for several hours. Observe the stem at intervals and note any movement of the dye through the shoot.
3. Remove the shoot and place it on a white tile. Use a sharp scalpel to cut very thin sections of the shoot at different points along the stem.
4. Observe the sections of stem through a hand lens and under the low-power lens of a microscope.
5. Compare the sections with Figure 15.3.
6. Make drawings of the sections. Label the drawings, and give them a title and a scale.

Uptake of water by roots

The regions just behind the growing tips of the roots of a plant are covered in thousands of tiny root hairs (Figure 15.6). These areas are the main sites of water absorption by the roots as the hairs greatly increase the surface area of the root epidermis.

Each hair is actually a single, specialised cell of the root epidermis. The long, thin outer projection of the root hair cell penetrates between the soil particles, reaching the soil water. The water in the soil has some solutes dissolved in it, such as mineral ions, but their concentrations are much lower than the concentrations of solutes inside the root hair cell. The soil water therefore has a higher water potential than the inside of the cell. This allows water to enter the root hair cell by osmosis. In turn, this water movement dilutes the contents of the cell, increasing its water potential. Water then moves out of the root hair cell into the outer tissue of the root (the root cortex). Continuing in this way, a gradient of water potential is set up across the root cortex, kept going by water being taken up by the xylem in the middle of the root (Figure 15.7).

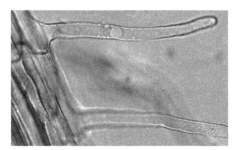

Figure 15.6 *These root hairs increase the surface area for water absorption.*

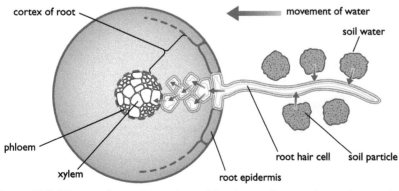

cortex of root

movement of water

soil water

phloem

xylem

root hair cell   soil particle

root epidermis

Figure 15.7 *Water is taken up by root hairs of the plant epidermis and carried across the root cortex by a water potential gradient. It then enters the xylem and is transported to all parts of the plant.*

Loss of water by the leaves – transpiration

Osmosis is also involved in the movement of water through leaves. The epidermis of leaves is covered by a waxy cuticle (see Chapter 9), which is impermeable to water. Most water passes out of the leaves as water vapour through pores called **stomata**. Water leaves the cells of the leaf mesophyll and evaporates into the air spaces between the spongy mesophyll cells. The water vapour then diffuses out through the stomatal pores (Figure 15.8).

Plants that live in dry habitats, such as cacti and succulents, usually have a very thick layer of waxy cuticle on their leaves, which reduces water loss as much as possible. Cacti have leaves reduced to spines, which have a low surface area and cut down water loss.

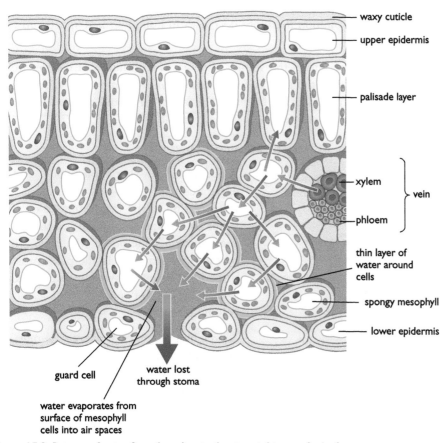

waxy cuticle
upper epidermis
palisade layer
xylem
vein
phloem
thin layer of water around cells
spongy mesophyll
lower epidermis
water lost through stoma
guard cell
water evaporates from surface of mesophyll cells into air spaces

Figure 15.8 *Passage of water from the xylem to the stomatal pores of a leaf.*

Loss of water from the mesophyll cells sets up a water potential gradient which 'draws' water by osmosis from surrounding mesophyll cells. In turn, the xylem vessels supply the leaf mesophyll tissues with water.

This loss of water vapour from the leaves is called **transpiration**. Transpiration causes water to be 'pulled up' the xylem in the stem and roots in a continuous flow known as the **transpiration stream**. The transpiration stream has more than one function. It:

- supplies water for the leaf cells to carry out photosynthesis
- carries mineral ions dissolved in the water
- provides water to keep the plant cells turgid (see Chapter 8)
- allows evaporation from the leaf surface, which cools the leaf, in a similar way to sweat cooling the human skin.

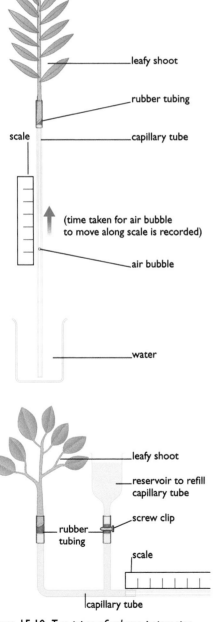

The concentration of water vapour inside the leaf is greater than the concentration in the air outside. We call this a diffusion (or concentration) gradient. A big difference in concentration means there is a steep gradient. The molecules of water vapour will diffuse out of the leaf more quickly if the gradient is steep.

leafy shoot

rubber tubing

scale ————— capillary tube

(time taken for air bubble to move along scale is recorded)

air bubble

water

leafy shoot

reservoir to refill capillary tube

screw clip

rubber tubing

scale

capillary tube

Figure 15.10 *Two types of volume potometer.*

Physical factors affecting the rate of transpiration

There are four main factors which affect the rate of transpiration:

- **Light intensity** – the rate of transpiration increases in the light, because the stomata in the leaves open.

- **Temperature** – high temperatures increase the rate of transpiration, by increasing the rate of evaporation of water from the mesophyll cells.

- **Humidity** – when the air around the plant is humid, this reduces the diffusion gradient between the air spaces in the leaf and the external air. The rate of transpiration therefore decreases in humid air and speeds up in dry air.

- **Wind speed** – the rate of transpiration increases with faster air movements across the surface of the leaf. The moving air removes any water vapour which might remain near the stomata. This moist air would otherwise reduce the diffusion gradient and slow down diffusion.

Measuring the rate of transpiration: potometers

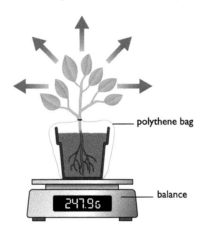

A potometer is a simple piece of apparatus which measures the rate of transpiration or the rate of uptake of water by a plant. (These are not the same thing – some of the water taken up by the plant may stay in the plant cells, or be used for photosynthesis.) There are two types: 'weight' and volume potometers.

polythene bag

balance

A 'weight' potometer measures the rate of loss of mass from a potted plant or leafy shoot over a relatively long period of time, usually several hours (Figure 15.9).

Figure 15.9 *A 'weight' potometer.*

The polythene bag around the pot prevents loss of moisture by evaporation from the soil. Most of the mass lost by the plant will be due to water evaporating from the leaves during transpiration (although there will be small changes in mass due to respiration and photosynthesis, since both of these processes exchange gases with the air).

A volume potometer is used to find the rate of uptake of water by a leafy shoot, by 'magnifying' this uptake in a capillary tube. The simplest is a straight vertical tube joined to the shoot by a piece of rubber tubing. More 'deluxe' versions have a horizontal capillary tube and a way of refilling the capillary to re-set the water at its starting position (Figure 15.10).

To set up a volume potometer, the whole apparatus is placed in a sink of water and any air in the tubing removed. A shoot is taken from a plant and the end of the stem cut at an angle. This makes it easier to push the stem into the rubber tubing, which is done under water to stop air entering. The apparatus is removed from the sink, and vaseline used to seal any joins. The movement of the water column in the capillary tube can be timed. If the water moves more quickly, this shows a faster rate of transpiration. The plant can then be exposed to different conditions to see how they affect the rate.

Activity 2: How does wind affect the rate of transpiration?

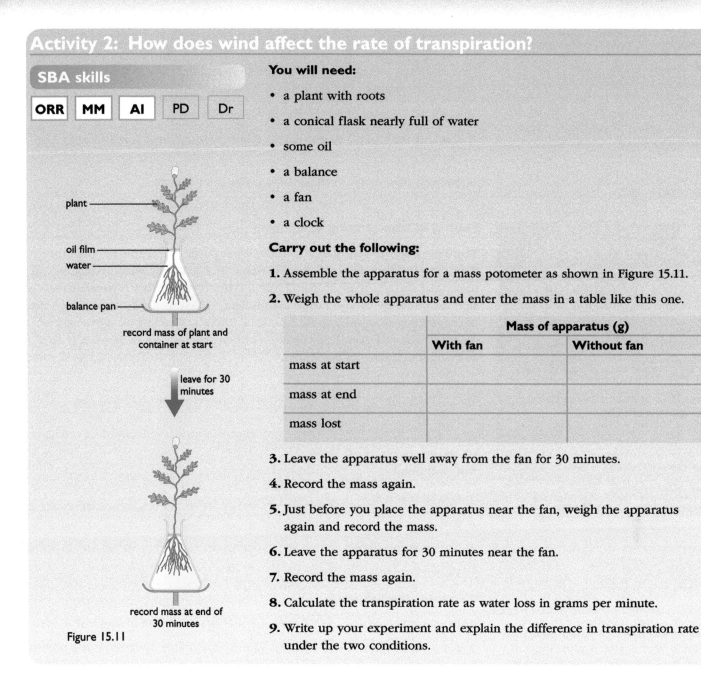

plant

oil film

water

balance pan

record mass of plant and container at start

leave for 30 minutes

record mass at end of 30 minutes

Figure 15.11

You will need:

- a plant with roots
- a conical flask nearly full of water
- some oil
- a balance
- a fan
- a clock

Carry out the following:

1. Assemble the apparatus for a mass potometer as shown in Figure 15.11.
2. Weigh the whole apparatus and enter the mass in a table like this one.

	Mass of apparatus (g)	
	With fan	**Without fan**
mass at start		
mass at end		
mass lost		

3. Leave the apparatus well away from the fan for 30 minutes.
4. Record the mass again.
5. Just before you place the apparatus near the fan, weigh the apparatus again and record the mass.
6. Leave the apparatus for 30 minutes near the fan.
7. Record the mass again.
8. Calculate the transpiration rate as water loss in grams per minute.
9. Write up your experiment and explain the difference in transpiration rate under the two conditions.

Activity 3: Investigating water loss from leaves

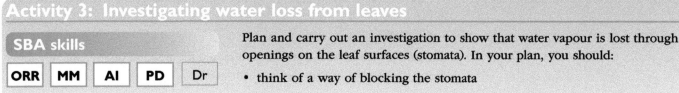

Plan and carry out an investigation to show that water vapour is lost through openings on the leaf surfaces (stomata). In your plan, you should:

- think of a way of blocking the stomata
- think of ways of making sure that the environmental conditions do not change.

Show your plan to your teacher before you carry it out.

SBA skills

ORR | MM | AI | PD | Dr

Plan and carry out an investigation to show that light intensity affects the rate of transpiration. In your plan, you should:

• think of a way of changing the light intensity

• think of ways of making sure that other environmental conditions do not change.

Show your plan to your teacher before you carry it out.

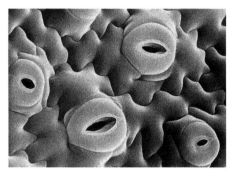

Control of transpiration by stomata

There are usually more stomata on the lower surface of the leaves than the upper surface in most plant species (Figure 15.12). If they were mainly on the upper leaf surface, the leaf would lose too much water. This is because the stomata would be exposed to direct sunlight, which would produce a high rate of evaporation from the exposed stomata. There is also less air movement on the underside of leaves. The evolution of this arrangement of stomata is an adaptation that reduces water loss.

Figure 15.12 *Stomata in the surface of the lower epidermis of a leaf.*

Activity 5: Looking at leaf surfaces

SBA skills

ORR | MM | AI | PD | Dr

When you paint nail varnish onto a leaf, it flows slowly, takes the shape of the leaf surface and hardens into that shape. When you peel it off, you have a mould of the leaf surface.

You will need:

• two leaves

• nail varnish

• a microscope slide

• a microscope

Carry out the following:

1. Paint a *thin layer* of nail varnish onto the upper surface of one leaf.

2. Paint a *thin layer* of nail varnish onto the lower surface of the other leaf.

3. Allow the nail varnish on both leaves to become dry and hard.

4. Peel off about 1 cm² nail varnish from each leaf.

5. Place the nail varnish on a microscope slide and observe under a microscope.

6. Count (or estimate if there are too many) the number of stomata you can see in the field of view on each surface.

7. Which surface has the most stomata?

8. Make a drawing of part of each slide.

9. Write up your experiment and suggest which side of the leaf will lose the most water vapour.

The stomata can open and close. The guard cells that surround each stoma have an unusual 'banana' shape, and the part of their cell wall nearest the stoma is particularly thick. In the light, water enters the guard cells by osmosis from the surrounding epidermis cells. This causes the guard cells to become turgid, and, as they swell up, their shape changes. They bend outwards, opening up the stoma. In the dark, the guard cells lose water again, they become flaccid and the stoma closes. No one knows for sure how this change is brought about, but it seems to be linked to the fact that the guard cells are the only cells in the lower epidermis that contain chloroplasts. In the light, the guard cells use energy to accumulate solutes in their vacuoles, causing water to be drawn in by osmosis (Figure 15.13).

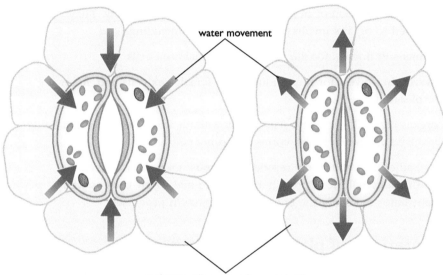

water enters guard cells by osmosis; guard cells become turgid, opening stoma

water leaves guard cells by osmosis; guard cells become flaccid, closing stoma

water movement

epidermis cells surrounding guard cells

Figure 15.13 *When the guard cells become turgid, the stoma opens. When they become flaccid, it closes.*

You can model the action of a guard cell using a long balloon and some sticky tape. Stick a few pieces of tape down one side of the balloon and then blow it up. As it inflates, the balloon will curve outwards like a turgid guard cell. The tape represents the thick inner cell wall.

Closure of stomata in the dark is a useful adaptation. Without the sun there is no need for loss of water vapour from the stomata to cool the leaves. In addition, leaves cannot photosynthesise in the dark, so they don't need water for this purpose. Therefore, it doesn't matter if the transpiration stream is shut down by closure of the stomata.

1. Describe three ways in which the structure of xylem cells adapts them to their function.
2. Explain how water passes from the soil to the xylem in a root.
3. Describe how water passes from the xylem in a leaf to the atmosphere outside the leaf.
4. Name three physical factors that affect the rate of transpiration and describe their effects.
5. Explain how stomata open and close.

Translocation – the movement of organic nutrients through plants

Tubes in the phloem are also formed by cells arranged end-to-end, but they have cell walls made of cellulose, and retain their cytoplasm. The end of each cell is formed by a cross-wall of cellulose with holes, called a **sieve plate**. The living cytoplasm extends through the holes in the sieve plates, linking each cell with the next, forming a long **sieve tube** (Figure 15.2, page 208). Sugars for energy, or amino acids for building proteins, are carried to young leaves and other growing points. Sugar may also be taken to the roots and converted into starch for storage. Despite being living cells, the phloem sieve tubes have no nucleus. They seem to be controlled by other cells that lie alongside the sieve tubes, called **companion cells** (Figure 15.2).

The sugars and amino acids transported to growing points are used for immediate production of new cells so that the plant can grow. They are also transported to organs involved in reproduction, including:

- stamens in flowers for the production of male sex cells

- ovaries in flowers for the production of female sex cells and, following fertilisation, seeds

- organs that will grow into new plants without sexual reproduction being involved, including bulbs, corms and runners.

Seeds and organs of asexual reproduction must grow into new plants. To do this they will need protein and lipid to make new cells as well as carbohydrates to supply the energy for the growth processes.

Adaptations of plants to conserve water

Plants lose water through their leaves by the process of transpiration. If water is freely available in the plant's habitat, this water loss is replaced by water absorbed through the roots. However, some plants live in habitats where the supply of water is limited. The most obvious example of this is a desert, where rainfall may be low all year round, or rain may only fall during one season.

Deserts are not the only habitats that limit the water available to a plant. Take for example sand dunes around the coast, or shingle beaches. There may be moderate or high rainfall in these situations, but any rain falling on the sand will rapidly drain through the sand particles, and not be available to the plants for very long.

Plants living in any of the above habitats, where water availability is low, have evolved adaptations which allow them to survive in these conditions. Such plants are called **xerophytes**. The adaptations allow the plants to maintain adequate levels of water in their cells, in other words this is an example of osmoregulation in plants. Adaptations for osmoregulation include:

- Increased root length or spread of the roots (Figure 15.14). Deep roots can reach far down into the soil to reach the water table. In other species the roots grow outwards to cover a large area, so that any rain falling on the soil is quickly absorbed before it can drain away.

(a)

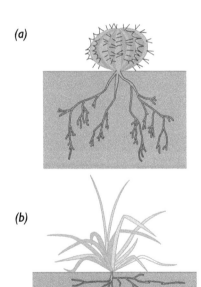

(b)

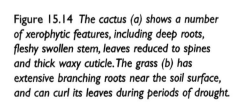

Figure 15.14 *The cactus (a) shows a number of xerophytic features, including deep roots, fleshy swollen stem, leaves reduced to spines and thick waxy cuticle. The grass (b) has extensive branching roots near the soil surface, and can curl its leaves during periods of drought.*

- Closure of stomata during very hot dry weather so that less water vapour is lost from the leaves.

- Increased thickness of cuticle on the leaves. The waxy cuticle on the epidermis of all leaves reduces water loss, but some water can still pass through the layer of cuticle. Xerophytes often have a much thicker layer of cuticle on their leaves, which conserves water.

- Water storage in swollen stems or leaves. Many xerophytes such as *Aloe* (a 'succulent') and cacti (Figure 15.14) have swollen fleshy leaves or stems which store water. This water can then be used by the rest of the plant during times of drought.

- Reduced leaf area. Since most transpiration takes place from the leaves of plants, reducing the area of the leaves reduces the rate of transpiration and conserves water. For example, cacti (Figure 15.14) have leaves reduced to spines on the fleshy stem. The spines also help protect the plant against predators which might attack the stem for its water content. Some cacti are covered with a coat of spines or hairs which are white. This reflects the sun off the plant, helping to prevent it from overheating.

- Stomata located in sunken pits. Plants such as oleander (*Nerium*) have all their stomata grouped together in sunken pits, protected by epidermal hairs (Figure 15.15). Water vapour builds up around the stomata, with the result that the air in the pits becomes saturated with water vapour, reducing the rate of evaporation from the stomata.

- Rolled leaves. Xerophytes such as marram grass (*Ammophila*) can roll up their leaves when water is in short supply (Figure 15.16). The stomata are mainly located on the inside surface of the rolled leaf, so that water vapour becomes trapped here, producing the same effect as with sunken stomata.

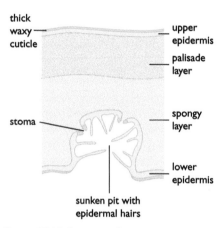

Figure 15.15 *Diagram of a section through a leaf of oleander, showing sunken stomata.*

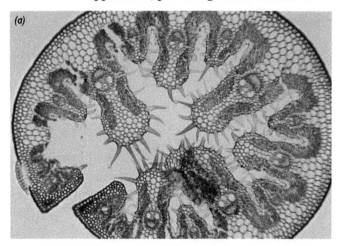

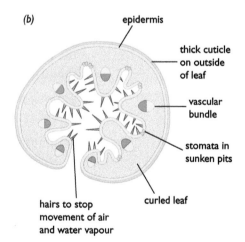

Figure 15.16 *(a) Photomicrograph of a cross-section through rolled leaf of marram grass, and (b) diagram of the cross section to show detail.*

6. What is a xerophyte?
7. Name two habitats in which xerophytes live.
8. Describe three ways that the leaves of xerophytes may be adapted to conserve water.

9. Give two important ways in which phloem cells are different from xylem cells.
10. Name two substances that are transported in the phloem.
11. Name three regions of a plant to which nutrients are transported in the phloem and explain why each region needs the supply of nutrients.
12. Why do seeds need a store of protein and starch?
13. Name two storage organs that are involved in asexual reproduction.

Chapter summary

In this chapter you have learnt that:

- water is transported through a plant in xylem cells that are dead, thick walled, hollow and tubular; the whole process is called transpiration

- water enters root hair cells and then passes to the xylem by osmosis

- water evaporates from cells inside a leaf and diffuses out through the stomata; most stomata are located on the lower epidermis

- the rate of transpiration is affected by temperature, wind speed and humidity of the air; water is lost more quickly from leaves when the environment is warm, windy and dry

- weight potometers and volume potometers can be used to measure the rate of transpiration

- organic nutrients, such as sugars and amino acids, are translocated in the living phloem cells; they can move both upwards and downwards though the plant

- growing points and organs involved in reproduction need a supply of nutrients to provide energy and the raw materials for growth

- xerophytes are plants which have adaptations that allow them to live in habitats where there is low water availability. These adaptations include deep roots, a thick waxy cuticle, sunken stomata and curled leaves.

Questions

1. In plants:

   A xylem transports water and dissolved ions, phloem transports only water

   B xylem transports water and sugars, phloem transports dissolved ions

   C xylem transports sugars and dissolved ions, phloem transports only water

   D xylem transports water and dissolved ions, phloem transports sugars

2. The rate of transpiration increases when it is:

   A warm, windy and dry
   B warm, windy and wet
   C cold, windy and dry
   D warm, still and dry

3. Water and sugars are transported through plants in the following directions:

   A water upwards and downwards, sugars downwards only
   B water downwards only, sugars upwards and downwards
   C water upwards only, sugars upwards and downwards
   D water upwards and downwards, sugars upwards only

4. Which of the following is **not** a feature of plants adapted to a habitat that is lacking in water?

   A thin cuticle
   B sunken stomata
   C curled leaves
   D deep roots

5. The diagram shows the pathway of water through a plant.

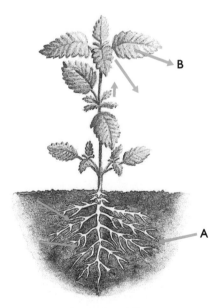

   The processes occurring at A and B are:

   A A – osmosis; B – transpiration
   B A – transpiration; B – evaporation
   C A – diffusion; B – osmosis
   D A – osmosis; B – evaporation

6. Having root hair cells and having most stomata on the lower epidermis are useful adaptations because:

   A there is a large surface area for absorption of water and exposure to the sun can make more water evaporate from leaves

   B there is a small surface area for absorption of water and exposure to the sun can make more water evaporate from leaves

   C there is a large surface area for absorption of water and reduced exposure to the sun reduces evaporation of water from leaves

   D there is a small surface area for absorption of water and reduced exposure to the sun reduces evaporation of water from leaves

7. Suggest reasons for each of the following observations:

   a) When transplanting a small plant from a pot to the garden, it is important to dig it up carefully, leaving the roots in a ball of soil. If this is not done, the plant may wilt after it has been transplanted. (2)

   b) A plant cutting is more likely to grow successfully if you remove some of its leaves before planting it in compost. (2)

   c) Plants that live in very dry habitats often have stomata located in sunken pits in their leaves. (2)

   d) Greenflies feed by sticking their hollow tube-like mouthparts into the phloem of a plant stem. (2)

8. A simple volume potometer was used to measure the uptake of water by a leafy shoot under four different conditions. During the experiment, the temperature, humidity and light intensity were kept constant. The conditions were:

   1. Leaves in still air with no vaseline applied to them.
   2. Leaves in moving air with no vaseline applied to them.
   3. Leaves in still air with the lower leaf surface covered in vaseline.
   4. Leaves in moving air with the lower surface covered in vaseline.

   The results are shown in the graph.

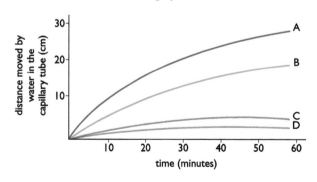

a) Describe how you would assemble a volume potometer. (3)

b) Complete the table below to show which condition (1, 2, 3 or 4) is most likely to produce curve A, B, C and D. One has been done for you.

Condition	Curve
1	B
2	
3	
4	

(3)

c) i) Explain why moving air affects the rate of uptake of water by the shoot. (2)

ii) Explain why applying vaseline to the leaves affects the rate of uptake of water by the shoot. (2)

d) Explain the value to the plant of the transpiration stream. (2)

9. The graph shows how the total diameter of the stomata changes over a 24-hour period.

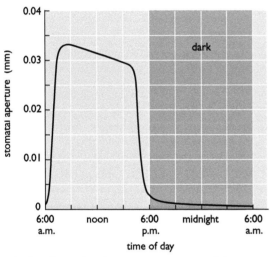

a) i) Describe the changes in stomatal diameter over this 24-hour period. (2)

ii) Copy the graph and draw a second line to show your estimation of the changes in total stomatal aperture in a plant growing in a desert. (2)

iii) Explain how stomata alter their diameter. (2)

b) i) Sketch a graph to show the changes in the rate of transpiration over the same 24-hour period. (2)

ii) Explain the shape of your graph. (2)

c) Describe how water passes out of leaves. (2)

10. The diagram shows the results of a 'ringing' experiment. In this procedure, a ring of phloem and bark is removed from a plant stem.

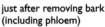

just after removing bark (including phloem) several weeks later

a) Describe the change in appearance of the tissue immediately above and below the ringed region during the period of the experiment. (3)

b) i) Explain the change above the ringed region. (1)

ii) Explain the change below the ringed region. (1)

c) Suggest why the roots of a ringed plant eventually die. (2)

11. Write a short description of how water moves from the soil through the plant. Your description must include these words: xylem, evaporation, root hair cells, water potential, stomata and osmosis.

Underline each of these words in your description. (6)

Chapter 16: Food Storage

> **When you have completed this chapter, you will be able to:**
>
> - identify the food substances stored in animals and plants and the sites where they are stored
>
> - discuss the importance of food storage in animals and plants.

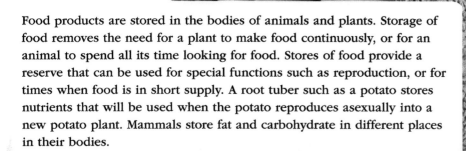

Food products are stored in the bodies of animals and plants. Storage of food removes the need for a plant to make food continuously, or for an animal to spend all its time looking for food. Stores of food provide a reserve that can be used for special functions such as reproduction, or for times when food is in short supply. A root tuber such as a potato stores nutrients that will be used when the potato reproduces asexually into a new potato plant. Mammals store fat and carbohydrate in different places in their bodies.

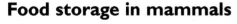

Food storage in mammals

There are two main types of food that are stored in the bodies of mammals:

- fat is stored under the skin, and around many of the internal organs

- carbohydrate is stored as glycogen in the liver and muscles.

Storage of fat

Fat is a useful long-term store of energy. When a mammal fasts (stops eating food for a lengthy period), fat is broken down into glycerol and fatty acids and used in respiration as an alternative to glucose. All mammals have a layer of fat under the skin (Figure 16.1, page 222). This **subcutaneous** fat is also a good insulating material, and prevents excessive heat loss from the body. This is especially important in aquatic mammals, which would lose heat easily to the water. For this reason, mammals such as whales and seals have a very thick layer of fat under their skin, called blubber (Figure 16.2, page 222).

There are also deposits of fat inside the body. Fat forms a layer under the muscles of the abdomen wall, and is found around organs such as the heart and kidneys (Figure 16.3, page 222). This fat also acts as an energy store and, in addition, helps to protect vital organs from mechanical damage.

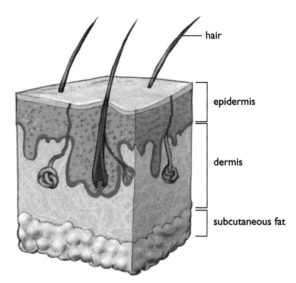

Figure 16.1 *A section through human skin, showing the subcutaneous fat layer.*

Figure 16.2 *Seals have a thick layer of subcutaneous fat which helps to prevent heat loss.*

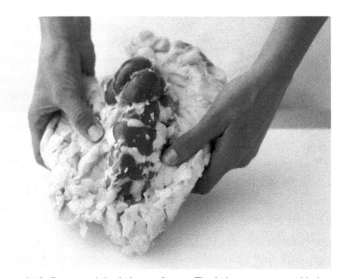

Figure 16.3 *Fat around the kidneys of an ox. The kidneys are just visible beneath the large mass of fat.*

Storage of glycogen

Carbohydrate is stored as a polysaccharide called glycogen. Glycogen is very similar in structure to starch in plants – they are both made of long chains of glucose molecules. They also have a similar function, as stores of carbohydrate.

Products of digestion are carried from the ileum of the gut to the liver through the hepatic portal vein (Figure 16.4). The liver acts as a metabolic 'processing plant', breaking down some substances and converting others into products that can be used by cells, or stored. One such conversion carried out by liver cells is to change glucose into glycogen for storage. The liver of a healthy person contains about 200 g of glycogen. When required, the glycogen is broken down into glucose again and released into the blood. Smaller amounts of glycogen are stored in muscle cells.

The conversion of glucose into glycogen and back to glucose is under the control of certain **hormones** in the body (see Chapter 18). The hormones **adrenaline** (from the adrenal gland) and **glucagon** (from the pancreas) both cause glycogen to be broken down into glucose, which increases the concentration of glucose in the blood. The hormone **insulin** (also from the pancreas) causes glucose to be taken up from the blood by the liver cells and stored as glycogen, thereby lowering the blood glucose levels.

1. Where are fats stored in the body?
2. How much glycogen is stored in the liver?
3. How does insulin cause a lowering of blood glucose?

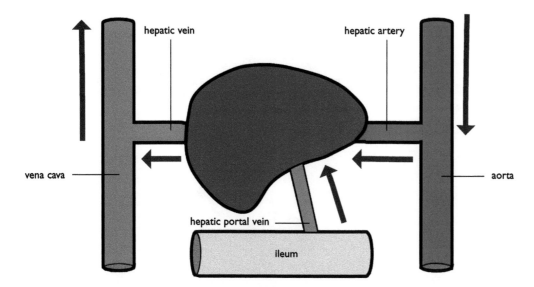

hepatic vein

hepatic artery

vena cava

aorta

hepatic portal vein

ileum

Figure 16.4 *The hepatic portal vein carries products of digestion to the liver.*

Food storage in plants

Plants make food in their leaves through the process of photosynthesis (see Chapter 9). The first storage product that is made is starch, which forms grains in chloroplasts.

Storage of starch

Starch is a good storage product, because it is insoluble and will remain in the cells until the plant needs to convert it into glucose. Starch is stored in many other plant organs, including seeds, roots, stems and fruits. Bananas contain much starch, as do seeds of cereal plants such as wheat, and tubers such as potatoes.

Asexual reproduction in animals is discussed in Chapter 20.

A potato is an example of a plant organ that is used for **asexual** or **vegetative** reproduction. This is where new organisms are formed without the involvement of specialised sex cells and fertilisation. Instead, cells in a part of the plant divide by **mitosis** (Chapter 24) and form a structure that breaks away from the parent plant and grows into a new plant. A potato **tuber** is a swollen underground stem. You can see evidence that it is a stem, because the 'eyes' on a potato are buds, which will grow into new leaves and roots (Figure 16.5a). The tuber will grow into a new potato plant using the energy stored as starch (Figure 16.5b).

(a)

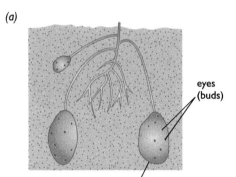

eyes
(buds)

Potato tubers form underground at the ends of branches from the main stem. Each potato can produce several new plants from the 'eyes' which are buds.

(b)

Figure 16.5 *Asexual reproduction in the potato plant.*

There are many other examples of vegetative reproduction in plants, such as overground stems called runners; and bulbs, which are swollen leaves full of food (Figure 16.6).

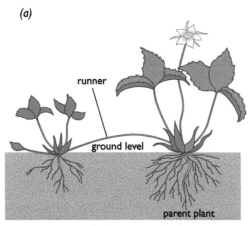

(a)

A new plant is produced where the runner touches the ground.

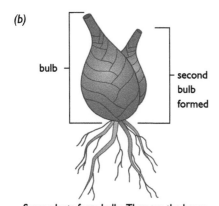

(b)

Some plants form bulbs. They are the bases of leaves which have become swollen with foo
Buds in them can develop into new plants.
Plants can form more than one bulb.

Figure 16.6 *Runners and bulbs – another example of vegetative reproduction in plants.*

Storage of other materials

Plants don't store only starch. Other materials that are stored include sugars, oils (lipids) and protein. Bulbs, such as onions (Figure 16.6b), are fleshy leaves that contain much stored sugar – mainly sucrose. Sugarcane is the stem of the sugarcane plant, and is our main commercial source of sugar. Many fruits also contain sugar; this is so they can attract animals to disperse the seeds inside the fruit (Chapter 21). Some plants store oils – the fruits of the olive tree are an example. Nuts, peas and beans contain much stored protein.

Seeds contain food reserves that are used when the seed germinates (starts to grow); this process will be described in Chapter 22.

4. Why is starch a good storage material?
5. What is the meaning of the term vegetative reproduction?
6. Why do fruits contain sugars?

SBA skills

ORR	**MM**	**AI**	PD	Dr

You will need:

- a variety of plant organs, including seeds, fruits, roots, tubers and bulbs
- a pestle and mortar
- iodine solution in a dropping bottle
- Benedict's solution in a dropping bottle
- ethanol
- biuret reagent in a dropping bottle
- a hot water bath (or a Bunsen burner, beaker, heatproof mat, tripod and gauze)
- dilute hydrochloric acid in a dropping bottle
- dilute sodium hydrogencarbonate in a dropping bottle
- boiling tubes
- test-tubes
- a spotting tile
- a spatula
- a test-tube holder
- a test-tube rack

Note that ethanol (alcohol) is *highly flammable.* You must not use it near a naked flame. *Turn off the Bunsen burner* before carrying out the ethanol test for lipid.

Carry out the following:

1. Construct a table like the one below to record your results.

Plant organ	Results of food tests				
	Test for starch	Test for reducing sugar	Test for non-reducing sugar	Test for oil (lipid)	Test for protein
potato tuber					
bean seed					
banana fruit					
onion bulb (etc.)					

2. Take a small amount of each plant organ and grind it up in a mortar.

3. Carry out the five food tests on each plant organ. The methods are given in Chapter 10, pages 133–135.

4. Summarise your results in the table. Explain why the different plant organs contain the food types that you have identified.

Chapter summary

In this chapter you have learnt that:

- the two main foods stored in animals are fat and glycogen
- fat is stored under the skin and around internal organs
- fat under the skin acts as insulation, and fat protects internal organs
- glycogen is stored in the liver
- the inter-conversion of glucose and glycogen is controlled by hormones
- plants store starch in their leaves and in other plant organs
- organs such as tubers and bulbs are used for vegetative reproduction
- plants can store other products, such as sugars, lipid and protein
- seeds use food stores for germination.

Questions

1. Animals store food as:

A   fat
B   carbohydrate and fat
C   fat and protein
D   carbohydrate, fat and protein

2. The hormone(s) that cause glycogen to be changed into glucose are:

A   glucagon
B   insulin
C   insulin and adrenaline
D   glucagon and adrenalin

3. The blood vessel leading from the ileum to the liver is called the:

A   hepatic portal vein
B   aorta
C   hepatic vein
D   hepatic artery

4. A tuber is:

A   a swollen root
B   an organ of sexual reproduction
C   an underground stem
D   a bulb containing starch

5. a) Where do we find large reserves of fat in the body of a mammal? (2)

b) Apart from storage, state two functions of fat in a mammal. (2)

c) Explain why glycogen makes a good compound for storage in the liver. (2)

6. The graph shows changes in the amounts of starch, sugars and protein in a seed as it begins to germinate.

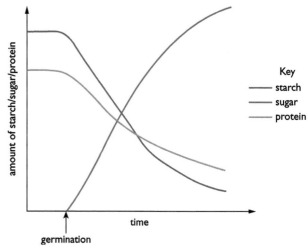

a) Describe how you could test a seed to see if it contained starch, reducing sugar and protein. (3)

b) Explain fully where the starch in the seed came from originally. (2)

c) i) Describe the changes in the levels of starch and glucose as the seed germinates. (2)

   ii) Explain the changes you described in (i). (2)

   iii) Describe *two* uses of glucose in plants. (2)

d) Explain the decrease in the amount of protein in the seed as it germinates. (2)

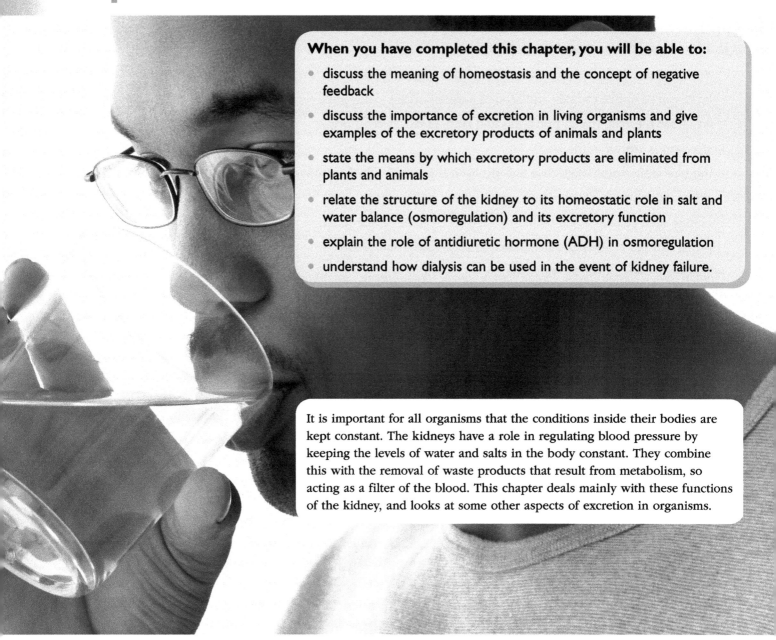

When you have completed this chapter, you will be able to:

- discuss the meaning of homeostasis and the concept of negative feedback

- discuss the importance of excretion in living organisms and give examples of the excretory products of animals and plants

- state the means by which excretory products are eliminated from plants and animals

- relate the structure of the kidney to its homeostatic role in salt and water balance (osmoregulation) and its excretory function

- explain the role of antidiuretic hormone (ADH) in osmoregulation

- understand how dialysis can be used in the event of kidney failure.

It is important for all organisms that the conditions inside their bodies are kept constant. The kidneys have a role in regulating blood pressure by keeping the levels of water and salts in the body constant. They combine this with the removal of waste products that result from metabolism, so acting as a filter of the blood. This chapter deals mainly with these functions of the kidney, and looks at some other aspects of excretion in organisms.

Homeostasis

If you were to drink a litre one water and wait for half an hour, your body would soon respond to this change by producing about the same volume of urine. In other words it would automatically balance your water input and water loss. Drinking is the main way that our bodies gain water, but there are other sources (Figure 17.1, page 229). Some water is present in the food that we eat, and a small amount is formed by cell respiration. The body also loses water, mostly in urine, and smaller volumes in sweat, faeces and exhaled air. Every day we gain and lose about the same volume of water, so that the total content of our bodies stays more or less the same. This is an example of **homeostasis**. The word homeostasis means 'steady state', and refers to keeping conditions inside the body relatively constant.

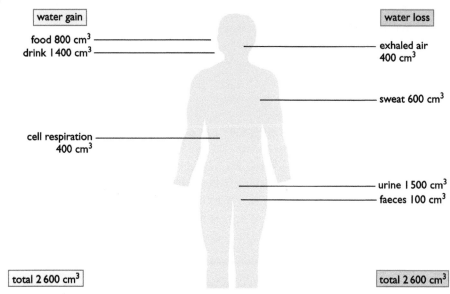

water gain		water loss
food 800 cm³		exhaled air 400 cm³
drink 1400 cm³		sweat 600 cm³
cell respiration 400 cm³		urine 1500 cm³
		faeces 100 cm³
total 2600 cm³		total 2600 cm³

Figure 17.1 *The daily water balance of an adult.*

Inside the body is known as the **internal environment**. You have probably heard of an environment, which means the surroundings of an organism. The internal environment is the surroundings of the cells inside a body. It particularly means the blood, together with another liquid called **tissue fluid**.

Tissue fluid is a watery solution of salts, glucose and other solutes. It surrounds all the cells of a body, forming a pathway for the transfer of nutrients between the blood and the cells. Tissue fluid is formed by leakage from blood capillaries. It is similar in composition to blood plasma, but lacks the plasma proteins.

It is not just water and salts that are kept constant in a body. Many other components of the internal environment are maintained. For example, the level of carbon dioxide in the blood is regulated, along with the blood pH, the concentration of dissolved glucose and the body temperature. Regulation of the glucose concentration of blood is controlled by hormones, and is covered in depth in Chapter 18.

Homeostasis is important because cells will only function properly if they are bathed in a tissue fluid which provides them with their optimum conditions. For instance, if the tissue fluid contains too many solutes, the cells will lose water by osmosis, and become dehydrated. If the tissue fluid is too dilute, the cells will swell up with water. Both conditions will prevent cells from working efficiently and might cause permanent damage. If the pH of the tissue fluid is not correct, it will affect the activity of the cells' enzymes, as will a body temperature much different from 37°C. It is important also that excretory products are removed. Substances such as urea must be prevented from building up in the blood and tissue fluid, where they would be toxic to cells.

The human kidneys have a very important role to play in homeostasis, because they produce **urine**. Urine contains a varying amount of water and salts, allowing us to lose as much as we need to maintain a balance of these substances in the body. This is called **osmoregulation**. However, the kidneys also excrete urea. Before we can look in more detail at the structure and function of the kidneys, we must consider the meaning of 'excretion'.

Homeostasis means 'keeping the conditions in the internal environment of the body constant'.

Excretion is the removal from the body of the waste products of metabolism.

Excretion

A large number of chemical reactions take place inside living cells. Some of these reactions break down substances; others build up large molecules from smaller ones. Together these reactions are called **metabolism**. Both types of reactions produce by-products, substances that might be harmful if they were allowed to build up in the cells. These harmful by-products of metabolism have to be removed from the body of an animal or plant. This is what is meant by **excretion**.

For example, most cells produce carbon dioxide and water from aerobic respiration (see Chapter 13):

glucose   +   oxygen $\rightarrow$ carbon dioxide   +   water   (+ energy)

$C_6H_{12}O_6$ + $6O_2$ $\rightarrow$ $6CO_2$ + $6H_2O$ (+ energy)

If carbon dioxide were to build up in the body (e.g. in the blood of a person) it would be toxic. It therefore has to be removed, which is one function of the lungs. In this respect, the lungs are acting as an excretory organ.

During the hours of sunlight, when a plant is photosynthesising, any carbon dioxide it produces by respiration is immediately used up in photosynthesis, so plants only excrete carbon dioxide from respiration at night, and during the day they excrete oxygen (Chapter 13).

Plants produce few other waste products. They have a lower rate of metabolism than animals, and can take just the amount of nutrients that they need from their environment in order to synthesise organic molecules. Plants do produce certain toxic wastes, however. For example, a number of plant species produce crystals of **calcium oxalate** which are deposited in their leaves. This substance is lost from the plants when they shed their leaves, and because it is stored as insoluble crystals, does not harm the plant itself. Many tree species store poisonous waste products such as **tannins** in their bark, and lose these when they shed the bark.

Animals produce a number of toxic waste products apart from carbon dioxide. For example, the blood pigment haemoglobin is broken down in the liver, forming coloured substances called **bile pigments**. These are passed out of the liver in the bile. They then pass into the intestine and are excreted with the faeces. The bile pigments are what gives bile its yellow-green colour.

However, the most obvious excretory organs in the human body are the kidneys. These produce a watery fluid called **urine**.

Urine

An adult human produces about 1.5 dm³ of urine every day, although this volume depends very much on the amount of water drunk and the volume lost in other forms, such as sweat. Every litre of urine contains about 40 g of waste products and salts (Table 17.1, page 231).

Cross-curricular links:
chemistry, oxygen, carbon dioxide, water

Cross-curricular links:
chemistry, oxygen, carbon dioxide, water

Sometimes plants use these toxic compounds as a means of defence against predators. Leaves containing poisonous compounds are less likely to be eaten by herbivores. A number of compounds have useful medicinal applications, such as quinine, used to treat malaria, and morphine, a powerful painkiller.

This illustrates the difference between **egestion** and **excretion**. Egestion is the removal of undigested food from the body, while excretion is the removal of the waste products of metabolism. Bile pigments are the result of the metabolism of haemoglobin by the liver cells, so they are an excretory product. However most faeces consist of undigested food and dead bacteria, which are not products of the cells, and their removal from the body is called egestion.

Substance	Amount (gm dm⁻³)
urea	23.3
ammonia	0.4
other nitrogenous waste	1.6
sodium chloride (salt)	10.0
potassium	1.3
phosphate	2.3

Table 17.1: *Some of the main dissolved substances in urine.*

Notice the words **nitrogenous waste**. Urea and ammonia are two examples of nitrogenous waste. It means that they contain the element **nitrogen**. All animals have to excrete a nitrogenous waste product.

The reason behind this is quite involved. Carbohydrates and fats only contain the elements carbon, hydrogen and oxygen. Proteins, on the other hand, also contain nitrogen. If the body has too much carbohydrate or fat, these substances can be stored, for example as glycogen in the liver, or as fat under the skin and around other organs. Excess proteins, or their building blocks (amino acids) cannot be stored. The amino acids are first broken down in the liver. They are converted into carbohydrate (which is stored as glycogen) and ammonia, in a process called **deamination**. The ammonia is then joined with carbon dioxide to make the main nitrogen-containing waste product, **urea** (Figure 17.2). The urea passes into the blood, to be filtered out by the kidneys during the formation of urine.

So the kidney is really carrying out two functions. It is a homeostatic organ, controlling the water and salt (ion) concentration in the body. It is also an excretory organ, concentrating nitrogenous waste in a form that can be eliminated. The urea passes into the blood, to be filtered out by the kidneys during the formation of urine.

The major excretory products of humans and plants are summarised in Table 17.2.

Salts in urine or in the blood are present as ions. For example, the sodium chloride in Table 17.1 will be in solution as sodium ions (Na^+) and chloride ions (Cl^-). Urine contains many other ions, such as potassium (K^+) phosphate (HPO_4^{2-}) and ammonium (NH_4^+), and removes excess ions from the blood.

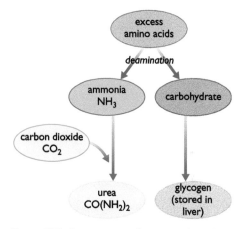

Figure 17.2 *Deamination of amino acids in the liver produces ammonia, which the liver cells convert into urea.*

Animals	Plants
• **carbon dioxide** and **water** from respiration	• **carbon dioxide** and **water** from respiration (used up in photosynthesis in the light)
• **urea** from deamination of excess amino acids	• **oxygen** from photosynthesis in the light
• **bile pigments** from the breakdown of haemoglobin	• many other toxic wastes, shed in leaves or bark, such as **calcium oxalate** crystals

Table 17.2: *The major excretory products of humans and plants.*

3. What is meant by 'excretion'?
4. What is a 'nitrogenous' waste?
5. Why do humans need to excrete urea?

The urinary system

The human urinary system is shown in Figure 17.3.

Each kidney is supplied with blood through a short **renal artery**. This leads straight from the body's main artery, the aorta, so the blood entering the kidney is at a high pressure. Inside each kidney, the blood is filtered, and the 'cleaned' blood passes out through each **renal vein** to the main vein, or vena cava. The urine passes out of the kidneys through two tubes, the **ureters**, and is stored in a muscular bag called the **bladder**.

The bladder has a tube leading to the outside, called the **urethra**. The wall of the urethra contains two ring-like muscles, called **sphincters**. They can contract to close the urethra and hold back the urine. The lower sphincter muscle is consciously controlled, or voluntary, while the upper one is involuntary – it automatically relaxes when the bladder is full.

A baby cannot control its voluntary sphincter. When its bladder is full, the baby's involuntary sphincter relaxes, releasing the urine. A toddler learns to control this muscle and hold back the urine.

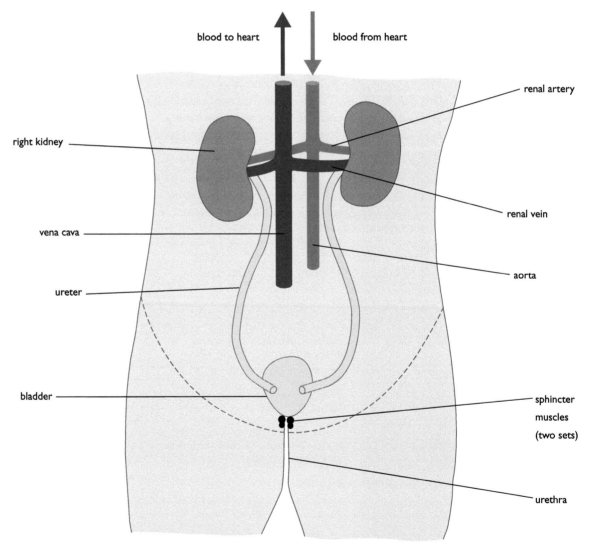

Figure 17.3 *The human urinary system.*

The kidneys

If you cut a kidney lengthwise as shown in Figure 17.4 on page 233, you should be able to find the structures shown.

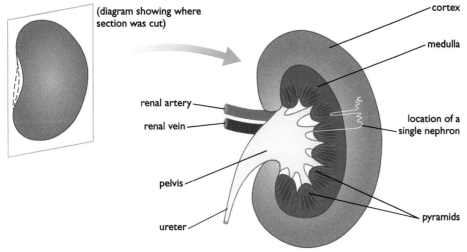

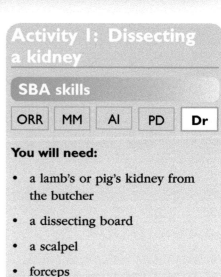

You will need:

- a lamb's or pig's kidney from the butcher
- a dissecting board
- a scalpel
- forceps
- seeker

Carry out the following:

1. Place the kidney flat on a board and cut through it longitudinally as shown in Figure 17.4.

2. Remove the top half and examine the parts of the kidney. How many of the structures shown in Figure 17.4 can you see?

3. Make a drawing of the kidney. Add a title, labels and annotations. Don't forget to include the magnification of your drawing.

Figure 17.4 *Section through a kidney cut along the plane shown.*

There is not much that you can make out without the help of a microscope. The darker outer region is called the **cortex**. This contains many tiny blood vessels that branch from the renal artery. It also contains microscopic tubes that are not blood vessels. They are the filtering units, called **kidney tubules** or **nephrons** (from the Greek word *nephros*, meaning kidney). The tubules then run down through the middle layer of the kidney, called the **medulla**. The medulla has bulges called **pyramids** pointing inwards towards the concave side of the kidney. The tubules in the medulla eventually join up and lead to the tips of these pyramids, where they empty urine into a space called the **pelvis**. The pelvis connects with the **ureter**, carrying the urine to the **bladder**.

By careful dissection, biologists have been able to find out the structure of a single tubule and its blood supply (Figure 17.5). There are about 1 million of these in each kidney.

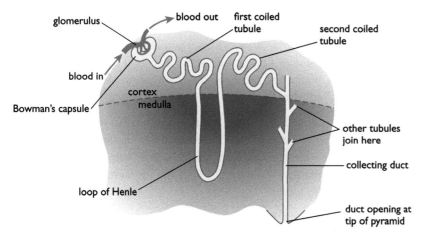

Figure 17.5 *A single nephron, showing its position in the kidney. Each kidney contains about 1 million of these filtering units.*

At the start of the nephron is a hollow cup of cells called the **Bowman's capsule**. It surrounds a ball of blood capillaries called a **glomerulus** (plural = glomeruli). It is here that the blood is filtered. Blood enters the kidney through the renal artery, which divides into smaller and smaller arteries. The smallest arteries, called **arterioles**, supply the capillaries of the glomerulus (Figure 17.6, page 234).

Chapter 17: Homeostasis and Excretion

233

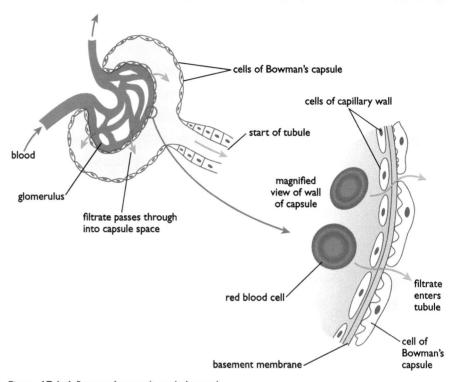

Figure 17.6 *A Bowman's capsule and glomerulus.*

The cells of the glomerulus capillaries do not fit together very tightly; there are spaces between them making the capillary walls much more permeable than others in the body. The cells of the Bowman's capsule also have gaps between them, so they only act as a coarse filter. It is the basement membrane that is the fine molecular filter.

Cross-curricular links:
chemistry, dialysis

A blood vessel with a smaller diameter carries blood away from the glomerulus, leading to capillary networks that surround the other parts of the nephron. Because of the resistance to flow caused by the glomerulus, the pressure of the blood in the arteriole leading to the glomerulus is very high. This pressure forces fluid from the blood through the walls of the capillaries and the Bowman's capsule, into the space in the middle of the capsule. Blood in the glomerulus and the space in the capsule are separated by two layers of cells, the capillary wall and the wall of the capsule. Between the two cell layers is a third layer called the **basement membrane**, which is not made of cells. These layers act like a filter, allowing water, ions and small molecules to pass through, but holding back blood cells and large molecules such as proteins. The fluid that enters the capsule space is called the **glomerular filtrate**. This process, where the filter separates different sized molecules under pressure, is called **ultrafiltration**.

The kidneys produce about 125 cm³ (0.125 dm³) of glomerular filtrate per minute. This works out at 180 dm³ per day. Remember though, only 1.5 dm³ of urine is lost from the body every day, which is less than 1% of the volume filtered through the capsules. The other 99% of the glomerular filtrate is *reabsorbed* back into the blood.

We know this because scientists have actually analysed samples of fluid from the space in the middle of the nephron. Despite the diameter of the space being only 20 µm (0.02 mm), it is possible to pierce the tubule with microscopic glass pipettes and extract the fluid for analysis. Figure 17.7 on page 235 shows the structure of the nephron and the surrounding blood vessels in more detail.

There are two **coiled regions** of the tubule in the cortex separated by a U-shaped loop that runs down into the medulla of the kidney, called the **loop of Henlé**. After the second coiled tubule, several nephrons join up to form a **collecting duct**, where the final urine passes out into the pelvis.

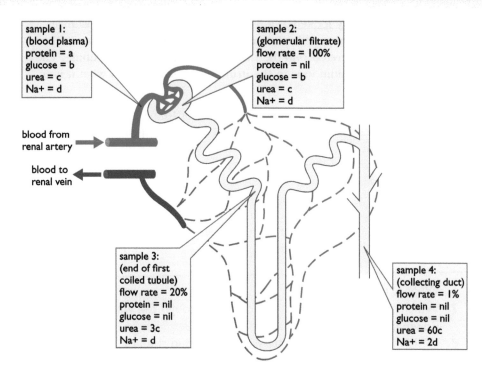

sample 1:
(blood plasma)
protein = a
glucose = b
urea = c
Na+ = d

sample 2:
(glomerular filtrate)
flow rate = 100%
protein = nil
glucose = b
urea = c
Na+ = d

blood from
renal artery

blood to
renal vein

sample 3:
(end of first
coiled tubule)
flow rate = 20%
protein = nil
glucose = nil
urea = 3c
Na+ = d

sample 4:
(collecting duct)
flow rate = 1%
protein = nil
glucose = nil
urea = 60c
Na+ = 2d

Figure 17.7 A nephron and its blood supply. Sample 1 is taken from the blood entering the glomerulus, and samples 2–4 show what is happening to the fluid as it travels along the nephron.

Samples 1–4 show the results of analysing the blood before it enters the glomerulus, and the fluid at three points inside the tubule. The flow rate is a measure of how much water is in the tubule. If the flow rate falls from 100% to 50%, this is because 50% of the water in the tubule has gone back into the blood. To make the explanation easier, the concentrations of dissolved protein, glucose, urea and sodium are shown by different letters (a to d). You can tell the relative concentration of one substance at different points along the tubule from this. For example, urea at a concentration '3c' is three times more concentrated than when it is 'c'.

In the blood (sample 1) the plasma contains many dissolved solutes, including protein, glucose, urea and salts (just sodium ions, Na+, are shown here). As we saw above, protein molecules are too big to pass through into the tubule, so the protein concentration in sample 2 is zero. The other substances are at the same concentration as in the blood.

Now look at sample 3, taken at the end of the first coiled part of the tubule. The flow rate that was 100% is now 20%. This means that 80% of the water in the tubule has been reabsorbed back into the blood. If no solutes were reabsorbed along with the water, their concentrations should be *five times* what they were in sample 2. Since the concentration of sodium hasn't changed, 80% of this substance must have been reabsorbed (and some of the urea too). However, the glucose concentration is now zero – *all* of the glucose is taken back into the blood in the first coiled tubule. This is necessary because glucose is a useful substance that is needed by the body.

Finally, look at sample 4. By the time the fluid passes through the collecting duct, its flow rate is only 1%. This is because 99% of the water has been reabsorbed. Protein and glucose are still zero, but most of the urea is still in the fluid. The level of sodium is only 2d, so not all of it has been reabsorbed, but it is still twice as concentrated as in the blood.

This is a summary of what happens in the kidney nephron:

Part of the plasma leaves the blood in the Bowman's capsule and enters the nephron. The filtrate consists of water and small molecules. As the fluid passes along the nephron, all the glucose is absorbed back into the blood in the first coiled part of the tubule, along with most of the sodium and chloride ions. In the rest of the tubule, more water and ions are reabsorbed, and some solutes like ammonium ions are secreted into the tubule. The final urine contains urea at a much higher concentration than in the blood. It also contains controlled quantities of water and ions.

This description has only looked at a few of the more important substances. Other solutes are concentrated in the urine by different amounts. Some, like ammonium ions, are secreted *into* the fluid as it passes along the tubule. The concentration of ammonium ions in the urine is about 150 times what it is in the blood.

You might be wondering what the role of the loop of Henlé is. The full answer to this is too complicated for this textbook, and a simple explanation will have to be sufficient for now. It is involved with concentrating the fluid in the tubule by causing more water to be reabsorbed into the blood. Mammals with long loops of Henlé can make a more concentrated urine than ones with short loops. Desert animals have many long loops of Henlé, so they are able to produce very concentrated urine, conserving water in their bodies. Animals that have easy access to water, such as otters or beavers, have short loops of Henlé. Humans have a mixture of long and short loops.

6. What is ultrafiltration?
7. Which substance in blood cannot pass through the wall of the Bowman's capsule?
8. Which substance is completely reabsorbed in the first coiled tubule?
9. Which substance is 60 times more concentrated in urine compared with blood?

Control of the body's water content

Not only can the kidney produce urine that is more concentrated than the blood, it can also *control* the concentration of the urine, and so can *regulate* the water content of the blood. Earlier in this chapter, you were asked to think about what would happen if you drank one litre of water. The kidneys respond to this 'upset' to the body's water balance by making a larger volume of more dilute urine. Conversely, if the blood becomes too concentrated, the kidneys produce a smaller volume of urine. These changes are controlled by a hormone produced by the pituitary gland, at the base of the brain. The hormone is called **anti-diuretic hormone**, or **ADH**.

'Diuresis' means the flow of urine from the body, so 'anti-diuresis' means producing less urine. ADH starts to work when your body loses too much water, for example, if you are sweating heavily and not replacing lost water by drinking.

The loss of water means that the concentration of the blood starts to increase. This is detected by special cells in a region of the brain called the **hypothalamus** (see Chapter 18). These cells are sensitive to the solute concentration of the blood, and cause the pituitary gland to release more ADH. The ADH travels in the bloodstream to the kidney. At the kidney tubules, it causes the collecting ducts to become more permeable to water, so that more water is reabsorbed back into the blood. This makes the urine more concentrated, so that the body loses less water and the blood becomes more dilute.

When the water content of the blood returns to normal, this acts as a signal to 'switch off' the release of ADH. The kidney tubules then reabsorb less water. Similarly, if someone drinks a large volume of water, the blood will become too dilute. This leads to lower levels of ADH secretion, the kidney tubules become less permeable to water, and more water passes out of the body in the urine. In this way, through the action of ADH, the level of water in the internal environment is kept constant.

The action of ADH illustrates the principle of **negative feedback**. A change in conditions in the body is detected, and starts a process that works to return conditions to normal. When the conditions are returned to normal, the corrective process is switched off (Figure 17.8). In the situation described, the blood becomes too concentrated. This switches on ADH release, which acts at the kidneys to correct the problem. The word 'negative' means that the process works to eliminate the change. When the blood returns to normal, ADH release is switched off. The feedback pathway forms a 'closed loop'. Many conditions in the body are regulated by negative feedback loops like this.

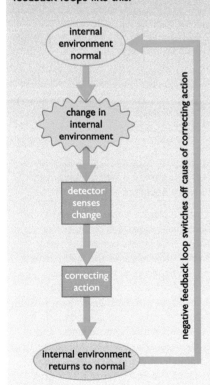

Figure 17.8 *In homeostasis, the extent of a correction is monitored by negative feedback.*

ORR	MM	AI	**PD**	Dr

In this experiment, you are provided with four samples of 'urine':

- The first sample is taken from a person with no medical problems.

- The second is from a patient suffering from diabetes, where glucose is present in the urine.

- The third is from a patient who has a kidney infection. She is excreting urine that contains protein.

- The fourth is from a patient who has an illness that results in the production of large volumes of very dilute urine.

The four samples have lost their original labels. Plan an investigation to find out which sample is from which patient. You should use suitable tests for organic substances (food tests) as well as observations, to decide which urine sample belongs to which patient. If your plan is suitable, you may be allowed to carry it out.

As well as causing the pituitry gland to release ADH, the receptor cells in the hypothalamus also stimulate a 'thirst centre' in the brain. This makes the person feel thirsty, so that he or she will drink water, diluting the blood.

Kidney failure – what happens when things go wrong

A kidney can stop working as a result of disease or an accident. We can live perfectly happily with only one kidney, but if both stop working we will die within a week or so, because poisonous waste builds up in the blood. If a person's kidneys fail, there are two ways they can be kept alive. One is to carry out a kidney **transplant**, and the other is for their blood to be filtered through an artificial kidney machine, in a process called **renal dialysis**.

Kidney dialysis machines

The artificial kidney, or renal dialysis machine, filters the patient's blood, removing urea and other waste, as well as excess water and salts. The filter is a special **dialysis membrane** called **Visking tubing**, which looks rather like cellophane. This thin material has millions of tiny holes in it. The holes will let small molecules like water, ions and urea pass through, but not larger molecules such as proteins, or blood cells.

Blood from the patient flows on one side of the membrane, and a watery liquid called **dialysis solution** flows past the other side, in the opposite direction (Figure 17.9). This is a solution of salts and glucose in exactly the concentrations that the body needs.

As the blood flows past the membrane, urea and unwanted water and salts diffuse through the holes in the membrane into the dialysis fluid. Cells and large molecules such

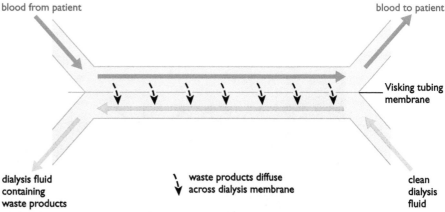

Figure 17.9 *The principle of renal dialysis. The dialysis membrane filters the blood, removing toxic waste.*

as proteins are kept back in the blood. The dialysis fluid is replaced with fresh solution all the time, so that after several hours, the patient's blood has been 'cleaned' of toxic waste and the correct balance of water and salts established.

The surface area of the dialysis membrane separating the blood and dialysis fluid must be large to filter the blood enough. To achieve this, the membrane can be arranged in different ways. In some dialysis machines it is in the form of many long narrow tubes, while in others it is arranged as a stack of flat sheets.

In order to carry out dialysis, it is easier to take blood from a vein than an artery, because veins are closer to the skin, and have a wider diameter than arteries (see Chapter 14). However, the blood pressure in veins is too low, so an operation is first carried out to join an artery to a vein, which raises the blood pressure. A tube is then permanently connected to the vein, so that the patient can be linked to the machine without having to use a needle each time. The purified blood returns to the patient through a second tube joined to the vein, and the 'used' dialysis fluid is discarded.

The kidney machine is a complex and expensive piece of apparatus (Figure 17.10). It has pumps to keep the blood and dialysis fluid flowing, traps to prevent air bubbles getting into the blood, and the oxygenation and temperature of the blood is controlled. Although dialysis will keep a person whose kidneys have failed alive, it is a time-consuming and unpleasant process. The patient has to have their blood 'cleaned' for many hours, two or three times a week. A transplant, if one becomes available, is a much better option.

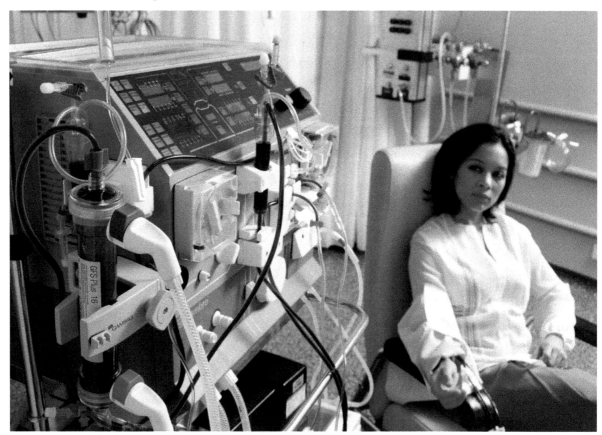

Figure 17.10 *A patient connected to a kidney dialysis machine.*

Chapter summary

In this chapter you have learnt that:

- homeostasis is the maintenance of constant conditions in the internal environment of the body

- factors such as body temperature, blood glucose concentration and water content are kept relatively constant

- excretion is the removal from the body of the waste products of metabolism, such as carbon dioxide from respiration and urea from deamination of proteins

- egestion is the removal from the body of undigested food (faeces)

- the kidneys are organs of excretion and homeostasis; they filter the blood and produce urine, which is a watery solution of urea and salts

- the functional units of a kidney are the nephrons or kidney tubules

- in the Bowman's capsule of a nephron, a solution of urea, salts, glucose and other solutes is forced out of the blood and into the tubule under pressure, by a process called ultrafiltration

- blood cells and proteins are too large to pass through the walls of the Bowman's capsule, so they remain in the blood

- in the first coiled tubule, 80% of water and salts and all the glucose is reabsorbed back into the blood

- in the last part of the tubule, a further 19% of the water and salts is reabsorbed

- water reabsorption is under the control of the hormone ADH (anti-diuretic hormone) and is an example of a negative feedback system, which keeps the concentration of the blood and tissue fluid constant in a process called osmoregulation

- kidney failure can be treated by renal dialysis, where the patient's blood is filtered in a kidney machine.

Questions

1. Which of the following is **not** regulated by homeostasis?

   A Blood glucose level
   B Body temperature
   C Body mass
   D Concentration of water in the blood

2. Which of these statements about the action of anti-diuretic hormone is/are true:

   I More ADH is released when the water content of the blood rises.

   II ADH increases the permeability of the collecting duct.
   III When more ADH is released, more water is reabsorbed.

   A I and II
   B I and III
   C II and III
   D I, II and III

3. In the diagram of a section through a kidney, which part contains the glomeruli?

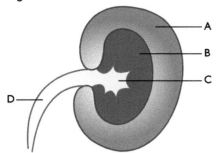

4. The diagram below shows some blood vessels entering and leaving the liver and the kidney.

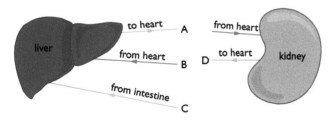

Which blood vessel (A, B, C or D) contains the highest concentration of urea?

5. The diagram below shows a simple diagram of a nephron (kidney tubule).

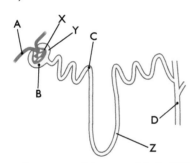

a) What are the names of the parts labelled X, Y and Z? (3)

b) Four places in the nephron and its blood supply are labelled A, B, C and D. Which of the following substances are found at each of these four places?

water   urea   protein   glucose   salt   (4)

c) Explain how the kidney carries out each of the following processes:

i)   Ultrafiltration   (3)

ii) Osmoregulation   (4)

d) What is meant by 'negative feedback'? (3)

6. The bar chart shows the volume of urine collected from a person before and after drinking 1 000 cm³ (1 dm³) of distilled water. The person's urine was collected immediately before the water was drunk and then at 30 minute intervals for four hours.

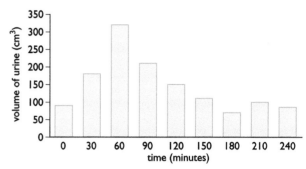

a) Describe how the output of urine changed during the course of the experiment. (4)

b) Explain the difference in urine produced at 60 minutes and at 90 minutes. (4)

c) The same experiment was repeated with the person sitting in a very hot room. How would you expect the volume of urine collected to differ from the first experiment? Explain your answer. (4)

d) Between 90 minutes and 120 minutes, the person produced 150 cm³ of urine. If the rate of filtration at the glomeruli during this time was 125 cm³ per minute, calculate the percentage of filtrate reabsorbed by the kidney tubules. (4)

7. a) By means of a fully labelled and annotated diagram, describe how urine is produced by a human kidney tubule (nephron). (6)

b) Suggest two ways in which the food a person eats may affect the composition of their urine. Explain your answers. (4)

c) Two symptoms of a malfunctioning kidney are (i) the presence of protein in the urine and (ii) high blood pressure (hypertension). Explain how each of these symptoms shows that the kidney is not functioning properly. (4)

8. a) What is an organ? (2)

b) The kidney is the major organ of excretion in humans. Name two other excretory organs and their products. (4)

c) Explain the meaning of the term homeostasis. (3)

d) The kidney is also a homeostatic organ. Explain fully how it is involved in the control of the body's water balance. (5)

Chapter 18: Sensitivity and Coordination in Animals and Plants

When you have completed this chapter, you will be able to:

- understand the meaning of the terms stimulus and response
- explain why response to stimuli is important for the survival of organisms
- define the terms receptor and effector
- explain the relationship between the receptor, the central nervous system and the effector
- identify the main sense organs and the stimuli to which they respond
- relate the structure of the human eye to its functions as a sense organ
- explain defects of vision and their correction
- understand the structure and functions of the human skin
- use simple flow diagrams to show the pathway along which an impulse travels in a reflex arc
- describe the functions of the main regions of the brain
- discuss the physiological, social and economic effects of drug abuse
- describe the responses of invertebrates to variations in light intensity, temperature and humidity, and how these responses can be investigated in a choice chamber
- understand the nature of the human endocrine system
- describe the role of hormones from the adrenal glands and pancreas
- describe the responses of plant stems and roots to light, touch and gravity.

In the body, 'coordination' means making things happen at the right time by linking up different body activities. Humans have two organ systems that do this – the nervous system and the hormone or endocrine system. Plants also show coordinated responses to stimuli.

Stimulus and response

Suppose when you are walking along you see a football coming at high speed towards your head. If your nerves are working properly, you will probably move or duck quickly to avoid contact. Imagine another situation where you are very hungry, and you smell food cooking. Your mouth might begin to 'water', in other words secrete saliva.

Each of these situations is an example of a **stimulus** and a **response**. A stimulus is a change in an organism's surroundings, and a response is a reaction to that change. In the first example, the approaching ball was the stimulus, and your movement to avoid it hitting you was the response. The change in your environment was detected by your eyes, which are an example of a **receptor** organ. The response was brought about by contraction of muscles, which are an **effector** organ (they produce an effect). Linking the two is the nervous system, an example of a coordination system. A summary of the sequence of events is:

$$\text{stimulus} \rightarrow \text{receptor} \rightarrow \text{coordination} \rightarrow \text{effector} \rightarrow \text{response}$$

In the second example, the receptor for the smell of food was the nose, and the response was secretion of saliva from glands. Glands secrete (release) chemical substances, and they are the second type of effector organ. Again, the link between the stimulus and the response is the nervous system. The information in the nerve cells is transmitted in the form of tiny electrical signals called nerve **impulses**.

Receptors

The role of any receptor is to detect the stimulus by changing its energy into the electrical energy of the nerve impulses. For example, the eye converts light energy into nerve impulses, and the ear converts sound energy into nerve impulses. When energy is changed from one form into another, this is called **transduction**. All receptors are **transducers** of energy (Table 18.1).

Receptor	Type of energy transduced
eye (retina)	light
ear (organ of hearing)	sound
ear (organ of balance)	movement (kinetic)
tongue (taste buds)	chemical
nose (organ of smell)	chemical
skin (touch/pressure/pain receptors)	movement (kinetic)
skin (temperature receptors)	heat
muscle (stretch receptors)	movement (kinetic)

Table 18.1: *Human receptors and the energy they transduce into electrical impulses.*

Notice how a 'sense' like touch is made up of several components. When we touch a warm surface, we will be stimulating several types of receptor, including touch and temperature receptors, as well as stretch receptors in the muscles. As well as this, each sense detects different aspects of the energy it receives. For example, the ears don't just detect sounds, but different loudness and frequencies of sound, while the eye not only forms an image, but also detects intensity of light and in humans can tell the difference between

Some animals can detect changes in their environment that are not sensed by humans. Insects such as bees can see ultraviolet (UV) light. The wavelengths of UV are invisible to humans (Figure 18.1).

(a)

(b)

Figure 18.1 *This yellow flower (a) looks very different to a bee, which sees patterns on the petals reflecting UV light (b).*

Some organisms can even detect the direction of magnetic fields. Many birds, such as pigeons, have a built-in compass in their brain, which they use for navigation. A species of bacterium can also do this, but as yet no one can explain why this might be an advantage to it!

different light wavelengths (colours). Senses tell us a great deal about changes in our environment.

> 1. What is a stimulus?
> 2. What are nerve impulses?
> 3. Name three types of sensory receptor found in the skin.

The central nervous system

The biological name for a nerve cell is a **neurone**. The impulses that travel along a neurone are not an electric current, as in a wire. They are caused by movements of charged particles (ions) in and out of the neurone. Impulses travel at speeds between about 10 m/s and 100 m/s, which is much slower than an electric current, but fast enough to produce a rapid response.

Impulses from receptors pass along nerves containing **sensory neurones**, until they reach the **brain** and **spinal cord**. These two organs are together known as the **central nervous system**, or **CNS** (Figure 18.2).

> The CNS is well protected by the skeleton. The brain is encased in the skull or **cranium** (nerves connected to the brain are **cranial** nerves) and the spinal cord runs down the middle of the spinal column, passing through a hole in each vertebra. Nerves connected to the **spinal** cord are called **spinal** nerves.

Figure 18.2 *The brain and spinal cord form the central nervous system. Cranial and spinal nerves lead to and from the CNS. The CNS sorts out information from the senses and sends messages to muscles.*

Other nerves contain **motor neurones**, transmitting impulses to the muscles and glands. Some nerves contain only sensory or motor cells, while other nerves contain both – they are 'mixed'. A typical nerve contains thousands of individual neurones.

Both sensory and motor neurones can be very long. For instance, a motor neurone leading from the CNS to the muscles in the finger has a fibre about 1 m in length, which is 100 000 times the length of the **cell body** (Figure 18.3).

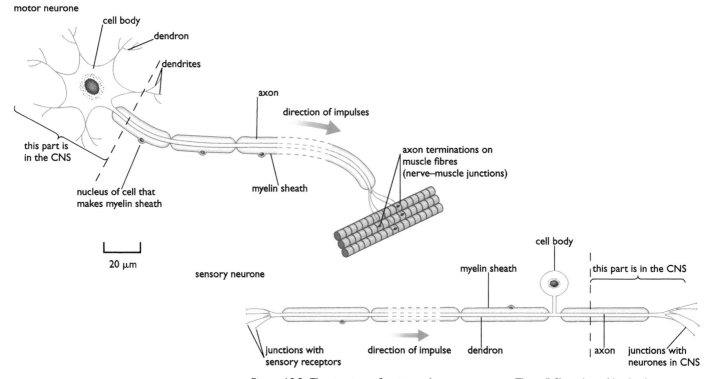

Figure 18.3 *The structure of motor and sensory neurones. The cell fibres (axon/dendron) are very long, which is indicated by the dashed sections.*

The cell body of a motor neurone is at one end of the fibre, in the CNS. The cell body has fine cytoplasmic extensions, called **dendrons**. These in turn form finer extensions, called **dendrites**. There can be junctions with other neurones on any part of the cell body, dendrons or dendrites. These junctions are called **synapses**. Later in this chapter, we will deal with the importance of synapses in nerve pathways. One of the extensions from the motor neurone cell body is much longer than the other dendrons. This is the fibre that carries impulses to the effector organ, and is called the **axon**. At the end of the axon furthest from the cell body, it divides into many nerve endings. These fine branches of the axon connect with a muscle at a special sort of synapse called a **nerve–muscle junction**. In this way, impulses are carried from the CNS out to the muscle. The signals from nerve impulses are transmitted across the nerve–muscle junction, causing the muscle fibres to contract. The axon is covered by a **sheath** made of a fatty material called **myelin**. The myelin sheath insulates the axon, preventing 'short circuits' with other axons, and also speeds up the conduction of the impulses. The sheath is formed by the membranes of special cells that wrap themselves around the axon as it develops.

A **sensory neurone** has a similar structure to the motor neurone, but the cell body is located on a side branch of the fibre, just outside the CNS. The fibre from the sensory receptor to the cell body is actually a dendron, while the fibre from the cell body to the CNS is a short axon. As with motor neurones, fibres of sensory neurones are often myelinated.

4. Explain the difference between a sensory neurone and a motor neurone.
5. What is the function of an axon?
6. What is the myelin sheath and what does it do?

The eye

Many animals have eyes, but few show the complexity of the human eye. Simpler animals, such as snails, use their eyes to detect light, but cannot form a proper image. Other animals, such as dogs, can form images but cannot distinguish colours. The human eye does all three. Of course it is not really the eye that 'sees' anything at all, but the brain that interprets the impulses from the eye. To find out how light from an object is converted into impulses representing an image, we need to look at the structure of this complex organ (Figure 18.4).

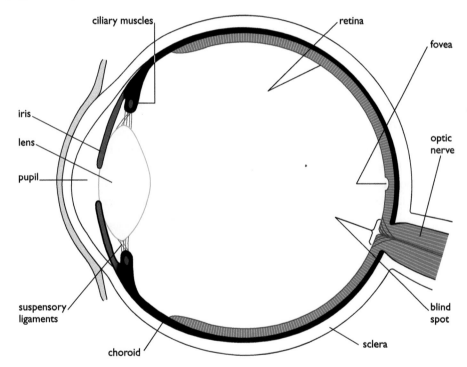

Figure 18.4 *A horizontal section through the human eye.*

The tough outer coat of the eye is called the **sclera**, which is the visible, white part of the eye. At the front of the eye, the sclera becomes a transparent 'window' called the **cornea**, which lets light into the eye. Behind the cornea is the coloured ring of tissue called the **iris**. In the middle of the iris is a hole called the **pupil**, which lets the light through. It is black because there is no light escaping from the inside of the eye.

Underneath the sclera is a dark layer called the **choroid**. It is dark because it contains many pigment cells, as well as blood vessels. The pigment stops light being reflected around inside the eye. In the same way, the inside of a camera is painted matt black to stop stray light bouncing around and fogging the image on the film.

The innermost layer of the back of the eye is the **retina**. This is the light-sensitive layer, the place where light energy is transduced into the electrical energy of nerve impulses. The retina contains cells called **rods** and **cones**. These cells react to light, producing impulses in sensory neurones. The sensory neurones then pass the impulses to the brain through the **optic nerve**.

Rod cells work well in dim light, but they cannot distinguish between different colours, so the brain 'sees' an image produced by the rods in black and white. This is why we can't see colours very well in dim light: only our rods are working properly. The cones, on the other hand, will only work in bright light, and there are three types that respond to different wavelengths or colours of light – red, green and blue. We can see all the colours of visible light as a result of these three types of cones being stimulated to a different degree. For example, if red, green and blue are stimulated equally, we see white. Both rods and cones are found throughout the retina, but cones are particularly concentrated at the centre of the retina, in an area called the **fovea**. Cones give a sharper image than rods, which is why we can only see objects clearly if we are looking directly at them, so that the image falls on the fovea.

To form an image on the retina, light needs to be bent or **refracted**. Refraction takes place when light passes from one medium to another of a different density. In the eye, this happens first at the air/cornea boundary, and again at the lens (Figure 18.5). In fact, the cornea acts as the first lens of the eye.

As a result of refraction at the cornea and lens, the image on the retina is upside down, or **inverted.** The brain interprets the image the right way up.

The role of the iris is to control the amount of light entering the eye by changing the size of the pupil. The iris contains two types of muscles. **Circular muscles** form a ring shape in the iris, and **radial muscles** lie like the spokes of a wheel. In bright light, the pupil is made smaller, or **constricted**.

The fact that the inverted image is seen the right way up by the brain makes the point that it is the brain that 'sees' things, not the eye. An interesting experiment was carried out to test this. Volunteers were made to wear special inverting goggles for long periods. These turned the view of their surroundings upside down. At first this completely disorientated them, and they found it difficult to make even simple coordinated movements. However, after a while, their brains adapted, until the view through the goggles looked normal. In fact, when the volunteers removed the goggles, the world then looked upside down!

Cross-curricular links:
physics, lenses

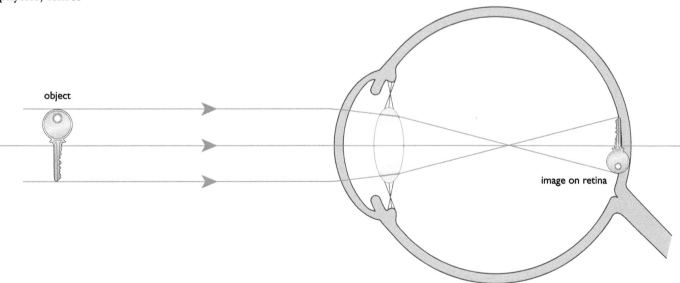

Figure 18.5 *How the eye forms an image. Refraction of light occurs at the cornea and lens, producing an inverted image on the retina.*

This happens because the circular muscles contract and the radial muscles relax. In dim light, the opposite happens. The radial muscles contract and the circular muscles relax, widening or **dilating** the pupil (Figure 18.6).

Try this for yourself! Working in pairs, shine a torch into your partner's eye and note the changes to the pupil.

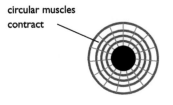

circular muscles contract

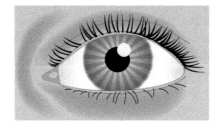

bright light
- circular muscles contract
- radial muscles relax
- pupil constricts

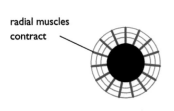

radial muscles contract

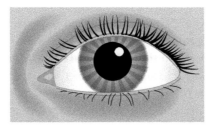

dim light
- circular muscles relax
- radial muscles contract
- pupil dilates

Figure 18.6 *The amount of light entering the eye is controlled by the iris, which alters the diameter of the pupil.*

Whenever our eyes look from a dim light to a bright one, the iris rapidly and automatically adjusts the pupil size. This is an example of a **reflex action**. You will find out more about reflexes later in this chapter. The purpose of the iris reflex is to allow the right intensity of light to fall on the retina. Light that is too bright could damage the rods and cones, and light that is too dim would not form an image. The intensity of light hitting the retina is the stimulus for this reflex. Impulses pass to the brain through the optic nerve, and straight back to the iris muscles, adjusting the diameter of the pupil. It all happens without the need for conscious thought – in fact we are not even aware of it happening.

There is one area of the retina where an image cannot be formed; this is where the optic nerve leaves the eye. At this position there are no rods or cones, so it is called the **blind spot**. The retina of each eye has a blind spot, but they do not cause a problem, because the brain puts the images from each eye together, cancelling out the blind spots of both eyes. As well as this, the optic nerve leaves the eye towards the edge of the retina, where vision is not very sharp anyway. To 'see' your own blind spot you can do a simple experiment. Cover or close your right eye. Hold this page about 30 cm from your eyes and look at the black dot below. Now, without moving the book or turning your head, read the numbers from left to right by moving your left eye slowly towards the right.

●   1   2   3   4   5   6   7   8   9   10   11   12   13   14   15

You should find that when the image of the dot falls on the blind spot it disappears. If you try doing this with both eyes open, the image of the dot will not disappear.

In the iris reflex, the route from stimulus to response is as follows:

stimulus (light intensity)
↓
retina (receptor)
↓
sensory neurones in optic nerve
↓
unconscious part of brain
↓
motor neurones in nerve to iris
↓
iris muscles (effector)
↓
response (change in size of pupil)

A way to prove to yourself that the eyes form two overlapping images is to try the 'sausage test'. Focus your eyes on a distant object. Place your two index fingers tip to tip, and bring them up in front of your eyes, about 30 cm from your face, while still focusing at a distance. You should see a finger 'sausage' between the two fingers. Now try this with one eye closed. What is the difference?

Accommodation

The changes that take place in the eye that allow us to see objects at different distances are called **accommodation**.

You have probably seen the results of a camera or projector that is not in focus – a blurred picture. In a camera, we can focus light from objects that are different distances away by moving the lens backwards or forwards, until the picture is sharp. In the eye, a different method is used. Rather than altering its position, the shape of the lens can be changed. A lens that is fatter in the middle (more convex) will refract light rays more than a thinner (less convex) lens. The lens in the eye can change shape because it is made of cells containing an elastic crystalline protein.

Figure 18.4 shows that the lens is held in place by a series of fibres called the **suspensory ligaments**. These are attached like the spokes of a wheel to a ring of muscle, called the **ciliary muscle**. The inside of the eye is filled with a transparent watery fluid that pushes outwards on the eye. In other words, there is a slight positive pressure within the eye. The changes to the eye that take place during accommodation are shown in Figure 18.7.

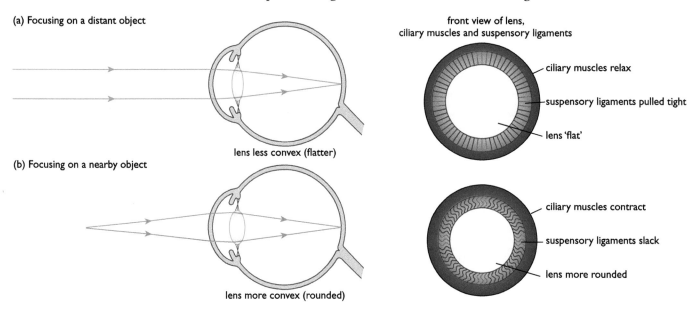

(a) Focusing on a distant object

front view of lens, ciliary muscles and suspensory ligaments

ciliary muscles relax

suspensory ligaments pulled tight

lens 'flat'

lens less convex (flatter)

(b) Focusing on a nearby object

ciliary muscles contract

suspensory ligaments slack

lens more rounded

lens more convex (rounded)

Figure 18.7 *Accommodation: how the eye focuses on objects at different distances.*

When the eye is focused on a distant object, the rays of light from the object are almost parallel when they reach the cornea (Figure 18.7a). The cornea refracts the rays, but the lens does not need to refract them much more to focus the light on the retina, so it does not need to be very convex. The ciliary muscles relax and the pressure in the eye pushes outwards on the lens, flattening it and stretching the suspensory ligaments. This is the condition when the eye is at rest – our eyes are focused for long distances.

When we focus on a nearby object, for example when reading a book, the light rays from the object are spreading out (diverging) when they enter the eye (Figure 18.7b). In this situation, the lens has to be more convex in order to refract the rays enough to focus them on the retina. The ciliary muscles now contract; the suspensory ligaments become slack and the elastic lens bulges outwards into a more convex shape.

7. Where is the choroid and what is its function?
8. List, in order, the parts of the eye that light passes through until it reaches the retina.
9. Why is the cornea the first lens of the eye?
10. What happens to the circular muscles of the iris when you look into a bright light?
11. What is the blind spot?
12. What shape is the lens of the eye when you look at a distant object?

Defects of vision

Long and short sight

In some people, the accommodation mechanism does not work properly, and they are unable to see clearly without the help of spectacles (glasses) or contact lenses. There are two main problems, called long sight and short sight.

In the case of **long sight**, either the lens is not convex enough (i.e. too flat) or the eyeball is too short from front to back, so that light rays from a nearby object are focused behind the retina (Figure 18.8a). This means that the image falling on the retina will be out of focus. A long-sighted person has difficulty focusing on nearby objects, for example, they will hold a book at arm's length to read it without glasses. Long sight can be corrected by using convex lenses or glasses that converge the light rays before they enter the eye (Figure 18.8b).

With **short sight**, either the lens is too convex or the eyeball is too long, so that the light rays from a distant object are focused in front of the retina (Figure 18.9a, page 250), again producing an out-of-focus image. A short-sighted person has problems focusing on distant objects. This fault can be corrected using concave lenses, which diverge (spread out) the light rays before they enter the eye (Figure 18.9b, page 250).

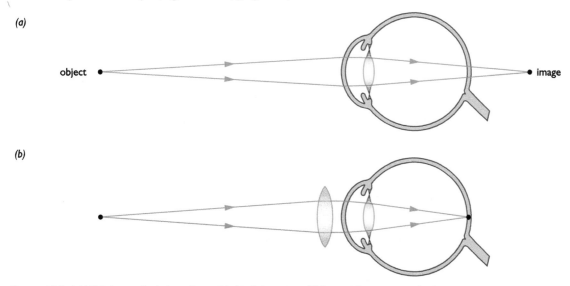

(a)

object

image

(b)

Figure 18.8 *(a) With long sight, light is focused behind the retina. (b) Long sight is corrected using a convex lens.*

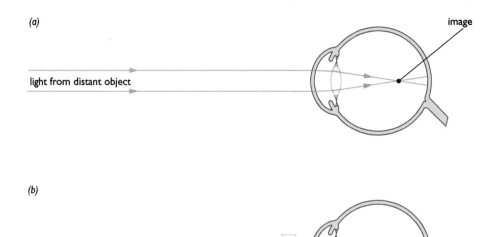

(a)

image

light from distant object

(b)

Figure 18.9 *(a) With short sight, light is focused in front of the retina. (b) Short sight is corrected using a concave lens.*

The causes and effects of long and short sight are summarised in Table 18.2.

	Long sight	Short sight
cause	lens not convex enough, or eyeball too short	lens too convex, or eyeball too long
result	light focused beyond retina	light focused short of retina
lens used for correction	convex lenses to diverge light before it enters the eye	concave lenses to converge light before it enters the eye

Table 18.2: *The causes and effects of long and short sight.*

Problems with vision in old age

Both long- and short-sightedness tend to occur more frequently in the elderly. However, a number of more serious defects of vision may also develop, including **cataracts** and **glaucoma**. A cataract is a condition where the lens of the eye becomes cloudy and opaque, so that the person is unable to see. This can be treated by surgery. The surgeon opens up the front of the eye and removes the affected lens. The patient will be able to see again, although they will need to wear glasses.

Glaucoma is caused by a build-up of fluid in the eye, producing excessive pressure. This pressure can damage the optic nerve, leading to poor vision and blindness. Both cataracts and glaucoma are more common in people with diabetes (see later in this chapter).

13. Why couldn't you use a short-sighted person's spectacles as a magnifying glass?

14. Older people sometimes need to use two pairs of glasses – one for distance vision and one for reading. Can you explain why?

The structure and functions of the skin

As you have seen, the human skin contains a number of different sense organs that detect touch, pain and temperature changes. In addition to its role in sensory reception, the skin has a number of functions that are related to the fact that it forms the outer surface of the body. These functions include:

- forming a tough outer layer able to resist mechanical damage

- acting as a barrier to the entry of disease-causing micro-organisms

- forming an impermeable surface, preventing loss of water

- controlling the loss of heat through the body surface

- protecting tissues from the harmful effects of ultraviolet (UV) radiation from the Sun.

Figure 18.10 shows the structure of human skin. It is made up of three layers – the epidermis, dermis and hypodermis.

> Sunscreens and sunblocks are creams, lotions or gels that are applied to the skin as extra defence against UV. They contain chemicals that absorb the UV radiation that damages the skin. The effectiveness of a sunscreen is measured by its sun protection factor, or SPF. The higher the value of SPF, the greater the protection given will be, for example, a SPF 50 sunscreen gives five times longer protection than an SPF 10 sunscreen.

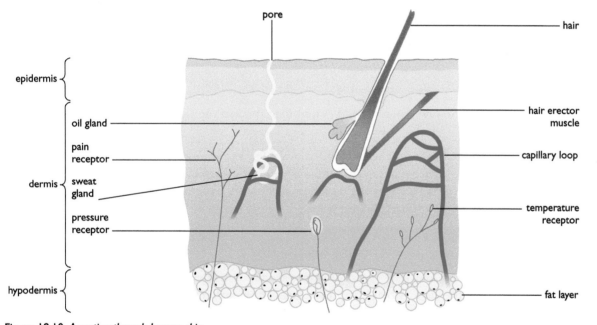

Figure 18.10 *A section through human skin.*

The outer **epidermis** consists of dead cells that stop water loss and protect the body against invasion by micro-organisms such as bacteria. The cells of the epidermis are continually being rubbed off and replaced by new cells from beneath, which divide and move upwards.

The living cells at the base of the epidermis contain a dark pigment called **melanin**, which gives your skin its characteristic colour. Melanin absorbs UV radiation, which can damage living cells by causing mutations to the DNA, sometimes resulting in skin cancers. Dark skins contain more melanin than lighter skins, and skin becomes darker on exposure to sunlight, due to more melanin being made in the cells.

The **hypodermis** contains fatty tissue, which insulates the body against heat loss and is a store of energy. The middle layer, the **dermis**, contains many sensory receptors. It is also the location of sweat glands, many small blood vessels and hair follicles. All of these structures are involved in control of the body's temperature (see page 252).

> Some people use chemicals in an attempt to lighten their skin. The 'skin-bleaching phenomenon' is a growing fashion among many people in Jamaica, who think it is fashionable and makes them look more attractive. Most skin bleachers use untested over-the-counter creams or ointments, illegal skin products or even homemade preparations. Many of these products contain harmful chemicals.

(Note: Read the instructions before you start, and design a suitable table to record your results before proceeding with the investigation.)

Different areas of skin have a different sensitivity to touch. Not surprisingly, areas such as the lips and fingertips have a very high density of touch receptors and so are more sensitive than other areas, such as the skin on the back of the arm. You can test this using a simple piece of apparatus consisting of two pins stuck in a cork. Work in pairs for this activity.

You will need:

- two pins and a cork
- a blindfold

Carry out the following:

1. Stick two small pins into a cork so that the pinheads are 2 cm apart.

2. Ask your partner to close his or her eyes, or use a blindfold.

3. *Gently* touch the back of his or her hand, using either one or two pins. Ask them to identify the number of pins that you were using and record whether their answer was right or wrong with a tick or cross.

4. Repeat this 10 times, each time randomly varying the number of pins used. Record the number of correct answers out of 10.

5. Repeat steps 1–4 with the pins 1 cm, 0.5 cm and 0.2 cm apart.

6. Now repeat steps 1–5 on other areas of the skin, such as the fingertips, palm of the hand or back of the arm.

7. Write up your method and results. You should present your results in a suitable graphical form, as well as in the table. What conclusions can be drawn about the distribution of touch receptors in the various areas of skin that you have tested?

The skin and temperature control

You may have heard of mammals and birds being described as 'warm blooded'. A better word for this is **homeothermic**. It means that these animals keep their body temperature constant, despite changes in the temperature of their surroundings (Figure 18.11, page 253). This is an example of **homeostasis** (see Chapter 17). In humans, body temperature is maintained at 37°C, give or take a few tenths of a degree.

The real difference between homeotherms and all other animals is that homeotherms can keep their temperatures constant by using **physiological** changes for generating or losing heat. For this reason, mammals and birds are also called **endotherms**, meaning 'heat from inside'.

An endotherm uses heat from the chemical reactions in its cells to warm its body. It then controls its heat loss by regulating processes like sweating and blood flow through the skin. Endotherms use behavioural ways to control their temperature too. For example, penguins 'huddle' to keep warm, and humans put on extra clothes in winter.

All other animals are 'cold blooded' – their body temperature changes with that of their surroundings. For example, a lizard placed in an aquarium at 20°C will have a body temperature of 20°C. If it is moved to an aquarium at 30°C, its body temperature will rise to 30°C too. (In the wild, lizards control their temperature to a certain extent by adapting their behaviour, e.g. by basking in sunlight to warm up, or hiding in holes to cool down.)

Physiology is a branch of biology that deals with how the bodies of animals or plants work, for example, how muscles contract, how nerves send impulses, or how xylem carries water through plants. In this chapter, you have read about kidney physiology.

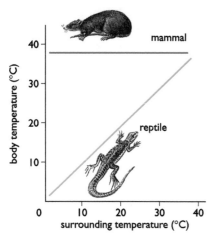

Figure 18.11 The temperature of a homeotherm such as a mammal is kept constant at different external temperatures, whereas the lizard's body temperature changes.

What is the advantage of a human maintaining a body temperature of 37°C? It means that all the chemical reactions taking place in the cells of the body can go on at a steady, predictable rate. The metabolism doesn't slow down in cold environments. If you watch goldfish in a garden pond, you will notice that in summer, when the pond water is warm, they are very active, swimming about quickly. In winter, when the temperature drops, the fish slow down and become very sluggish in their actions. This would happen to mammals too, if their body temperature was not kept steady.

It is also important that the body does not become *too* hot. The cells' enzymes work best at 37°C. At higher temperatures, enzymes, like all proteins, are destroyed by **denaturing** (see Chapter 11). Endotherms have all evolved a body temperature around 40°C (Table 18.3) and enzymes that work best at this temperature.

Monitoring body temperature

In humans and other mammals, the core body temperature is monitored by a part of the brain called the **thermoregulatory centre**. This is located in the hypothalamus of the brain. It acts as the body's thermostat.

If a person goes into a warm or cold environment, the first thing that happens is that temperature receptors in the skin send electrical impulses to the hypothalamus, which stimulates the brain to alter behaviour. We start to feel hot or cold, and usually do something about it, such as finding shade or having a cold drink.

If changes to our behaviour are not enough to keep our body temperature constant, the thermoregulatory centre in the hypothalamus detects a change in the temperature of the blood flowing through it. It then sends signals via nerves to other organs of the body, which regulate the temperature by physiological means.

Imagine that the hypothalamus detects a rise in the central (core) body temperature. Immediately, it sends nerve impulses to the skin. These bring about changes to correct the rise in temperature.

First of all, the **sweat glands** produce greater amounts of sweat. This liquid is secreted onto the surface of the skin. When a liquid evaporates, it turns into a gas.

Species	Average and normal range of body temperature (°C)
brown bear	38.0 ± 1.0
camel	37.5 ± 0.5
elephant	36.2 ± 0.5
fox	38.8 ± 1.3
human	36.9 ± 0.7
mouse	39.3 ± 1.3
polar bear	37.5 ± 0.4
shrew	35.7 ± 1.2
whale	35.7 ± 0.1
duck	43.1 ± 0.3
ostrich	39.2 ± 0.7
penguin	39.0 ± 0.2
thrush	40.0 ± 1.7
wren	41.0 ± 1.0

Table 18.3: The body temperatures of a range of mammals and birds.

A **thermostat** is a switch that is turned on or off by a change in temperature. It is used in electrical appliances to keep their temperature steady. For example, a thermostat in an iron can be set to 'hot' or 'cool' to set the temperature of the iron for ironing different materials.

This change needs energy, called the **latent heat of vaporisation**. When sweat evaporates, the energy is supplied by the body's heat, cooling the body down. It is not that the sweat is cool – it is secreted at body temperature. It only has a cooling action when it evaporates. In very humid atmospheres (e.g. a tropical rainforest) the sweat stays on the skin and doesn't evaporate. It then has very little cooling effect.

Secondly, hairs on the surface of the skin lie flat against the skin's surface. This happens because of the relaxation of tiny muscles called **hair erector muscles** attached to the base of each hair. In cold conditions, these contract and the hairs are pulled upright. The hairs trap a layer of air next to the skin, and since air is a poor conductor of heat, this acts as insulation. In warm conditions, the thinner layer of trapped air means that more heat will be lost. This is not very effective in humans, because the hairs over most of our body do not grow very large. It is very effective in hairy mammals like cats or dogs. The same principle is used by birds, which 'fluff out' their feathers in cold weather.

Lastly, there are tiny blood vessels called capillary loops in the dermis. Blood flows through these loops, radiating heat to the outside and cooling the body down. If the body is too hot, small arteries (arterioles) leading to the capillary loops dilate (widen). This increases the blood flow to the skin's surface. At the same time, blood flow through deeper capillaries is reduced by arterioles narrowing (Figure 18.12). This is called **vasodilation**.

In cold conditions, the opposite happens. The surface blood vessels undergo **vasoconstriction**, so that less heat is lost. Vasoconstriction and vasodilation are brought about by tiny rings of muscles in the walls of the arterioles, called sphincter muscles, like the sphincters you met earlier in this chapter, at the outlet of the bladder.

There are other ways that the body can control heat loss and heat gain. In cold conditions, the body's **metabolism** speeds up, generating more heat. The liver, a large organ, can produce a lot of metabolic heat in this way. The hormone **adrenaline** stimulates the increase in metabolism (see later in this chapter). **Shivering** also takes place, where the muscles contract and relax rapidly. This also generates a large amount of heat.

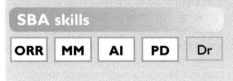

Figure 18.12 *Blood flow through the surface of the skin is controlled by vasodilation or vasoconstriction.*

Activity 2: Does insulation reduce the loss of heat from a hot 'body'?

SBA skills

ORR	MM	AI	PD	Dr

You will need:

- two 100-cm³ beakers or cans with lids – the lids must have a hole for a thermometer to fit through
- hot water (a temperature of about 60–70°C is hot enough)
- two thermometers
- elastic bands
- some cotton wool

Carry out the following:

1. Wrap cotton wool around one of the beakers to act as insulation. Secure it in place with elastic bands.

2. Place hot water in each of the beakers. The water should be at the same starting temperature in each beaker.

3. Place a lid on each beaker and insert thermometers through the holes in the lids.

4. Measure the starting temperature of the water in each beaker.

5. Record the temperature of the water at 1 minute intervals for 20 minutes.

6. Plot a graph of the temperature against time, using the same axes for both sets of results.

7. Write up your experiment and explain your results.

8. Is the apparatus a good 'model' of the effect of insulation in an animal? Explain your answer.

Sweating, vasodilation and vasoconstriction, hair erection, shivering, and changes to the metabolism, along with behavioural actions, work together to keep the body temperature to within a few tenths of a degree of the 'normal' 37°C. If the difference is any bigger than this, it shows that something is wrong. For instance, a temperature might be due to an illness.

> Students often describe vasodilation incorrectly. They talk about the blood vessels 'moving nearer the surface of the skin'. They don't *move* at all, it's just that more blood flows through the surface vessels.

15. List five functions of the skin.
16. What is the normal human body temperature?
17. How does sweat cool the body?
18. What is vasodilation in the skin and what does it do?

Activity 3: Does a wet insulating surface increase heat loss from a 'body'?

SBA skills

| ORR | MM | AI | PD | Dr |

If the insulating fur of a small animal such as a mouse becomes wet, its body temperature may fall so much that it may die. From this observation, plan an investigation that you could carry out using the apparatus in Activity 2. You must make a prediction and say how you would carry out the investigation, including details of how your experiment will be controlled. If there is time, you may be able to try out your method.

Reflex actions

You saw on page 247 that the dilation and constriction of the pupil by the iris is an example of a reflex action. You now need to understand a little more about the nerves involved in a reflex. The nerve pathway of a reflex is called the **reflex arc**. The 'arc' part means that the pathway goes into the CNS and then straight back out again, in a sort of curve or arc (Figure 18.13, page 256).

> A reflex action is a *rapid, automatic* (or *involuntary*) response to a stimulus. The action often (but not always) protects the body. Involuntary means that it is not started by impulses from the brain.

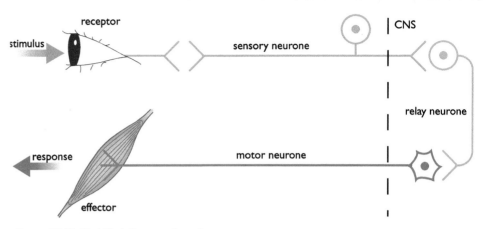

Figure 18.13 *Simplified diagram of a reflex arc.*

The iris–pupil reflex is brought about by impulses travelling through **cranial** nerves (those leading to and from the brain), so it is known as a **cranial reflex**. Others, such as the withdrawal reflex and the knee-jerk reflex are brought about by **spinal** nerves (leading to and from the spinal cord). They are therefore called **spinal reflexes**.

The iris–pupil reflex protects the eye against damage by bright light. Other reflexes are protective too, preventing serious harm to the body. Take for example the reflex response to a painful stimulus. This happens when part of your body, such as your hand, touches a sharp or hot object. The reflex results in your hand being quickly withdrawn. Figure 18.14 shows the nerve pathway of this reflex in more detail.

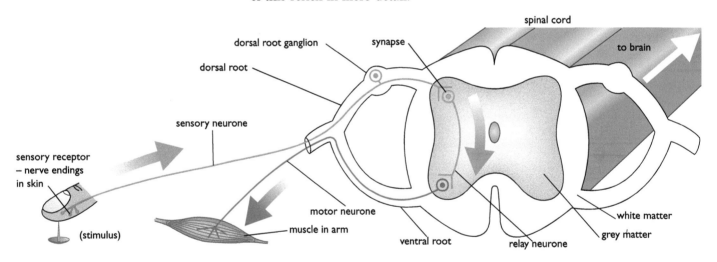

Figure 18.14 *A reflex arc in more detail.*

'Dorsal' and 'ventral' are words describing the back and front of the body. The dorsal roots of spinal nerves emerge from the spinal cord towards the back of the person, while the ventral roots emerge towards the front. Notice that the cell bodies of the sensory neurones are all located in a swelling in the dorsal root, called the **dorsal root ganglion**.

The stimulus is detected by temperature or pain receptors in the skin. These generate impulses in sensory neurones. The impulses enter the CNS through a part of the spinal nerve called the **dorsal root**. In the spinal cord, the sensory neurones connect by synapses with short **relay neurones**, which in turn connect with motor neurones. The motor neurones emerge from the spinal cord through the **ventral root**, and send impulses back out to the muscles of the arm. These muscles then contract, pulling the arm away from the harmful stimulus.

The middle part of the spinal cord consists mainly of nerve cell bodies, which gives it a grey colour. This is why it is known as **grey matter**. The outer part of the spinal cord is called **white matter**, and has a whiter appearance because it contains many axons with their fatty myelin sheaths.

Impulses travel through the reflex arc in a fraction of a second, so that the reflex action is very fast, and doesn't need to be started off by impulses from the brain. However, this doesn't mean that the brain is unaware of what is going on. This is because in the spinal cord, the reflex arc neurones also form synapses with nerve cells leading to and from the brain. The brain therefore receives information about the stimulus. This is how we feel the pain.

Movements are sometimes a result of reflex actions, but we can also contract our muscles as a **voluntary action**, using nerve cell pathways from the brain linked to the same motor neurones. A voluntary action is under conscious control.

The knee-jerk reflex

You can demonstrate a spinal reflex on yourself quite easily. It is the well-known **knee-jerk reflex**. Sit down and cross your legs, so that the upper leg hangs freely over the lower one. Grip the muscles of the top of the upper thigh with one hand and tap the area below the kneecap with a rubber hammer or the edge of the other hand (Figure 18.15).

This may need a little practice, but you should eventually see the lower leg jerk forward as the muscles at the front of the thigh contract.

The reflex arc that brings this about is very similar to the withdrawal reflex (Figure 18.14), but in this case, the stimulus is not detected in the skin, but in stretch receptors in the tendon below the knee. Tapping the tendon causes these receptors to be stretched. They react by sending nervous impulses towards the spinal cord through sensory neurones. The impulses then pass out again to the thigh muscles, through motor neurones, resulting in contraction of the muscle.

Of course, you would not normally experience a tap on the knee from a rubber hammer in everyday life! Where this reflex normally acts is in situations where the knee joint is unexpectedly flexed. For example, if you stumble, the stretch receptors will be stimulated in the same way, and the contraction of the thigh muscle will help to correct the stumble.

Synapses

Synapses are critical to the working of the nervous system. The CNS is made of many billions of nerve cells, and these have links with many others through synapses. In the brain, each neurone may form synapses with thousands of other neurones, so that there are an almost infinite number of possible pathways through the system.

A synapse is actually a gap between two nerve cells. The gap is not crossed by the electrical impulses passing through the neurones, but by chemicals. Impulses arriving at a synapse cause the ends of the fine branches of the axon to secrete a chemical called a **neurotransmitter**. This chemical diffuses across the gap and attaches to the membrane of the second neurone. It then starts off impulses in the second cell (Figure 18.16, page 258). After the neurotransmitter has 'passed on the message', it is broken down by an enzyme.

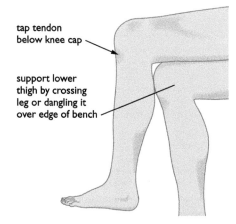

tap tendon below knee cap

support lower thigh by crossing leg or dangling it over edge of bench

Figure 18.15 *Demonstration of the knee-jerk reflex.*

In the knee-jerk reflex, the route from stimulus to response is as follows:

stimulus (tap below the knee)
↓
stretch receptor
↓
sensory neurones in leg
↓
central nervous system
↓
motor neurones in the leg
↓
thigh muscle
↓
response
(knee jerk)

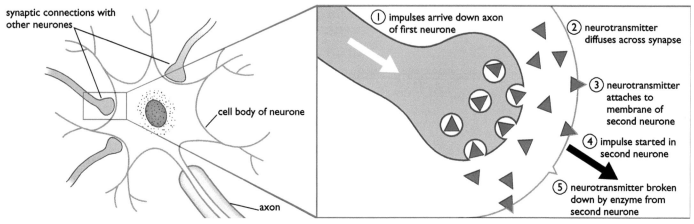

synaptic connections with other neurones

cell body of neurone

axon

① impulses arrive down axon of first neurone

② neurotransmitter diffuses across synapse

③ neurotransmitter attaches to membrane of second neurone

④ impulse started in second neurone

⑤ neurotransmitter broken down by enzyme from second neurone

Figure 18.16 *The sequence of events happening at a synapse.*

Remember that many nerve cells, particularly those in the brain, have thousands of synapses with other neurones. The output of one cell may depend on the inputs from many cells adding together. In this way, synapses are important for integrating information in the central nervous system (Figure 18.17).

Because synapses are crossed by chemicals, it is easy for other chemicals to interfere with the working of the synapse. They may imitate the neurotransmitter, or block its action. This is the way that many well-known drugs, both useful and harmful, work. We will return to this topic later.

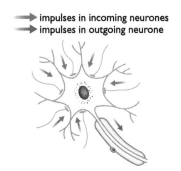

➡ impulses in incoming neurones
➡ impulses in outgoing neurone

Figure 18.17 *Synapses allow the output of one nerve cell to be a result of integration of information from many other cells.*

19. What is a reflex action?
20. Name the stimulus, the receptor, the effector and the response in the 'withdrawal' reflex.
21. What is a synapse? How does it work?

Activity 4: Testing reaction times

SBA skills

| ORR | MM | AI | PD | Dr |

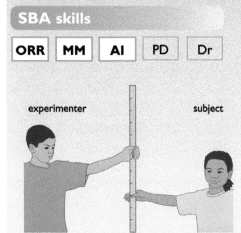

experimenter   subject

Figure 18.18 *Testing reaction times using a falling ruler.*

This investigation tests a subject's reaction time by measuring how quickly he or she can catch a falling metre ruler. The person can use two senses to help them do this – sight and touch.

You will need:

• a metre ruler

Carry out the following:

1. Hold a metre ruler vertically with the lower end touching the subject's open hand (Figure 18.18). Here the subject can both see and feel the ruler. It is best if you hold the ruler so that the 20 cm mark is level with the top of the hand.

2. Tell the person that you will be letting the ruler fall soon, and that he or she must catch it as quickly as they can.

3. Wait for a few seconds, then let go of the ruler.

4. Record the distance the ruler has fallen (subtract 20 cm from the measurement on the ruler at the top of the subject's hand where he/she caught it).

5. Repeat steps 1–4 10 times, waiting variable lengths of time before you let the ruler fall.

6. Now repeat the whole exercise, this time with the ruler visible to the subject but not touching the hand.

7. Finally, repeat the exercise again with the ruler touching the subject's hand, but ask the subject to close his or her eyes (or use a blindfold). This time the subject can feel the ruler against the hand but cannot see it.

8. Calculate the mean distance the ruler has fallen under the three different conditions. Plot a bar graph of your results.

9. Are reaction times faster using sight only, touch only, or sight and touch?

10. Write up your experiment and explain your results.

The brain

The functions of different parts of the brain were first worked out by studying people who had suffered brain damage through accident or disease. These days, we have very sophisticated electronic equipment that can record the activity in a normal living brain, but we are still relatively ignorant about the workings of this most complex organ of the body.

Your brain is sometimes called your 'grey matter'. This is because the positions of the grey and white matter are reversed in the brain compared with the spinal cord. The grey matter, mainly made of nerve cell bodies, is on the outside of the brain, and the axons that form the white matter are in the middle of the brain. The brain is made up of different parts, each with a particular function (Figure 18.19 on page 260).

The largest part of the brain is the **cerebrum**, which is made of two **cerebral hemispheres**. The cerebrum is the source of all our conscious thoughts. It has an outer layer called the **cerebral cortex**, which has many folds all over its surface (Figure 18.20, page 260 on page 260).

The cerebrum has three main functions:

- It contains **sensory areas** that receive and process information from all our sense organs.

- It has **motor areas**, which are where all our voluntary actions originate.

- It is the origin of 'higher' activities, such as memory, reasoning, emotions and personality.

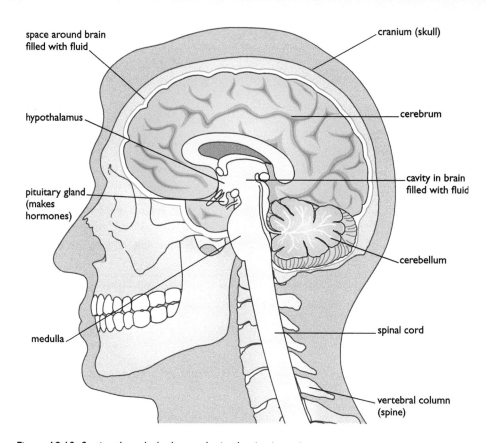

Figure 18.19 *Section through the human brain, showing its main parts.*

Different parts of the cerebrum carry out particular functions. For example, the sensory and motor areas are always situated in the same place in the cortex (Figure 18.19). Some parts of these areas deal with more information than others. Large parts of the sensory area deal with impulses from the fingers and lips, for example (Figure 18.21).

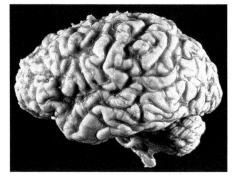

Figure 18.20 *A side view of a human brain. Notice the folded surface of the cerebral cortex.*

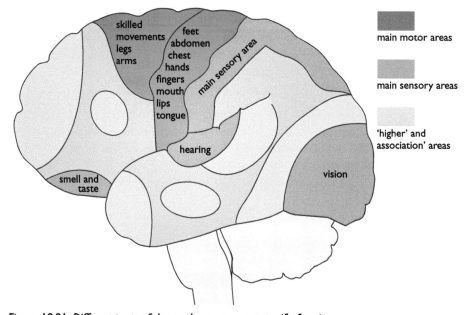

Figure 18.21 *Different parts of the cerebrum carry out specific functions.*

Behind the cerebrum is the **cerebellum**. This region is concerned with coordinating the contraction of sets of muscles, as well as maintaining balance. This is important when you are carrying out complicated muscular activities, such as running or riding a bike. Underneath the cerebrum, connecting the spinal cord with the rest of the brain, is the brain stem or **medulla**. This controls basic body activities such as heartbeat and breathing rate.

The **pituitary gland** is located at the base of the brain, just below a part of the brain called the **hypothalamus**. The pituitary gland secretes a number of chemical 'messengers' called hormones, into the blood. The pituitary and hypothalamus are both discussed later in this chapter.

The autonomic nervous system

The medulla controls the heart rate and breathing through nerves that are part of the **autonomic nervous system**. The autonomic nervous system consists of a series of nerves outside the CNS, leading to many organs. They control activities that are normally involuntary. As well as heart rate and breathing, these include activities such as:

- constriction and dilation of the pupil in the iris–pupil reflex
- relaxation of the involuntary sphincter muscle of the bladder
- contraction of the muscle layers in the wall of the intestine during peristalsis
- secretion of sweat by the sweat glands in the skin
- dilation of the bronchioles leading to the lungs.

The autonomic nervous system consists of two sets of nerves, which tend to produce opposite effects. One set, called the **sympathetic nerves**, usually works in times when the body is active or stressed. These nerves prepare the body for action by, for example, increasing the heart rate, dilating the bronchioles of the lungs and slowing down digestion. The other set, called the **parasympathetic nerves**, produces the opposite effects. We have no conscious control over these processes.

Drugs and the nervous system

Drugs are chemicals that affect processes in a person's body. Many drugs are useful. For example, penicillin is an antibiotic that kills many of the bacteria that cause disease, and aspirin is an effective painkiller. However, a number of drugs act by interfering with the nervous system, and some of these can have very harmful side effects. A good example of this is the drug nicotine, which is present in tobacco. When a person smokes a cigarette, especially the first time, they get a 'buzz' from smoking. Their heart beats faster, their blood pressure rises and they feel excited. This is because nicotine is a **stimulant**, meaning that it increases brain activity. It does this by mimicking the action of neurotransmitters at the synapses of nerve cells in the brain. Some other stimulants also affect synapses. **Caffeine**, the drug in tea and coffee, causes more neurotransmitter to be released than normal.

Nicotine is a highly addictive drug. A person who smokes develops a craving for the drug. On top of this, their body 'gets used to it' and so more must be taken into the body to satisfy the craving. When the person tries to stop smoking, they suffer **withdrawal symptoms**, such as becoming irritable.

Activity 5: Testing the effect of caffeine on the heart rate

SBA skills

| ORR | MM | AI | PD | Dr |

Caffeine is present in coffee, tea and many types of cola. Design an investigation to test the hypothesis that drinking caffeine increases the pulse rate. You will need to measure the pulse rate of yourself and other volunteers before and after consuming drinks containing caffeine. Make sure that your plan contains proper controls, and include ways to ensure that your findings are reliable. When you have had your plan checked by your teacher, you may be allowed to carry out the investigation.

Nicotine, along with the other components of tobacco smoke, has a number of harmful effects on the lungs, heart and blood system.

The opposite of a stimulant is a **depressant** drug. One example is **alcohol** (ethanol). The alcohol in beer, wine and spirits slows down the nervous system, even when drunk in small quantities, and increases the time a person takes to react to a stimulus. That is why driving after drinking alcohol is so dangerous. The driver will not react quickly to sudden danger, such as a person walking into the road. Larger amounts of alcohol in the body interfere with the drinker's balance and muscular control, and lead to blurred vision and slurred speech. High concentrations of alcohol in the blood can even cause coma and death.

Some legal drugs are prescribed by doctors to their patients because the drugs have mind-altering (psychoactive) properties. These include medications for pain relief, depressants for anxiety and sleep disorders, and stimulants. However, prescription drugs may be taken:

- for reasons not intended by the doctor

- in larger amounts than were intended

- by someone other than the patient.

This is called 'prescription drug abuse'. Abuse of prescription drugs is one of the fastest-growing drug problems in many countries across the world. For example, in the United States of America, abuse of prescription and over-the-counter drugs is the third most common type of drug misuse after alcohol and marijuana. Another type of prescription drug abuse is the use of steroids in sport (see page 267 in this chapter).

> The use of illegal drugs and the abuse of legal drugs have serious consequences beyond their physiological effects on a person's body. There are social and economic effects too. Drug abuse often prevents people working or generally contributing to society, and leads them into a life of crime to pay for their drug addiction.

> **22.** Explain the difference between the sensory and motor areas of the cerebral cortex.
> **23.** What is the function of the cerebellum?
> **24.** Alcohol is a depressant drug. What does this mean?

Can other animals detect changes in their environment?

You have seen how humans can detect and respond to changes in their environment, which are called stimuli. All animals have this ability, which is essential for survival in their habitat. However, the particular stimuli that they respond to depend on the nature of their habitat and the factors that are important for their survival. For example, woodlice live in leaf litter, under dead logs and in other similar habitats where it is damp and cool. They feed on dead organic matter, such as decaying leaves. If they are exposed to the sun for too long, they will dry up and die. From these observations, your hypothesis might be that if given a choice, woodlice would select dark, moist and cool conditions.

You can test this hypothesis using a piece of apparatus called a **choice chamber**. There are a number of designs of choice chambers, but they work on the same principle. They allow small invertebrates such as woodlice to 'choose' between two contrasting conditions, such as light and dark, or dry and damp. One design is made from two Petri dishes glued together (Figure 18.22, page 263).

> **Did you know?**
> Woodlice are actually crustaceans, and are related to crabs and shrimps. They breathe through gills located in cavities on the underside of their bodies. They can only live in damp habitats, where their gills remain moist enough for them to breathe.

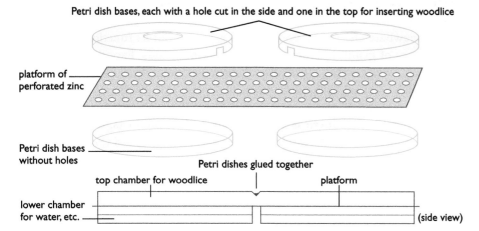

Petri dish bases, each with a hole cut in the side and one in the top for inserting woodlice

platform of perforated zinc

Petri dish bases without holes

Petri dishes glued together

top chamber for woodlice

platform

lower chamber for water, etc.

(side view)

Figure 18.22 *A choice chamber made from two Petri dishes.*

To keep the air humid around the woodlice, you can put water in both bottom chambers of the apparatus. To give the animals a choice between light and dark, you can then cover one side with black paper and leave the other side in the light. To compare their response to humid versus dry air, you can put water in the lower chamber on one side, and a chemical drying agent, such as silica gel, in the other side.

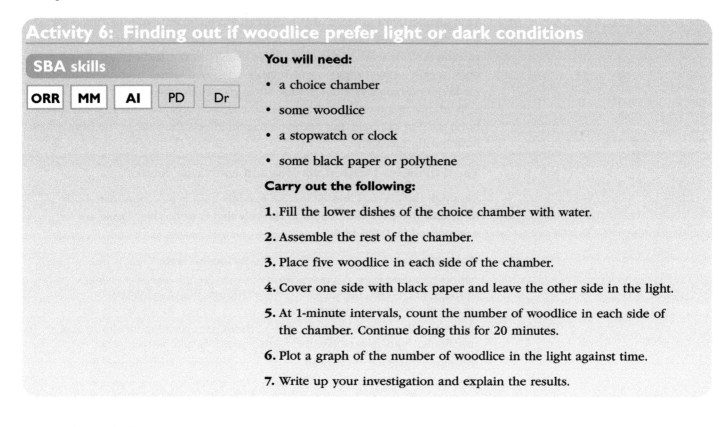

Activity 6: Finding out if woodlice prefer light or dark conditions

SBA skills

| ORR | MM | AI | PD | Dr |

You will need:

- a choice chamber
- some woodlice
- a stopwatch or clock
- some black paper or polythene

Carry out the following:

1. Fill the lower dishes of the choice chamber with water.
2. Assemble the rest of the chamber.
3. Place five woodlice in each side of the chamber.
4. Cover one side with black paper and leave the other side in the light.
5. At 1-minute intervals, count the number of woodlice in each side of the chamber. Continue doing this for 20 minutes.
6. Plot a graph of the number of woodlice in the light against time.
7. Write up your investigation and explain the results.

Chemical coordination – glands and hormones

The nervous system is a coordination system forming a link between stimulus and response. The body has a second coordination system, which does not involve nerves. This is the **endocrine** system. It consists of organs called endocrine **glands**, which make chemical messenger substances called hormones.

A gland is an organ that releases or **secretes** a substance. This means that cells in the gland make a chemical that passes out of the cells. The chemical then travels somewhere else in the body, where it carries out its function. There are two types of glands – **exocrine** and **endocrine** glands. Exocrine glands secrete their products through a tube or **duct**. For example, salivary glands in your mouth secrete saliva down salivary ducts, and tear glands secrete tears through ducts that lead to the surface of the eye. Endocrine glands have no duct, and so are called **ductless** glands. Instead, their products, the hormones, are secreted into the blood vessels that pass through the gland (Figure 18.23).

> The receptors for some hormones are located in the cell membrane of the target cell. Other hormones have receptors in the cytoplasm and some in the nucleus. Without specific receptors, a cell will not respond to a hormone at all.

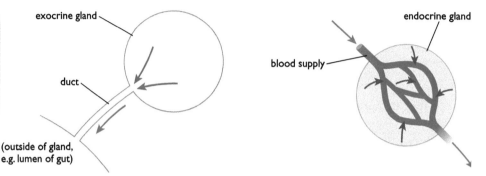

Figure 18.23 *Exocrine glands secrete their products though a duct, while endocrine glands secrete hormones into the blood.*

This section looks at some of the main endocrine glands and the functions of the hormones they produce. Because hormones are carried in the blood, they can travel to all areas of the body. They usually only affect certain tissues or organs, called target organs, which can be a long distance from the gland that made the hormone. Hormones only affect particular tissues or organs if the cells of that tissue or organ have special chemical receptors for that particular hormone. For example, the hormone insulin affects the cells of the liver, which have insulin receptors.

The differences between nervous and endocrine control

Although the nervous and endocrine systems both act to coordinate body functions, there are differences in the way that they do this. These are summarised in Table 18.4.

Nervous system	Endocrine system
works by nerve impulses transmitted through nerve cells (although chemicals are used at synapses)	works by hormones transmitted through the bloodstream
nerve impulses travel quickly and usually have an 'instant' effect	hormones travel more slowly and generally take longer to act
response is usually short-lived	response is usually longer-lasting
impulses act on individual cells such as muscle fibres, so have a very localised effect	hormones can have widespread effects on different organs (although they only act on particular tissues or organs if the cells have the correct receptors)

Table 18.4: *The nervous and endocrine systems compared.*

The positions of the endocrine glands

The main endocrine glands are shown in Figure 18.24, and a summary of some of the hormones that they make and their functions, is given in Table 18.5.

Gland	Hormone	Some functions of the hormones
pituitary	follicle stimulating hormone (FSH)*	stimulates egg development and oestrogen secretion in females, and sperm production in males
	luteinising hormone (LH)*	stimulates egg release (ovulation) in females and testosterone production in males
	anti-diuretic hormone (ADH)	controls the water content of the blood (see Chapter 17)
	growth hormone (GH)	speeds up the rate of growth and development in children
thyroid	thyroxin	controls the body's metabolic rate (how fast chemical reactions take place in cells)
pancreas	insulin	lowers blood glucose
	glucagon	raises blood glucose
adrenals	adrenaline	prepares body for physical activity
testes	testosterone*	controls development of male secondary sexual characteristics
ovaries	oestrogen*	controls development of female secondary sexual characteristics; regulates menstrual cycle
	progesterone*	regulates menstrual cycle

* The hormones involved in reproduction are dealt with in Chapter 20.

Table 18.5: *Some of the main endocrine glands, the hormones they produce and their functions.*

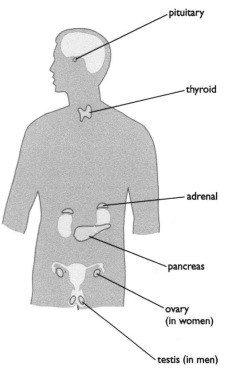

Figure 18.24 *The main endocrine glands of the body.*

The pituitary gland is found at the base of the brain (see Figure 18.19, page 260). It produces a number of hormones, including those that regulate reproduction. The pituitary contains neurones linking it to a part of the brain called the hypothalamus, and some of its hormones are produced under the control of the brain.

> The pituitary is a link between the nervous and endocrine coordination systems.

The **pancreas** is both an endocrine *and* an exocrine gland. It secretes two hormones involved in the regulation of blood glucose, and is also a gland of the digestive system, secreting enzymes through the pancreatic duct into the small intestine (see Chapter 12). The sex organs of males and females, as well as producing sex cells or gametes, are also endocrine organs.

Both the **testes** and **ovaries** make hormones that are involved in controlling reproduction. We will look at the functions of some hormones in more detail.

25. Explain the difference between an exocrine and an endocrine gland.
26. What is the function of the hormone thyroxin?
27. Name two hormones made by the ovaries.

Adrenal means 'next to the kidneys', which describes where the adrenal glands are located, i.e. on top of these organs (see Figure 18.24 on page 265).

Adrenaline – the 'fight or flight' hormone

When you are frightened, excited or angry, your **adrenal** glands secrete the hormone **adrenaline**.

Adrenaline acts at a number of target organs and tissues, preparing the body for action. In animals other than humans, this action usually means dealing with an attack by an enemy, where the animal can stay and fight or run away – hence 'fight or flight'. This is not often a problem with humans, but there are plenty of other times when adrenaline is released.

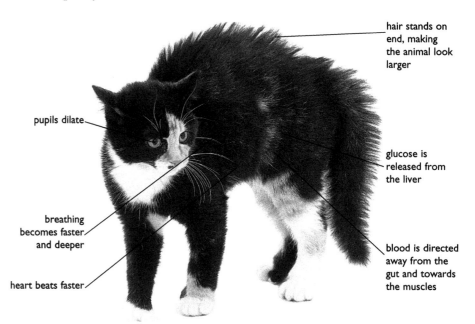

hair stands on end, making the animal look larger

pupils dilate

glucose is released from the liver

breathing becomes faster and deeper

blood is directed away from the gut and towards the muscles

heart beats faster

Figure 18.25 *Adrenaline affects the body of an animal in many ways.*

If an animal's body is going to be prepared for action, the muscles need a good supply of oxygen and glucose for respiration. Adrenaline produces several changes in the body that cause this to happen (Figure 18.25), as well as other changes to prepare for fight or flight:

- The breathing rate increases and breaths become deeper, taking more oxygen into the body.

- The heart beats faster, sending more blood to the muscles, so that they receive more glucose and oxygen for respiration.

- Blood is diverted away from the intestine and into the muscles.

- In the liver, stored carbohydrate is changed into glucose and released into the blood. The muscle cells absorb more glucose and use it for respiration.

- The pupils dilate, increasing visual sensitivity to movement.

- Body hair stands on end, making the animal look larger to an enemy.

- Mental awareness is increased, so reactions are faster.

In humans, adrenaline is not just released in a 'fight or flight' situation, but in many other stressful activities too, such as preparing for a race, going for a job interview or taking an exam.

Controlling blood glucose

You saw earlier that adrenaline can raise blood glucose from stores in the liver. The liver cells contain carbohydrate in the form of glycogen. Glycogen is made from long chains of glucose sub-units joined together (see Chapter 10), producing a large insoluble molecule. Being insoluble makes glycogen a good storage product. When the body is short of glucose, the glycogen can be broken down into glucose, which then passes into the bloodstream.

Adrenaline raises blood glucose concentration in an emergency, but two other hormones act all the time to control the level, keeping it fairly constant. Both of these hormones are made by the pancreas. **Insulin** stimulates removal of glucose from the bloodstream into cells and causes the liver cells to convert glucose into glycogen. This lowers the glucose concentration in the blood when it is too high.

The other hormone is glucagon. This stimulates the liver cells to break down glycogen into glucose, raising the concentration of glucose in the blood if it is too low. Together, they work to keep the blood glucose approximately constant, at a little less than 1 g of glucose in every 1 dm³ of blood. Both hormones are released by special cells in the pancreas in direct response to the level of glucose in the blood passing through this organ. In other words:

$$\text{glucose} \underset{\text{glucagon}}{\overset{\text{insulin}}{\rightleftharpoons}} \text{glycogen}$$

The concentration of glucose in your blood will start to rise after you have had a meal. Sugars from digested carbohydrate pass into the blood and are carried to the liver in the hepatic portal vein (Chapter 12). Here, the glucose is converted to glycogen, so the blood leaving the liver in the hepatic vein will have a lower concentration of glucose.

Sometimes the pancreas does not produce enough insulin, or the target organs do not respond correctly to the insulin. These problems result in a disease called **diabetes** (see Chapter 23).

Hormones and sport

One of the effects of the male sex hormone, testosterone, is that it stimulates the growth of muscles. Because of this, testosterone is known as an **anabolic steroid**. Anabolic means that it causes reactions that build up proteins in the cells (anabolism). A steroid is a type of lipid molecule. Other sex hormones, such as oestrogen, are also steroids.

In the 1950s, some athletes started to inject themselves with testosterone to develop their muscles and gain extra strength. The testosterone also made them more aggressive and determined to win. The use of testosterone and other steroids to improve performance was soon banned by sporting authorities. Not only is it unfair, it is also highly dangerous. The high level of steroid drugs used by some athletes can lead to sterility, heart disease, kidney damage and liver cancer. When used by female athletes, steroids have caused loss of periods and even development of male characteristics.

New kinds of illegal drugs are being developed all the time to give athlete's an 'edge' over their competitors. Many are anabolic steroids; others are based on other types of natural hormone, such as growth hormone (Table 18.5).

Most of the cells of the pancreas are concerned with making digestive enzymes. However, in the pancreas tissue, there are small groups of cells called the **Islets of Langerhans**. These contain two types of cell. Larger α (alpha) cells secrete glucagon, and smaller β (beta) cells secrete insulin.

Insulin is a protein, and if it were to be taken by mouth in tablet form, it would be broken down in the gut. Instead, it is injected into muscle tissue, where it is slowly absorbed into the bloodstream.

They are illegal, and people who are caught using any of them are usually banned from taking part in their sport.

> **28.** What makes glycogen a good storage carbohydrate?
> **29.** Which cells in the pancreas secrete insulin?
> **30.** Explain what happens after you eat a carbohydrate-rich meal.
> **31.** What is an anabolic steroid?

Plant responses

Like animals, plants sense and respond to their environment, but the responses are usually much slower than those of animals because their movements are due to changes in the plant's growth.

Animals usually respond very quickly to changes in their environment – for example, the reflex action resulting from a painful stimulus (page 256) is over in a fraction of a second.

As in animals, some species of plant can respond rapidly to a stimulus, for example, the Venus' flytrap (Figure 18.26). This plant has modified leaves, which close quickly around their 'prey', trapping it. The plant then secretes enzymes to digest the insect. The movement is brought about by rapid changes in turgor of specialised cells at the base of their leaves.

Tropisms

Most plants do not respond to stimuli as quickly as this, because their response normally involves changing their rate of growth. Different parts of plants may grow at different rates, and a plant may respond to a stimulus by increasing growth near the tip of its shoot or roots. Imagine a plant growing normally in a pot. Usually, most light will be falling on the plant from above. If you turn the plant on its side and leave it for a day or two, you will see that its shoot starts to grow upwards (Figure 18.27).

There are two stimuli acting on the plant in Figure 18.7. One is the direction of the light that falls on the plant. The other stimulus is gravity. Both light and gravity are **directional stimuli** (they act in a particular direction). The growth response of a plant to a directional stimulus is called a **tropism**. If the growth response is *towards* the direction of the stimulus, it is a *positive* tropism, and if it is *away* from the direction of the stimulus, it is a *negative* tropism. The stem of the plant in Figure 18.27 is showing a positive **phototropism** and a negative **geotropism**, which both make the stem grow upwards.

The aerial part of a plant (the 'shoot') needs light to carry out photosynthesis. This means that in most species, a positive phototropism is the strongest tropic response of the shoot. If a shoot grows towards the light, it ensures that the leaves, held out at an angle to the stem, will receive the maximum amount of sunlight. This response is easily seen in any plant placed near a window, or another source of 'one-way' or *unidirectional* light (Figure 18.28, page 269).

In darkness or uniform light, the shoot shows a negative geotropism. As you might expect, the roots of plants are strongly positively geotropic.

Figure 18.26 *The Venus' flytrap catches and digests insects to gain extra nutrients. The plant responds very quickly to a fly landing on one of its leaves.*

Figure 18.27 *This bean has responded to being placed horizontally. The growing shoot has started to bend upwards.*

Cross-curricular links:
physics, light and gravity

This response makes sure that the roots grow down into the soil, where they can reach water and mineral ions, and obtain anchorage.

The roots of some species that have been studied are also negatively phototropic, but most roots don't respond to directional light at all. In the same way, some experiments have shown that roots of a few species show positive **hydrotropism** (attraction to water).

Thigmotropism is a growth response to touch (or contact). This is most obvious in climbing plants, such as vines, peas and beans, where the shoot of the plant grows towards any object it touches. This tropism results in the stem coiling around a stick or another plant, which provides it with support (Figure 18.29). Stems are positively thigmotropic, whereas roots are negatively thigmotropic. When roots touch an object, such as a stone in the soil, they grow away from it. This makes it easier for a root to grow through the soil.

The common tropisms are shown in Table 18.6.

Stimulus	Name of response	Response of shoots	Response of roots
light	phototropism	grow towards light source (positive phototropism)	most species show no response, but a few grow away from light (negative phototropism)
gravity	geotropism	grow away from direction of gravity (negative geotropism)	grow towards direction of gravity (positive geotropism)
water	hydrotropism	none	some species may grow towards water (positive hydrotropism)
touch	thigmotropism	grow towards direction of touch (positive thigmotropism)	grow away from direction of touch (negative thigmotropism)

Table 18.6: *Common responses of plants to directional stimuli (tropisms).*

32. What is a tropism?
33. Shoots show positive phototropism. Why is this response an advantage to the plant?
34. Roots show positive geotropism. Why is this response an advantage to the plant?

Detecting the light stimulus – plant hormones

Plants do not have the obvious sense organs and nervous system of animals, but since they respond to stimuli such as light and gravity, they must have some way of detecting them and coordinating the response. The detection system of phototropism was first investigated by the great English biologist Charles Darwin (see Chapter 27) in the late 19th century. Instead of using stems, Darwin (and later scientists) used cereal **coleoptiles**, which are easier to grow and use in experiments (Figure 18.30 on page 270).

Phototropisms are growth responses to light from one direction. Geotropisms are growth responses to the direction of gravity ('geo' refers to the Earth).

Light from one direction is called unidirectional. If light shines on the plant evenly from all directions, this is called uniform light.

Figure 18.28 *The shoots of these cress seedlings are showing a positive phototropism.*

Figure 18.29 *Thigmotropism in a plant stem.*

Charles Darwin is best known for his theory of evolution by natural selection. However, he carried out research in other areas of biology, including a study of plant hormones.

A coleoptile is a protective sheath that covers the first leaves of a cereal seedling. It protects the delicate leaves as the shoot emerges through the soil (Figure 18.30).

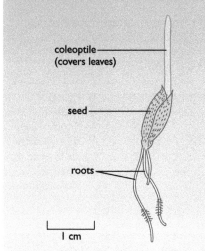

Figure 18.30 *Germinating oat seedling.*

Darwin showed that the stimulus of unidirectional light was detected by the tip of the coleoptile and transmitted to a growth zone, just behind the tip (Figure 18.31).

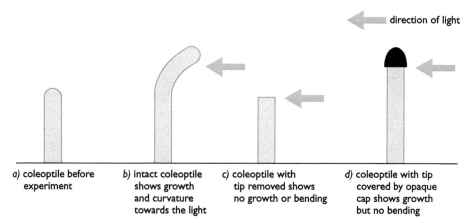

Figure 18.31 *Darwin's experiments with phototropism (1880).*

Since plants don't have a nervous system, biologists began to look for a chemical messenger, or hormone, that might be the cause of phototropism in coleoptiles. Between 1910 and 1926, several scientists investigated this problem. Some of their results are summarised in Figure 18.32 on page 271.

- In experiment 1, the stimulus for growth was found to pass through materials such as gelatin, which absorbs water-soluble chemicals, but not through materials such as mica (a mineral), which is impermeable to water. This made biologists think that the stimulus was a chemical that was soluble in water.

- In experiment 2, it was shown that the phototropic response could be brought about even *without* unidirectional light, by removing a coleoptile tip ('decapitating' the coleoptile) and placing the tip on one side of the decapitated stalk.

- In experiment 3, it was found that the hormone could be collected in another water-absorbing material (a block of agar jelly). Placing the agar block on one side of the decapitated coleoptile stalk caused it to bend.

Experiments like 2 and 3 led scientists to believe that the hormone caused bending by stimulating growth on the side of the coleoptile furthest from the light. The theory is that the hormone is produced in the tip of the shoot, and diffuses back down the shoot. If the shoot is in the dark, or if light is all around the shoot, the hormone diffuses at equal rates on each side of the shoot, so it stimulates the shoot equally on all sides. However, if the shoot is receiving light from one direction, the hormone moves away from the light as it diffuses downwards. The higher concentration of hormone on the 'dark' side of the shoot stimulates cells there to grow, making the shoot bend towards the light (Figure 18.33, page 271).

Since these experiments were carried out, scientists have identified the hormone responsible. It is called **auxin**. Several other types of plant hormone have been found. Like auxin, they all influence growth and development of plants in one way or another, so that many scientists prefer to call them **plant growth substances** rather than plant hormones.

experiment 1

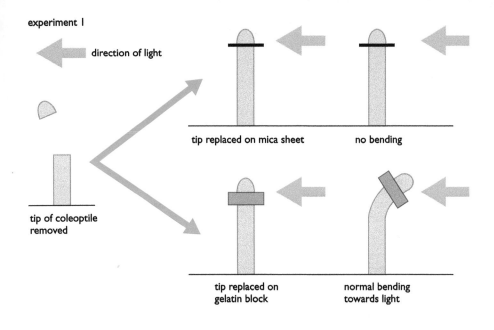

direction of light

tip of coleoptile removed

tip replaced on mica sheet — no bending

tip replaced on gelatin block — normal bending towards light

experiment 2

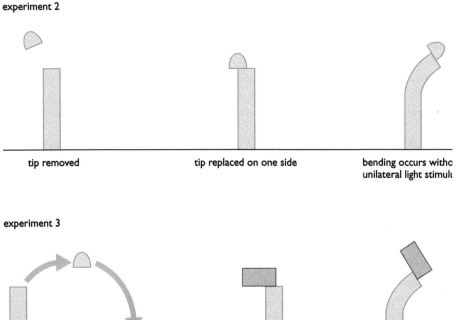

tip removed — tip replaced on one side — bending occurs without unilateral light stimuli

experiment 3

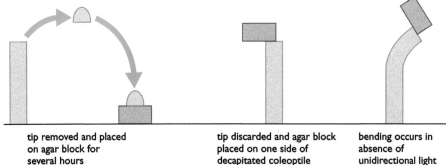

tip removed and placed on agar block for several hours

tip discarded and agar block placed on one side of decapitated coleoptile

bending occurs in absence of unidirectional light

Figure 18.32 *Experiments on coleoptiles that helped to explain the mechanism of phototropism.*

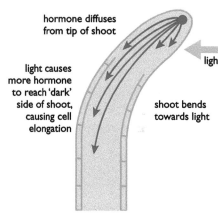

hormone diffuses from tip of shoot

light

light causes more hormone to reach 'dark' side of shoot, causing cell elongation

shoot bends towards light

Figure 18.33 *How movement of a plant hormone causes phototropism.*

'Auxin' should really be auxins, since there are a number of chemicals with very similar structures making up a group of closely related plant hormones.

Plant hormones and geotropism

Bending of the root and shoot during geotropism is also thought to be due to plant hormones. If a broad bean seedling is placed in the dark in a horizontal position, its shoot will bend upwards and its root downwards (Figure 18.34, page 272).

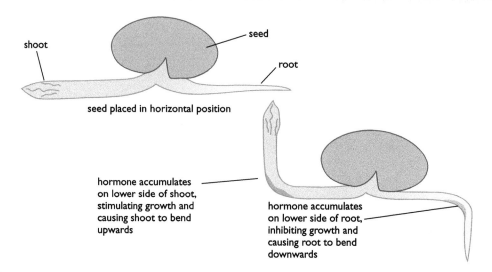

shoot

seed

root

seed placed in horizontal position

hormone accumulates
on lower side of shoot,
stimulating growth and
causing shoot to bend
upwards

hormone accumulates
on lower side of root,
inhibiting growth and
causing root to bend
downwards

Figure 18.34 *Geotropism in a broad bean seedling. It was once thought that movement of auxin caused this response. We now know that the true explanation is not as simple as this.*

As well as auxins, there are four other main groups of plant hormones called gibberellins, cytokinins, abscisic acid and ethene. They control many aspects of plant growth and development, other than from tropisms. These include growth of buds, leaves and fruit, fruit ripening, seed germination, leaf fall and opening of stomata, to name just a few. In addition to auxins, abscisic acid, ethene and gibberellins have all been shown to be involved in geotropisms.

It was once thought that these geotropic responses of the shoot and root were due to auxin. The auxin was supposed to be produced at the tip of the shoot, and to sink under the influence of gravity as it diffused back from the tip. A high concentration of auxin would increase growth on the lower side of the shoot, causing it to bend upwards. The downward growth of the root was supposed to be caused in a similar way, except that auxin *inhibited* growth on the lower side. We know now that these explanations are not the whole story. Although some movement of auxin happens due to the effect of gravity, it is not enough to explain geotropisms, where other hormones seem to be involved.

35. What is a coleoptile?
36. What is auxin?
37. Explain how auxin brings about the growth of a shoot towards the light.

Activity 7: Investigating which part of a shoot is sensitive to light

SBA skills

ORR	MM	AI	PD	Dr

You will need:

- some cereal seeds such as maize or wheat
- three boxes that are lightproof except for a 'window' on one side (you can make these from cardboard)
- three dishes containing soil or cotton wool
- light from a bench lamp (or window)

Carry out the following:

1. Soak a number of seeds overnight, then allow them to germinate in the dishes. Leave room between the seeds so that you can measure them when they start to grow.

2. When the coleoptiles have grown to a length of about 10–15 mm, treat them as follows (Figure 18.35):

  • Cut off about 2 mm from the tip of the coleoptiles in the first dish.

  • Cover the tips of the coleoptiles in the second dish with aluminium foil.

  • Leave the coleoptiles in the third dish untreated as a control.

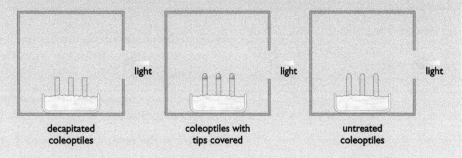

decapitated coleoptiles   coleoptiles with tips covered   untreated coleoptiles

Figure 18.35 *Apparatus to find the effect of unidirectional light on the growth of three groups of cereal coleoptiles.*

3. For each dish, measure the length of the coleoptiles and calculate the mean coleoptile length for each treatment group.

4. Place the dishes in the boxes with a light shining from one side (Figure 18.35). Leave them for 24–48 hours.

5. Find the new mean length of the coleoptiles in each dish and calculate the percentage change in length. Record any change in direction of growth of the coleoptiles in the three treatment groups.

6. Write up the experiment and explain your results.

Activity 8: Use of a clinostat to show geotropism in roots

SBA skills

| ORR | MM | AI | PD | Dr |

A **clinostat** is a piece of apparatus consisting of an electric motor turning a cork disc. Germinating seeds can be attached to the disc. The motor turns the disc and seeds around very slowly, so that the movement eliminates any directional stimulus that may be acting on the seeds (Figure 18.36 on page 274). The clinostat can be turned through 90°, so that the disc rotates either horizontally or vertically.

You will need:

• two clinostats

• several bean seeds

• some cotton wool

• some pins

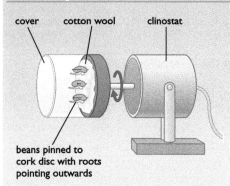

cover   cotton wool   clinostat

beans pinned to
cork disc with roots
pointing outwards

Figure 18.36 *A clinostat being used to show geotropism in roots.*

Carry out the following:

1. Soak a few bean seeds in water overnight. Place the seeds on wet cotton wool for a day or two until the first root of each seed (called the radicle) has grown to about 2 cm in length.

2. Attach wet cotton wool to the cork disc of two clinostats. Pin three or four of the germinating bean seeds onto the discs, with their radicles pointing outwards. Place the covers over the discs to keep the air around the beans moist.

3. Turn the clinostats on their sides, as shown in Figure 18.36. Switch on one clinostat, but leave the other switched off. What is the purpose of this second clinostat?

4. Leave both clinostats set up for a few days. Make sure that you keep the cotton wool damp.

5. Observe the results. Which way do the radicles of each set of beans grow?

6. Write up your experiment and explain the results.

Activity 9: Using a clinostat to show phototropism in shoots

SBA skills

ORR	MM	AI	**PD**	Dr

Plan an investigation using two clinostats to find out if the growing shoot of a plant responds to unidirectional light. State the hypothesis you will test, and explain how you will make sure that your experiment is controlled.

Chapter summary

In this chapter you have learnt that:

- a stimulus is a change in an organism's surroundings, which produces a reaction called a response

- stimuli are detected by receptors and the response is brought about by effectors

- in animals, the links between the receptor and the effector can be the nervous system or the endocrine system

- there are several types of receptor in the body, detecting different kinds of energy in the environment

- the central nervous system (CNS) consists of the brain and spinal cord; nerves composed of neurones lead from the CNS

- nerve cells carrying impulses into the CNS are called sensory neurones, and nerves leading out to the effectors are called motor neurones

- a neurone is adapted to carry impulses quickly over long distances via a fibre called an axon

- many neurones are surrounded by an insulating layer called the myelin sheath

- the eye is a complex organ with a structure that allows it to form an image on a light-sensitive layer at the back of the eye, called the retina; the image is formed by refraction of light at the cornea and lens

Chapter summary (continued)

- cells in the retina called rods and cones convert light into nerve impulses, which travel through the optic nerve to the brain

- the iris contains two sets of muscles that can alter the diameter of the pupil, controlling the amount of light entering the eye by a reflex action

- focusing in the eye is carried out by the ciliary muscles altering the shape of the lens; the lens is made less convex when focusing on a distant object, and more convex if the object is nearby

- long sight is due to a lens that is too flat or an eyeball that is too short; it is corrected by wearing glasses with convex lenses

- short sight is due to a lens that is too convex or an eyeball that is too long; it is corrected by wearing glasses with concave lenses

- cataracts and glaucoma are problems with the eye that develop in old age

- the skin has a number of functions, including acting as a sense organ, protecting against UV light and helping to maintain a constant body temperature

- the mechanisms that the skin uses to control body temperature are sweating, vasodilation and vasoconstriction, and erection of hairs on the skin

- a reflex action is a rapid, automatic response to a stimulus that is not initiated by the brain; reflexes include the iris/pupil reflex, withdrawal reflex and knee-jerk reflex

- many legal and illegal drugs act upon the nervous system, which means that drug abuse can have harmful physiological, social and economic effects

- the pathway involved in a reflex is called a reflex arc. The stimulus is detected by receptors, which generate impulses in a sensory nerve. The sensory nerve passes the impulses into the CNS. Motor nerves carry the impulses out from the CNS to the effector organ

- nerve cells in the CNS are linked by synapses, which are gaps between two neurones; synapses are crossed by a chemical called a neurotransmitter

- the brain consists of a number of distinct regions with different functions, such as the cerebrum, with its sensory and motor areas; the cerebellum, which coordinates muscle contractions during movement; and the medulla, which controls some automatic activities like the heart rate

- the autonomic nervous system controls many activities in the body that are involuntary and automatic, such as breathing and peristalsis

- invertebrates, such as woodlice, respond to stimuli, allowing them to survive in their habitat; their responses can be investigated in a choice chamber

- the endocrine system is another coordinating system between stimulus and response; it consists of glands, which secrete chemicals called hormones into the bloodstream., and which then travel in the blood to target organs, where they bring about a response

- the adrenal glands secrete adrenaline, which controls the 'flight or fight' response of an animal to danger

- the concentration of glucose in the blood is detected by the pancreas and controlled by two hormones released by the pancreas – glucose and glucagon – insulin lowers blood glucose and glucagon raises it

- hormones such as anabolic steroids and growth hormone are used illegally by some men and women to improve their performance in competitive sports

- plants respond to directional stimuli such as light, gravity and touch by a change in their growth, which is called a tropism

- plant stems grow towards a unidirectional light source (a positive phototropism), which is important for a plant to be able to carry out photosynthesis

- stems show a negative geotropism, and some stems show a positive thigmotropism

- roots grow towards the direction of gravity (a positive geotropism), which is important for the plant's survival, because roots need to reach down into the soil for water and minerals

- roots show a negative thigmotropism, and some roots show a positive hydrotropism

- tropisms are controlled by plant hormones called auxins.

- the response of a shoot to unidirectional light is brought about by auxin moving away from the light and causing the 'dark' side of the shoot to grow, bending it towards the light

- auxins in roots inhibit growth on the side nearest gravity, so that the root grows downwards.

Questions

1. A girl is sitting under a shady tree reading a book. She looks up into the sunny sky at an aeroplane. Which of the following changes will take place in her eyes?

   A The pupils dilate and the lens becomes less convex.
   B The pupils dilate and the lens becomes more convex.
   C The pupils constrict and the lens becomes less convex.
   D The pupils constrict and the lens becomes more convex.

2. Which of these statements about how nerve cells work is/ are true?

   I   Neurotransmitters carry a nerve impulse along a neurone.
   II An electrical charge carries a nerve impulse across a synapse.

A I
B II
C I and II
D neither

The diagram below shows a section through the eye. Use the diagram to answer questions 3 and 4.

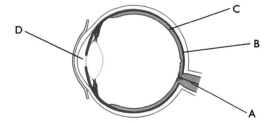

3. Which part is the choroid?

4. Which part transmits nerve impulses to the brain?

5. Which of the following is a hormone controlling blood glucose levels?

   A   Glucagon
   B   Glycogen
   C   Glycine
   D   Galactose

6. Which of the following will **not** happen when the hormone adrenaline is released?

   A   Increase in heart rate
   B   Increase in blood flow to the gut
   C   Dilation of the pupils
   D   Increase in breathing rate

7. During the knee-jerk reflex, which of the following is the correct order of events?

   A   effector → motor neurone → CNS
       → sensory neurone → receptor
   B   effector → sensory neurone → CNS
       → motor neurone → receptor
   C   receptor → motor neurone → CNS
       → sensory neurone → effector
   D   receptor → sensory neurone → CNS
       → motor neurone → effector

8. Which of the following happens when the human body temperature rises above normal?

   A   Blood vessels in the skin constrict and blood flow decreases.
   B   Blood vessels in the skin constrict and blood flow increases.
   C   Blood vessels in the skin dilate and blood flow decreases.
   D   Blood vessels in the skin dilate and blood flow increases.

9. The diagram below shows a bean seedling that has been growing in the dark.

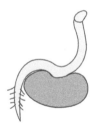

The response of the shoot can be described as:

   A   a positive phototropism
   B   a negative phototropism
   C   a positive geotropism
   D   a negative geotropism

10. Which of the following is correct?

   A   Roots show negative geotropism and positive phototropism.
   B   Roots show positive geotropism and positive hydrotropism.
   C   Stems show negative phototropism and negative geotropism.
   D   Stems show positive phototropism and positive geotropism.

11. Which of these statements about the action of auxin in the growth of a coleoptile towards a light source is/are true?

   I   Auxin is made in the tip of the coleoptile.
   II   Auxin diffuses towards the light.
   III   A high concentration of auxin stimulates growth of the coleoptile towards the light.

   A   I and II
   B   I and III
   C   II and III
   D   I, II and III

12. The diagram shows a section through a human eye.

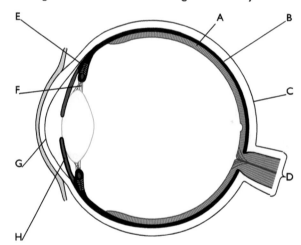

   a)   The table lists the functions of some of parts A to H. Copy the table and write the letters of the correct parts in the boxes.   (5)

Function	Letter
refracts light rays	
converts light into nerve impulses	
contains pigment to stop internal reflection	
contracts to change the shape of the lens	
takes nerve impulses to the brain	

   b)   i)   Which label shows the iris?   (1)
        ii) Explain how the iris controls the amount of light entering the eye.   (4)
        iii) Why is this important?   (2)

13. The diagram shows some parts of the nervous system involved in a simple reflex action that happens when a finger touches a hot object.

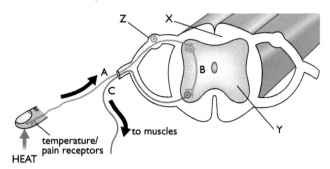

a) What type of neurone is:

   i) neurone A

   ii) neurone B

   iii) neurone C? (3)

b) Describe the function of each of these types of neurone. (6)

c) Which parts of the nervous system are shown by the labels X, Y and Z? (3)

d) In what form is information passed along neurones? (1)

e) Explain how information passes from one neurone to another. (2)

14. a) Which part of the human brain is responsible for controlling:

   i) keeping your balance when you walk (1)

   ii) maintaining your breathing when you are asleep (1)

   iii) making your leg muscles contract when you kick a ball? (1)

b) A stroke is caused by a blood clot blocking the blood supply to part of the brain.

   i) One patient, after suffering a stroke, was unable to move his left arm. Which part of his brain was affected? (1)

   ii) Another patient lost her sense of smell following a stroke. Which part of her brain was affected? (1)

15. a) List *five* examples of stimuli that affect the body and state the response produced by each stimulus. (10)

b) For *one* of your five examples, explain:

   i) the nature and role of the receptor (2)

   ii) the nature and role of the effector organ. (2)

c) For the same example as in (b) describe the chain of events
from stimulus to response. (5)

16. The graph shows the changes in blood glucose in a healthy woman over a 12-hour period.

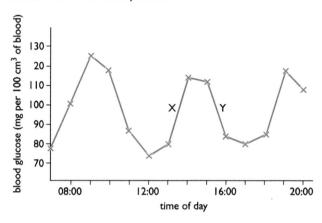

a) Explain why there was a rise in blood glucose at X. (2)

b) How does the body bring about a decrease in blood glucose at Y? Your answer should include the words insulin, liver and pancreas. (5)

c) Diabetes is a disease where the body cannot control the concentration of glucose in the blood.

   i) Why is this dangerous? (2)

   ii) Describe *two* ways a person with diabetes can monitor their blood glucose level. (4)

   iii) Explain *two* ways that a person with diabetes can help to control their blood glucose level. (4)

17. Blowfly larvae (maggots) normally live in the rotting flesh of dead animals. A student used a choice chamber to investigate the preference of maggots for light or dark conditions. She placed 10 maggots in one side of the choice chamber and 10 in the other side. One side was illuminated by using a laboratory lamp with a 100-W bulb, and the other was covered with black plastic, to keep out the light. She counted the number in the light every minute for 10 minutes. Here are her results.

Time (min)	Number of maggots in the light
1	10
2	11
3	13
4	8
5	7
6	4
7	6
8	3
9	3
10	1

a) Plot a graph of the number of maggots in the light against time. *(5)*

b) What conclusion should the student make from the results? Explain your answer. *(3)*

c) Could the maggots have been responding to any stimulus other than light intensity? Explain your answer. *(2)*

d) Assuming the maggots were responding to light intensity, how would this help them to survive in their natural habitat? *(3)*

e) How could you use a choice chamber to test the hypothesis that maggots are attracted by the smell of meat? Give full experimental details. *(5)*

18. **a)** What are the main stimuli affecting the growth of:

    i)   the shoot *(2)*

    ii) the root? *(2)*

b) How does a plant benefit from a positive phototropism in its stem? *(2)*

19. Draw a labelled diagram to show how auxin brings about phototropism in a coleoptile that has light shining on it from one direction. *(4)*

20. An experiment was carried out to investigate phototropism in a coleoptile. The diagram shows what was done. Predict the results you would expect to get in each of the experiments *(a)* to *(c)*. Explain your answers.

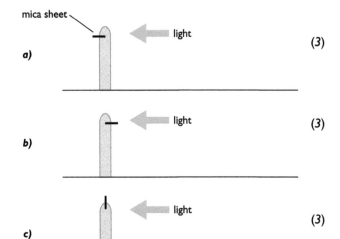

a) *(3)*

b) *(3)*

c) *(3)*

21. Plants can respond to a range of stimuli.

a) Plant shoots detect and grow towards light.

    i)   What is this process called? *(1)*

    ii) Explain how a plant bends towards the light. *(3)*

    iii) Explain the advantage to the plant of this response. *(1)*

b) In an investigation, young plant shoots were exposed to light from one side. The wavelength of the light was varied. The graph summarises the results of the investigation.

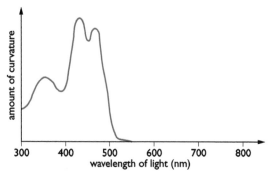

    i)   Describe the results shown in the graph. *(2)*

    ii) Suggest why the results show this pattern. *(2)*

c) *i)*   Name *two* other stimuli that produce growth responses in plants. *(2)*

    ii) For each stimulus you name, describe the way that both roots and shoots respond. *(4)*

    iii) What is the benefit to the plant of these responses? *(2)*

22. A cataract is an eye problem suffered by some people, especially the elderly. The lens of the eye becomes opaque (cloudy), which blocks the passage of light. It can lead to blindness. Cataracts can be treated by a simple eye operation, where a surgeon removes the lens. After the operation, the patient is able to see again, but the eye is unable to carry out accommodation, and the patient will probably need to wear glasses.

a) Explain why the eye can still form an image after the lens has been removed. *(3)*

b) Explain what is meant by accommodation and how it is brought about in the eye. *(6)*

c) Why is accommodation not possible after a cataract operation? *(3)*

d) Suggest the type of lenses that the patient's glasses will need to have. Explain your answer. *(4)*

e) Name *one* other type of eye problem that is more common in elderly people and describe its symptoms. *(3)*

23. **a)** Describe the functions of the following parts of the brain:

    i)   The cerebral hemispheres

    ii) The pituitary gland

    iii) The medulla oblongata *(6)*

b) The brain contains billions of nerve cells (neurones), each of which forms synapses with thousands of other neurones.

   i)   Draw a labelled diagram of a synapse. *(3)*

   ii) Explain how information is transmitted along a nerve cell and then across a synapse. *(3)*

   iii) Alcohol is a *depressant* drug, while caffeine is a *stimulant*. Explain what these terms mean, referring to synapses in your answer. *(4)*

24. a) What is a reflex action? *(3)*

b) Give *three* examples of reflex actions, and in each case, state the advantage to a person of the action. *(6)*

c) A boy accidently put his hand onto a drawing pin, and quickly removed it again. Describe the pathway involved in this reflex from stimulus to response. *(6)*

d) What is the difference between a cranial reflex and a spinal reflex? *(2)*

e) Explain how coughing can be either a reflex or a voluntary action. *(3)*

25. a) Briefly describe the mechanism by which:

   i)   the stem (shoot) of a plant responds to a directional light source

   ii) the root of a plant responds to the force of gravity. *(8)*

b) How are each of these responses advantageous to the plant? *(4)*

c) Many small invertebrate animals are rarely seen during daylight, and spend most of the time hiding under stones or in leaf litter. Suggest *three* ways that this may increase their chance of survival. *(3)*

d) Describe how you could use a choice chamber to test the hypothesis that millipedes prefer moist rather than dry conditions. *(5)*

Chapter 19: Support and Movement

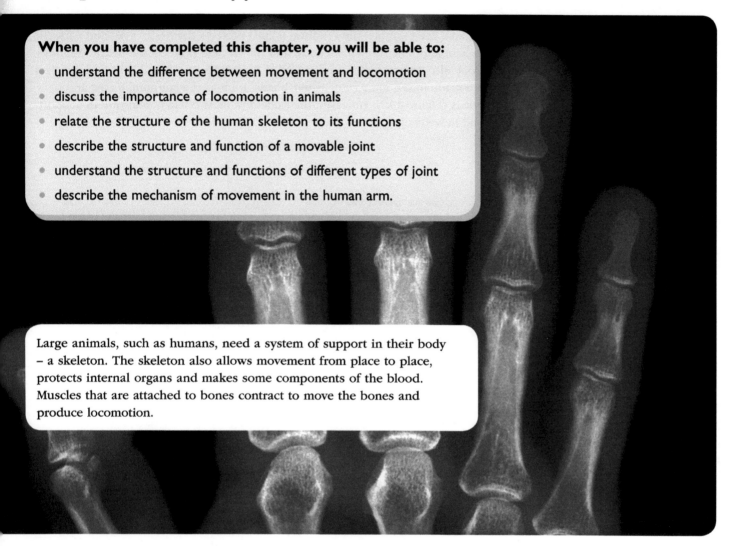

When you have completed this chapter, you will be able to:

- understand the difference between movement and locomotion
- discuss the importance of locomotion in animals
- relate the structure of the human skeleton to its functions
- describe the structure and function of a movable joint
- understand the structure and functions of different types of joint
- describe the mechanism of movement in the human arm.

Large animals, such as humans, need a system of support in their body – a skeleton. The skeleton also allows movement from place to place, protects internal organs and makes some components of the blood. Muscles that are attached to bones contract to move the bones and produce locomotion.

Movement and locomotion

Movement and locomotion mean different things. Both plants and animals can move, but only animals carry out locomotion, that is, move their whole body from place to place.

Plants carry out growth movements. For example, plants show tropisms, which are slow growth movements that occur in response to a directional stimulus, such as light (see Chapter 18). Seeds also show growth movements when they germinate, producing a root and shoot (see Chapter 22).

There are several reasons why it is important for animals to move from place to place, while plants do not carry out locomotion. These are mainly due to the differences between the methods of feeding and reproduction in plants and animals. Plants can stay in one place to make food by photosynthesis, while animals have to move about to obtain their food. Animals also have to carry out locomotion to seek shelter, and to find a mate and reproduce. Plants do not need shelter in the same way that animals do, and they do not need to move to reproduce, because their pollen grains are moved from one flower to another, either on the wind or by insects (see Chapter 21).

The skeleton

The human skeleton (Figure 19.1) has several functions. It protects many vital organs. For example, the **cranium** (skull) protects the brain, eyes and ears, the vertebrae protect the spinal cord, and the ribcage protects the heart and lungs. Bone also makes some components of the blood. Red blood cells and platelets are made in the marrow or larger bones, such as the sternum, femur and pelvis (Figure 19.2). However, the skeleton's most obvious roles are in support and movement.

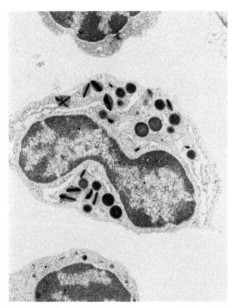

Figure 19.2 *Bone marrow contains cells that divide to produce all our red blood cells and platelets.*

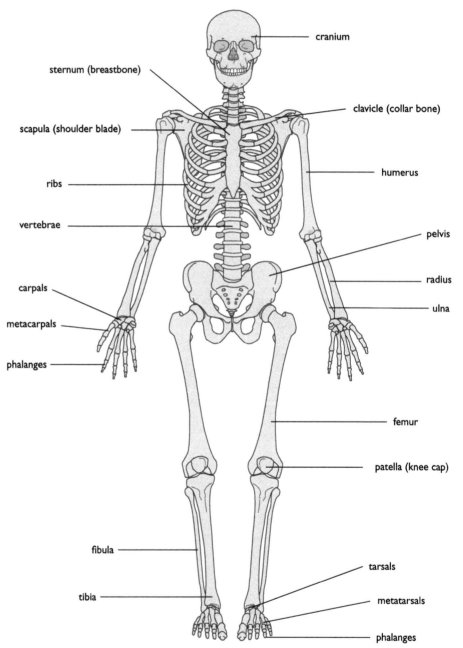

cranium

sternum (breastbone)

clavicle (collar bone)

scapula (shoulder blade)

humerus

ribs

vertebrae

pelvis

radius

carpals

ulna

metacarpals

phalanges

femur

patella (knee cap)

fibula

tarsals

tibia

metatarsals

phalanges

Figure 19.1 *The human skeleton.*

The vertebral column, cranium and ribcage make up the **axial skeleton**. The other bones are attached to this axis – the scapulas, clavicles, pelvis and limbs bones, forming the **appendicular skeleton**.

Bone

Bone is a hard substance because it contains calcium salts, mainly calcium phosphate. This results in a rigid material that can resist bending and compression (squashing) forces. Although out of the body a bone looks dead, in the body it is a living organ, made of cells called **osteocytes**. These cells, along with protein fibres, stop the bone being too brittle. The presence of living cells and blood vessels in the bone also means that bones can repair themselves if they are broken. There are many different shapes and sizes of bones in the human body, from the tiny ear ossicles to long bones like the femur. Figure 19.3 shows a section through a long bone, and Figure 19.4 is a photomicrograph of a section through the outer part of a bone.

> If a long bone such as the femur from a chicken is left in a beaker of acid, the acid will dissolve the calcium salts from the bone. The fibres that are left are tough but flexible, so afterwards the bone will bend, and can even be tied in a knot!

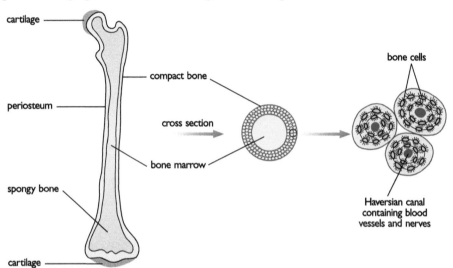

Figure 19.3 *Section through a long bone.*

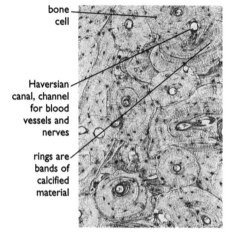

Figure 19.4 *This is a cross-section through a normal bone. The rings are calcified material.*

The middle of a bone is composed of **spongy bone**, with fewer calcium salts, and containing spaces like a sponge, making it less hard. The spaces are filled with bone marrow. Some larger bones have a hollow central cavity containing bone marrow, which makes them lighter. Marrow stores fats and produces blood cells.

The outside of the bone is made of harder material, called **compact bone**. In the embryo, bones start off made of cartilage, but as the embryo grows, the cartilage is gradually replaced by bone, a process called **ossification**. This process is nearly complete by the time a baby is born. Osteocytes arrange themselves in rings called **Haversian systems**, around canals containing blood vessels and nerves. The osteocytes secrete calcium phosphate salts, which, along with protein fibres, make up the bone **matrix**.

> The arrangement of the bone matrix in concentric rings makes a long bone very good at resisting compression forces due to the weight of the body, while still allowing the bone a certain amount of side-to-side flexibility.

Cartilage remains present at the ends of long bones, where it acts as a cushion between two bones at a joint (see below). Cartilage is a tough but flexible tissue containing cells called **chondrocytes**. They secrete a matrix containing various types of protein fibres.

Covering the outer surface of the bone is a tough membrane called the **periosteum**.

> 1. Explain the difference between movement and locomotion.
> 2. State three functions of the skeleton.
> 3. What is the main inorganic component of bone?

Joints

When humans move, the bones move relative to each other. The point where two bones meet is called a **joint**. We say that the bones **articulate** at joints. A movable joint such as the hip or elbow needs to have certain features. These include:

- a way to keep the ends of the bones held together, so that they don't separate (**dislocate**)

- a means of reducing friction between the ends of the moving bones

- a shock-absorbing surface between the two bones.

You can see the structures that do these things in Figure 19.5.

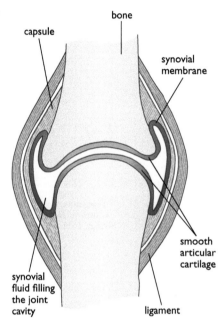

Figure 19.5 *The structure of a synovial joint. The size of the space filled with synovial fluid has been exaggerated in the diagram.*

Movable joints are called **synovial** joints. They contain a liquid called **synovial fluid**, which is secreted by the **synovial membrane**, lining the space in the middle of the joint. Synovial fluid is oily and acts as a lubricant, reducing the friction between the ends of the bones. The end of each bone has an articulating surface covered with a smooth layer of **cartilage**. Cartilage is a strong material, but it is not brittle. It acts as a shock absorber between the ends of the bones, rather like a rubber gym mat compresses to absorb the shock when you fall over.

The joint is surrounded by a tough fibrous **capsule**, and held together by **ligaments**, which run from one bone to the other across the joint. Ligaments are composed of fibres that make them very tough. They have great strength to resist stretching, called **tensile** strength. However, ligaments have some elasticity, so that they allow joints to bend without the bones becoming dislocated.

There are various kinds of joints in the body. They are classified into three groups by how much movement they allow:

- freely movable joints

- partially movable joints

- immovable (fixed) joints.

> The correct definition of an 'elastic' material is one which, when you bend or stretch it, will return to its original shape. In this chapter we use the word 'elastic' to mean 'easily stretched'.

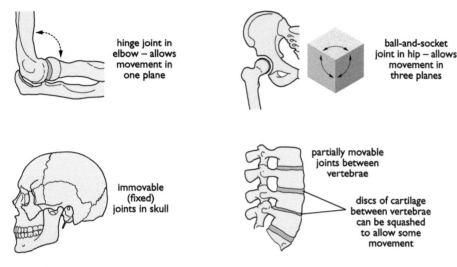

Figure 19.6 *Different types of joint.*

Freely movable joints include **ball and socket** joints, such as the shoulder and hip, and **hinge** joints, such as the elbow.

A ball and socket joint allows movement any direction, in all three planes. Imagine your shoulder joint is at the corner of a box. You can move your upper arm in any of the three planes formed by the sides of the box: side, back or top. Compare this with a hinge joint, such as your elbow – due to the shape of the joint, you can only move this in one plane, as when bending your arm (Figure 19.6).

Partially movable joints allow a slight degree of movement. The joints between vertebrae are like this. They allow a little movement when the spinal column bends (Figure 19.6).

Some joints are **immovable** or **fixed** joints, for example those between the bones of the skull. The cranium consists of 22 bones, most of them fused together, allowing no movement.

4. What is the function of each of the following in a joint?
   (a) Synovial fluid
   (b) Cartilage
   (c) Ligaments
5. Which part of a joint makes synovial fluid?
6. Name an example of the following types of joint in the body:
   (a) A ball-and-socket joint
   (b) A hinge joint

Muscles

Muscles are organs that are attached to bones and move them by contracting, pulling on the bone. At the end of a muscle there are **tendons**. A tendon attaches the muscle to the bone. Tendons have very high tensile strength, like ligaments, but unlike ligaments they are not very elastic. This means that they don't stretch when the muscle contracts.

> The word we use to mean 'not very elastic' is **inelastic**. Both ligaments and tendons have a high tensile strength, but ligaments are fairly elastic, while tendons are inelastic. Ligaments join bone to bone across a joint, while tendons join muscle to bone.

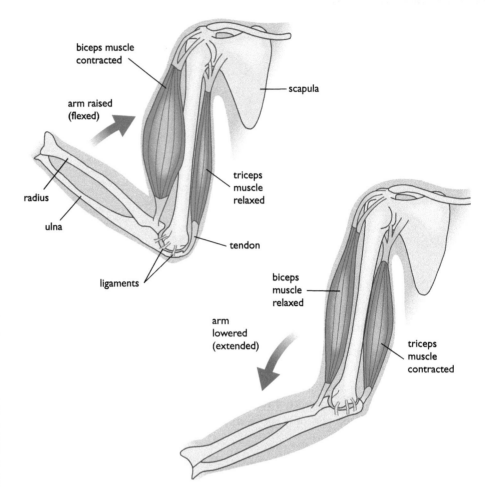

Figure 19.7 *The biceps and triceps muscles contract to move the arm at the elbow joint. The biceps flexes the arm, while its antagonistic partner, the triceps, extends the arm.*

Figure 19.8 *A body-builder performing a biceps curl – you can see clearly the contraction of the biceps.*

Muscles cannot push, they can only pull – in other words they are not able to expand actively. When muscles get longer, it is because they are stretched by the contraction of another muscle. When a muscle is being stretched it is relaxed (the opposite of contracted). Because of this, muscles usually work in pairs; one contracting while the other relaxes. These are called **antagonistic pairs**. One of the simplest examples of an antagonistic pair of muscles is the arrangement of the **biceps** and **triceps** muscles in the arm (Figure 19.7).

When the biceps muscle contracts it bends, or **flexes** the arm at the elbow joint. Contraction of the triceps straightens, or **extends** the arm. Of course, there are other muscles in the arm which produce movement in other directions.

When a muscle contracts, the bone at one end of the muscle moves and the bone at the other end stays still. The place where the muscle is attached to the stationary bone is called the **origin**. The place where it is attached to the moving bone is called the **insertion**. When a muscle contracts, the insertion moves towards the origin.

You can identify other antagonistic pairs of muscles in the body. For example, when we run we use several sets of muscles that cause bending at the hip, knee and ankle joints (Figure 19.9, page 287).

Figure 19.9 *Antagonistic muscles are used in running. The main muscle that flexes the knee joint is the biceps femoris, while its antagonistic partner, extending the knee, is the quadriceps. Can you work out which muscles flex and extend at the hip and ankle?*

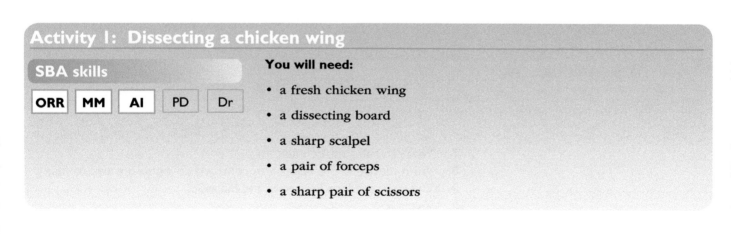

Activity 1: Dissecting a chicken wing

SBA skills

| ORR | MM | AI | PD | Dr |

You will need:

- a fresh chicken wing
- a dissecting board
- a sharp scalpel
- a pair of forceps
- a sharp pair of scissors

Activity 1: Dissecting a chicken wing

SBA skills

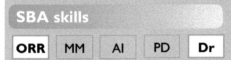

| ORR | MM | AI | PD | Dr |

Carry out the following:

1. Place a chicken wing on the dissecting board and observe it closely. In many ways, it looks similar to a human arm, in that it comprises three parts – an 'upper arm', a 'forearm' and a 'hand' (Figure 19.10(a)). The wing even has an alula, which is the equivalent of the human thumb.

2. Feel the flesh of the wing. You should be able to feel a long bone in the upper arm (the humerus) and two in the lower arm (the radius and ulna).

3. Carefully remove the skin from the wing, starting from the top of the upper arm and proceeding towards the wing tip. Pull up the loose skin and cut it away from the underlying tissue. This takes time, but if you are careful you can expose most of the underlying muscles (Figure 19.10(b)).

4. The 'upper arm' has two muscles, just like the biceps and triceps in the human arm. If you pull on these in turn, you can flex and straighten the 'elbow' joint, mimicking what the muscles do when they contract. Look for the tendons that attach the muscles to the bones. Compare the arrangement of these muscles with that shown in Figure 7.8. The 'forearm' has another bundle of muscles passing down into the 'hand'.

5. Make a drawing of the arrangement of bones, muscles and tendons in the 'upper arm' of the chicken wing.

6. Cut away the muscles of the wing to reveal the bones of the wing skeleton (Figure 19.10(c)).

7. Break open the 'elbow joint' of the wing. Why is this difficult to do? (Look at the structure of a synovial joint in Figure 7.6). When you open the joint, you should be able to see the smooth articular cartilage covering the ends of the bones.

(a)

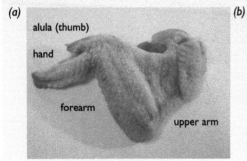

alula (thumb)
hand
forearm
upper arm

(b)

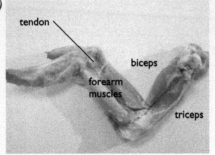

tendon
biceps
forearm muscles
triceps

(c)

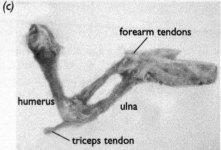

forearm tendons
humerus
ulna
triceps tendon

Figure 19.10 *Dissection of a chicken wing.*

7. What are antagonistic muscles?
8. Which muscle in the upper arm contracts to cause the arm to flex?
9. Why must tendons be strong and inelastic?
10. Where is:
    (a) the origin of the triceps muscle
    (b) the insertion of the triceps muscle?

Chapter summary

In this chapter you have learnt that:

- plants carry out growth movements, while animals can move from place to place (locomotion)

- locomotion is necessary in animals in order for them to obtain food, seek shelter and reproduce

- the main functions of the skeleton are support, protection, movement and locomotion, and making blood cells

- bone is a living tissue made from specialised cells and protein fibres hardened with calcium salts

- synovial joints have a specialised structure that allows movement to occur

- there are several types of joint, allowing different degrees of movement, such as ball-and-socket, hinge, partially movable and fixed joints

- a muscle moves a bone by contracting and pulling on the bone

- muscles are present in antagonistic pairs working against each other, for example, the biceps and triceps in the upper arm.

Questions

1. Which of the following are all functions of the human skeleton?

   A   Movement, support and circulation
   B   Movement, breathing and protection against infection
   C   Support, production of red blood cells and protection from injury
   D   Support, protection from injury and storing iron

2. Synovial joints:

   A   always allow movement in all directions
   B   do not allow any movement
   C   always allow some movement
   D   always allow movement in one plane

3. Muscles move bones by:

   A   relaxing and pulling
   B   relaxing and pushing
   C   contracting and pushing
   D   contracting and pulling

4. Which of the following is **not** true of movements in plants?

   A   They are brought about by hormones.
   B   They often involve the whole organism changing position.
   C   They often involve growth.
   D   Muscles are not involved.

5. *a)* Which parts of the body are protected by:

   i)   the cranium

   ii) the vertebral column

   iii) the ribs?   (3)

   b) Between the vertebrae are discs of cartilage. From your knowledge of the properties of cartilage, suggest what their function is.   (2)

   c) Name *three* components of bone.   (3)

   d) The synovial membrane, synovial fluid and ligamants are parts of a synovial joint. Explain the function of each.   (6)

6. The diagram shows the bones and some of the muscles of the back leg of a rabbit.

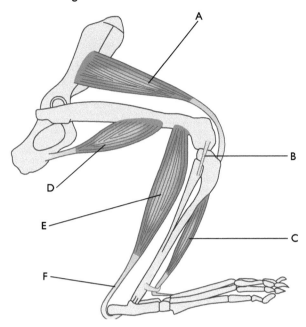

a) Which label (A to F) is a muscle that::

    i)   straightens the knee joint

    ii)   flexes the ankle joint?   *(2)*

b) Which label shows a tendon?   *(1)*

c) Is this tendon at the origin or the insertion of the muscle? Explain your answer.   *(2)*

d) Tendons are *inelastic*. Why is this an important property for them to have?   *(2)*

7. The diagram shows the muscles and bones in a human forearm when the forearm is being raised.

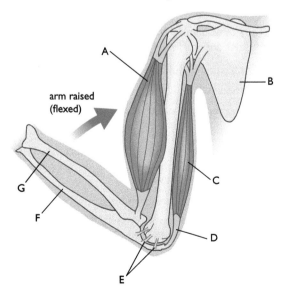

arm raised (flexed)

a) Name the parts labelled A, B, C, D, E, F and G.   *(7)*

b) A and C are antagonistic muscles. What does this mean?   *(2)*

c) *i)*   Describe the function of the structures labelled E.

    ii)   What properties of structures E adapt them to this function?   *(2)*

d) What must happen for the forearm to be lowered? Explain your answer.   *(2)*

8. Movement of the forearm is brought about by contraction of antagonistic muscles. Contraction of the biceps muscle flexes the arm, while contraction of the triceps muscle extends the arm at the elbow. The elbow is an example of a hinge joint, whereas the shoulder joint is a ball and socket joint. Both are examples of synovial joints, allowing free movement between bones.

a) Draw a labelled diagram of the arrangement of the arm bones and muscles from the shoulder blade (scapula) to the wrist.   *(6)*

b) Explain the meaning of antagonistic muscles, with reference to the biceps and triceps.   *(5)*

c) Explain the difference between the movement allowed by a hinge joint and a ball and socket joint.   *(2)*

d) What adaptations are shown by a synovial joint that allow free movements between bones?   *(4)*

e) Osteoarthritis is a disease that causes damage to the cartilage of joints, leaving the ends of the bones to rub against each other. This causes great pain and stiffness in the joint, and eventually prevents movement altogether. It can be treated by surgery, where the joint is replaced with an artificial one. In the case of an artificial hip joint, it consists of a metal ball inserted into a plastic socket.

Suggest *three* features that an artificial hip joint will need in order to work efficiently.   *(3)*

Chapter 20: Reproduction in Animals

When you have completed this chapter, you will be able to:

- understand the differences between sexual and asexual reproduction
- describe the structure and function of the reproductive system in humans
- understand the stages of sexual reproduction in humans:
  - production of sex cells
  - transfer of the male sex cell to the female sex cell
  - fertilisation
  - development of the embryo
- describe the functions of the placenta during pregnancy
- describe the stages of birth
- describe the menstrual cycle and its control by hormones
- discuss the advantages and disadvantages of different methods of birth control
- discuss the transmission and control of sexually transmitted infections.

The plural of sperm is sperm.
The plural of ovum is ova.

One of the characteristics of living organisms that sets them apart from non-living things is their ability to reproduce. Reproduction is all about an organism passing on its genes. This can be through special sex cells, or asexually, without the production of these cells. In this chapter, you will look at the differences between sexual and asexual reproduction, and study in detail the process of human reproduction.

Sexual and asexual reproduction compared

In any method of reproduction, the end result is the production of more organisms of the same species. Humans produce more humans, jellyfish produce more jellyfish and salmonella bacteria produce more salmonella bacteria. However, the way in which they reproduce differs. There are two main types of reproduction: **sexual reproduction** and **asexual reproduction**.

In sexual reproduction specialised **sex cells** are produced. There are usually two types, a mobile male sex cell (a **sperm**) and a stationary female sex cell (an **ovum**).

Figure 20.1 *A sperm fertilising an ovum.*

The sperm must move to the ovum and fuse (join) with it. This is called **fertilisation**. The single cell formed by fertilisation is called a **zygote**. This cell will divide many times by mitosis to form all the cells of the new animal.

In asexual reproduction, there are no specialised sex cells and there is no fertilisation. Instead, cells in one part of the body divide by mitosis to form a structure that breaks away from the parent body and grows into a new organism. Not many animals reproduce in this way. Figure 20.2 (page 292) shows *Hydra* (a small animal similar to jellyfish) reproducing by budding. Cells in the body wall divide to form a small version of the adult. This eventually breaks off and becomes a free-living *Hydra*. One animal may produce several 'buds' in a short space of time.

Asexual reproduction produces identical offspring

All the offspring produced when *Hydra* buds, are genetically identical – they have exactly the same **genes**. This is because all the cells of the new individual are produced by mitosis from just one cell in the body of the adult. When cells divide by mitosis, the new cells that are produced are exact copies of the original cell (see Chapter 24). As a result, all the cells of an organism that is produced asexually have the same genes as the cell that produced them – the original adult cell. So *all* asexually produced offspring from one adult will have the same genes as the cells of the adult. They will *all* be genetic copies of that adult and so will be identical to each other.

Asexual reproduction is useful to a species when the environment in which it lives is relatively stable. If an organism is well adapted to this stable environment, asexual reproduction will produce offspring that are also well adapted. However, if the environment changes significantly, then *all* the individuals will be affected equally by the change. It may be such a dramatic change that none of the individuals are adapted well enough to survive. The species will die out in that area.

Sexual reproduction produces offspring that show genetic variation

There are four key stages in any method of sexual reproduction:

1 Sex cells (sperm and ova) are produced.

2 The male sex cell (sperm) is transferred to the female sex cell (ovum).

3 Fertilisation occurs – the sperm fuses with the ovum.

4 The zygote formed develops into a new individual.

Production of sex cells

Sperm are produced in the male sex organs – the **testes**. Ova are produced in the female sex organs – the **ovaries**. Both are produced when cells inside these organs divide. These cells do not divide by mitosis but by **meiosis** (Chapter 24). Meiosis produces cells that are not genetically identical and have only half the number of chromosomes as the original cell.

Transfer of the sperm to the ovum

Sperm are specialised for swimming. They have a tail-like **flagellum** that moves them through water or a water-based liquid. Figure 20.3 shows the structure of a sperm.

A gene is a section of DNA that determines a particular characteristic or feature. Genes are found in the nucleus of a cell on the chromosomes.

Individuals produced asexually from the same adult organism are called **clones**.

buds – young *Hydra* growing from the parent – they will eventually break off and become independent

parent *Hydra*

Figure 20.2 *Hydra reproducing asexually by budding.*

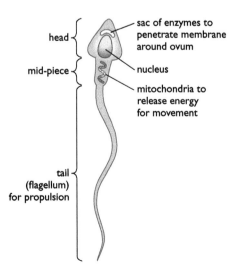

head

sac of enzymes to penetrate membrane around ovum

mid-piece

nucleus

mitochondria to release energy for movement

tail (flagellum) for propulsion

Figure 20.3 *The structure of a sperm.*

Some male animals, such as those of most fish, release their sperm into the water in which they live. The female animals release their ova into the water and the sperm then swim through the water to fertilise the ova. This is **external fertilisation** as it takes place *outside* the body. Before the release takes place, there is usually some mating behaviour to ensure that the male and female are in the same place at the same time. This gives the best chance of fertilisation occurring before water currents sweep the sex cells away.

Other male animals, such as those of birds and mammals, **ejaculate** their sperm in a special fluid into the bodies of the females. **Internal fertilisation** then takes place inside the female's body. Fertilisation is much more likely as there are no external factors to prevent the sperm from reaching the ova. Some form of **sexual intercourse** precedes ejaculation.

Fertilisation

Once the sperm has reached the ovum, its nucleus must enter the ovum and fuse with the ovum's, nucleus. As each sex cell has only half the normal number of chromosomes, the zygote formed by fertilisation will have the full number of chromosomes. In humans, sperm and ovum each have only 23 chromosomes. The zygote has 46 chromosomes like all other cells in the body. Figure 20.4 shows the main stages in fertilisation.

> Red blood cells are exceptions. They have no nucleus, so have no chromosomes.

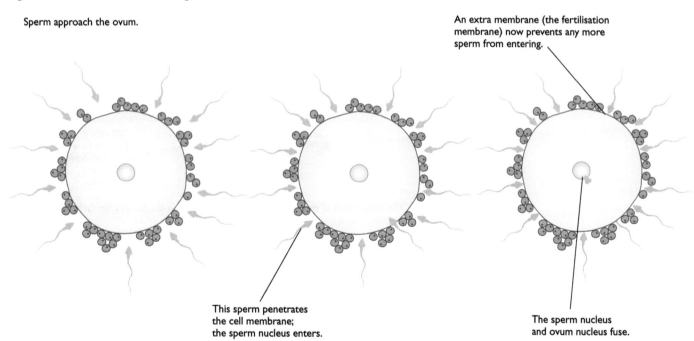

Sperm approach the ovum.

An extra membrane (the fertilisation membrane) now prevents any more sperm from entering.

This sperm penetrates the cell membrane; the sperm nucleus enters.

The sperm nucleus and ovum nucleus fuse.

Figure 20.4 *The main stages in fertilisation.*

Fertilisation does more than just restore the full chromosome number, it provides an additional source of genetic variation. The sperm and ova are all genetically different because they are formed by meiosis. Therefore, each time fertilisation takes place, it brings together a different combination of genes.

Development of the zygote

Each zygote that is formed must divide to produce all the cells that will make up the adult. All these cells must have the full number of chromosomes, so the zygote divides repeatedly by mitosis. Figure 20.5 on page 294, shows the importance of meiosis, mitosis and fertilisation in the human life cycle.

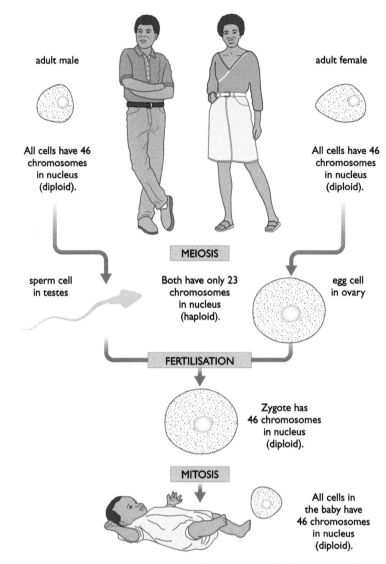

Figure 20.5 *The importance of meiosis, mitosis and fertilisation in the human life cycle.*

However, mitosis is not the only process involved in development, otherwise all that would be produced would be a ball of cells. During the process, cells move around and different shaped structures are formed. Also, different cells specialise to become bone cells, nerve cells, muscle cells, and so on.

1. Describe two differences between sexual and asexual reproduction.
2. Name the organs that produce sperm and eggs.
3. What type of cell division produces sperm and eggs?
4. Explain the difference between internal fertilisation and external fertilisation.
5. Describe (a) two features of a sperm that allow it to move, and (b) one feature of a sperm that allows it to penetrate an ovum.
6. Describe two processes, other than mitosis, that must occur to allow a zygote to develop into a new individual.

Reproduction in humans

Humans reproduce sexually and fertilisation is internal. Figures 20.6 and 20.7 show the structure of the human female and male reproductive systems.

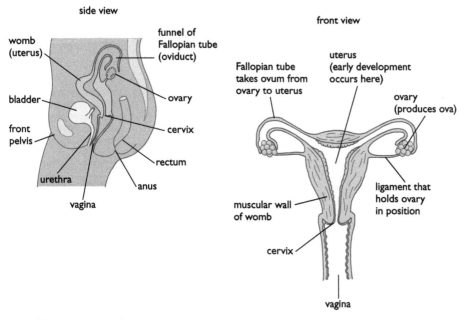

Figure 20.6 *The human female reproductive system.*

The sperm are produced in the testes by meiosis. During sexual intercourse, they pass along the sperm duct and are mixed with a fluid from the seminal vesicles. This mixture, called **semen**, is ejaculated into the vagina of the female. The sperm then begin to swim towards the Fallopian tubes.

One ovum is released into a Fallopian tube each month from an ovary. If an ovum is present in the Fallopian tubes, then it may be fertilised by sperm introduced during intercourse. The zygote formed will begin to develop into an **embryo**, which will **implant** in the lining of the uterus. Here, the embryo will develop a **placenta**, which will allow the embryo to obtain materials such as oxygen and nutrients from the mother's blood. It also allows the embryo to get rid of waste products such as urea and carbon dioxide, as well as anchoring the embryo in the uterus. The placenta secretes female hormones, in particular progesterone, that maintain the pregnancy and prevent the embryo from aborting. Figure 20.8 on page 296 shows the structure and position of the placenta.

During pregnancy, a membrane called the **amnion** encloses the developing embryo. The amnion secretes a fluid called **amniotic fluid**, which protects the developing embryo against jolts and bumps. As the embryo develops, it becomes more and more complex. When it becomes recognisably human, we no longer call it an embryo but a **fetus**. At the end of nine months of development, there just isn't any room left for the fetus to grow and it sends a hormonal 'signal' to the mother to initiate birth. This is called 'going into labour'. Figure 20.8 also shows the position of a human fetus just before birth.

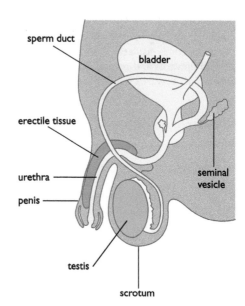

Figure 20.7 *The human male reproductive system.*

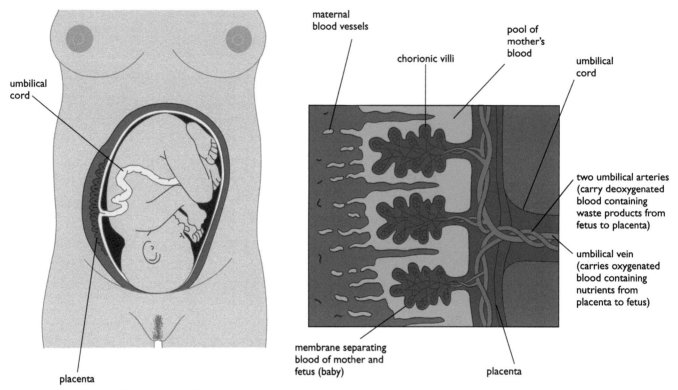

Figure 20.8 *The position of the fetus just before birth, and the structure of the placenta.*

Just before birth, the fetus takes up so much room that many of the mother's organs become displaced. The heart is pushed upwards and rotates so that the base points towards the left breast.

There are three stages to the birth of a child:

1 **Dilation of the cervix** – the cervix gets wider to allow the baby to pass through. The muscles of the uterus contract quite strongly and rupture the amnion, allowing the amniotic fluid to escape. This is called the breaking of the waters.

2 **Delivery of the baby** – strong contractions of the muscles of the uterus push the baby head first through the cervix and vagina to the outside world.

3 **Delivery of the afterbirth** – after the baby has been born, the uterus continues to contract and pushes the placenta out, together with the membranes that surrounded the baby. These are known as the afterbirth.

Figure 20.9 shows the stages of birth.

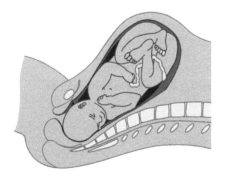

I Baby's head pushes cervix; mucous plug dislodges and waters break.

2 Uterus contracts to push baby out through the vagina.

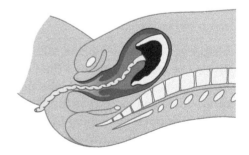

3 The placenta becomes detached from the wall of the uterus and is expelled through the vagina as the afterbirth.

Figure 20.9 *The stages of birth.*

7. Describe the functions of the testes, the sperm duct and the seminal vesicles.
8. Describe the functions of the ovaries, the uterus, the cervix and the Fallopian tubes.
9. What is meant by implantation?
10. Describe three functions of the placenta.
11. Describe how birth is initiated, and name and describe the three stages of birth.

Hormonal control of the female reproductive cycle

The **menstrual cycle** is a cycle of events that, each month, prepares a woman for possible pregnancy. It is controlled by sex hormones and involves the maturation of an ovum in an ovary, as well as changes to the structure of the uterine lining.

Approximately every 28 days, cells in the ovary begin to produce and secrete the hormone **oestrogen**. This circulates in the bloodstream to the uterus, where it causes the lining of the uterus to become thicker. This extra thickness makes implantation possible if fertilisation occurs. At the same time, an ovum is maturing inside a group of cells called a **follicle** in one of the ovaries and, about half-way through the month, the mature ovum is released into one of the Fallopian tubes. This is **ovulation**. The remains of the follicle, called the **corpus luteum**, begins to secrete the hormone **progesterone**. This hormone **vascularises** the newly thickened uterus lining, that is it causes many blood vessels to develop in it. This means that, should implantation occur, the embryo will be able to receive nutrients immediately from the mother's blood. If fertilisation and implantation do not occur, the ovary reduces its production of the two hormones. Without these hormones, the new uterus lining cannot remain in place. It breaks away and lost through the vagina. This loss of uterus lining and blood is called **menstruation**. Figures 20.10 and 20.11 summarise the main events of the menstrual cycle.

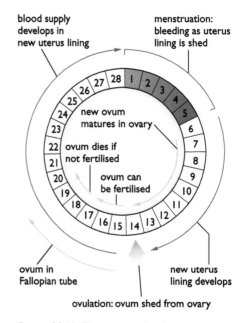

Figure 20.10 *The menstrual cycle.*

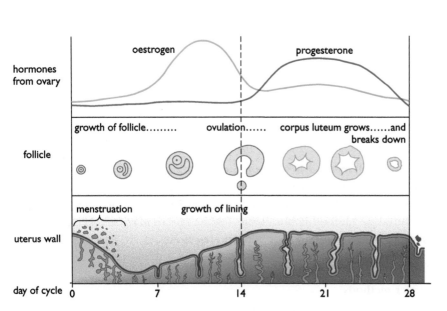

Figure 20.11 *How the changes in hormone levels influence the menstrual cycle.*

Contraception

Contraception means avoiding conception – that is, avoiding becoming pregnant. Not having sexual intercourse is the most certain method of contraception, but many couples do want to have sex without the woman becoming pregnant. One way of doing this is the so-called 'natural' method. This uses knowledge of the menstrual cycle to avoid intercourse at times when sperm could possibly fertilise an ovum. Fertilisation is most likely to take place in the middle of the menstrual cycle, either side of the day of ovulation. Sperm and ova live for only a few days. Sperm released into the vagina 4 days before ovulation will have died by the time ovulation occurs. Two or 3 days after ovulation, an unfertilised ovum will have been lost through the vagina. By avoiding intercourse during this period, and limiting it to the rest of the menstrual cycle (the 'safe period'), a woman can usually avoid becoming pregnant (Figure 20.12).

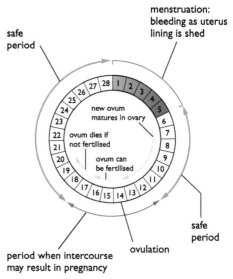

Figure 20.12 *The 'safe period'.*

For this method to be successful, the woman's menstrual cycle must be regular, and she must know when ovulation occurs in her cycle. The only disadvantage of this method is that it is very unreliable. There is usually a slight increase in a woman's body temperature around the time of ovulation, so she can take her temperature daily to try to judge when she is ovulating. Even if she does this, the 'safe period' method results in many unplanned pregnancies. However, some couples rely on the 'safe period' method, usually because they have moral or religious reasons for not using other means of contraception.

Another 'natural' method is where the man withdraws his penis from the woman's vagina before ejaculation. As with the 'safe period' method, this has the advantage that it does not involve using artificial contraception. But again, the 'withdrawal' method is unreliable – some sperm may be released before ejaculation, or the man may not withdraw in time.

There are four other categories of contraception:

- barrier methods
- intrauterine devices (IUDs)

Figure 20.13 *Condoms are easily obtained from clinics or pharmacists.*

- hormonal methods

- sterilisation.

Barrier methods use some kind of barrier to prevent sperm reaching the ovum. The **condom** is a sheath of very fine rubber that fits over the penis, and catches the semen before it enters the vagina (Figure 20.13, page 298). Another version, called the **femidom,** is similar, but inserted into the vagina before intercourse. Condoms and femidoms are easy to obtain and use, and have the added advantage that they are the only methods of contraception which give protection against sexually transmitted diseases.

Another barrier method is the **diaphragm** or **cap** (Figure 20.14). This is a dome-shaped piece of rubber that a woman inserts into her vagina before intercourse. The cap covers the cervix, preventing sperm from entering the uterus. A **spermicidal cream** is used with the cap as extra protection against pregnancy (spermicidal creams also increase the reliability of condoms). The cap is left in position for 6 hours after intercourse. Caps differ in size, and the woman needs a medical examination by a doctor or nurse to select the correct size of cap to use.

Intrauterine devices (IUD or coil) are small pieces of plastic or copper of various shapes. An IUD is inserted through the cervix into the uterus (Figure 20.15). The copper IUD works by preventing a fertilised egg from implanting in the lining of the uterus. Some IUDs contain the hormone progesterone, which thickens the mucus in the cervix, stopping sperm from getting through. One disadvantage of the IUD is that it has to be fitted by a doctor.

Hormonal methods mean taking the **oral contraceptive pill**. The **combined pill** contains a mixture of oestrogen and progesterone. These hormones prevent the production of FSH and LH from the pituitary gland (see page 127). This means that the follicles inside the ovary do not develop, and ovulation does not take place. Without an egg being released, a woman cannot become pregnant. She takes the pill for 21 days of the menstrual cycle (Figure 20.16) then stops taking it for the last 7 days. During this time she will have a period.

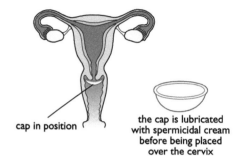

cap in position

the cap is lubricated with spermicidal cream before being placed over the cervix

Figure 20.14 *The contraceptive cap (diaphragm).*

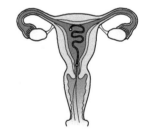

IUD in place in uterus

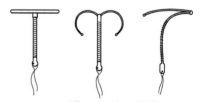

some different styles of IUD

Figure 20.15 *Intrauterine devices.*

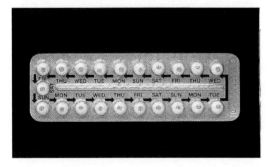

Figure 20.16 *Synthetic hormones can be used to prevent pregnancy.*

The **mini-pill** contains only progesterone. It works by causing a thickening of mucus in the cervix, which acts as a barrier to sperm.

Sterilisation is a surgical operation that can be carried out on men to prevent sperm passing to the penis, or on women to prevent eggs passing to the uterus. This makes them unable to have children (sterile). Sterilisation is a method that is normally only used by couples who have produced the number of children that they want, and do not want to use other methods

The combined pill has been linked to a number of health problems in women, the main one being an increased risk of blood clots (thromboses) forming in blood vessels, which can be fatal. The increased risk is only very slight for most women, but is higher if they are smokers. The mini-pill is not quite as reliable as the combined pill that contains oestrogen, but it does not increase the chances of blood clots developing.

of contraception to prevent further pregnancies. The operation is usually irreversible, so the decision to opt for sterilisation must be considered carefully.

Male sterilisation is called **vasectomy**. Under local anaesthetic, the sperm ducts are cut and tied so that no sperm can get through. After the operation, the man can still ejaculate but the semen will contain no sperm. In women, the Fallopian tubes are cut by a similar operation, called **tubal ligation**.

Table 20.1 compares the main methods of contraception in order of their rates of failure, and lists some of the advantages and disadvantages of each method.

Method	Failure rate (%)*		Some advantages and disadvantages
	Typical use	Careful use	
none – intercourse without contraception	85	85	
withdrawal	27	4	High failure rate.
'safe period'	25	9	High failure rate. Woman needs to have a regular cycle, and to keep records of the cycle.
diaphragm with spermicidal cream	16	6	Medical examination needed to select correct size. Not simple to use – must be inserted before intercourse and left in place for 6 hours afterwards.
condom	15	2	Easy to obtain and use. Gives protection against sexually transmitted diseases. May slip off during intercourse.
mini-pill	8	0.5	Low failure rate if used carefully. Must be taken every day, at the same time each day, to be effective.
combined pill	8	0.3	Low failure rate if used carefully. Must be taken every day. Links to some health problems.
intrauterine device	0.8	0.6	Must be fitted by a doctor. Can cause heavier periods.
tubal ligation	0.5	0.5	Very low failure rate. Operation usually not reversible.
vasectomy	0.1	0.1	Very low failure rate. Operation usually not reversible.

* The values for 'failure rate' are the number of pregnancies that result per 100 women per year. There are two sets of numbers. The 'careful use' column shows the failure rate when the contraceptives are used carefully, exactly as they should be. These values are from medical trials under controlled conditions. However, people are human, and make mistakes. The rates in the 'typical use' column are from surveys of couples who use the contraceptive methods normally, sometimes without taking the same degree of care as in the medical trials. This results in a higher rate of failure.

Table 20.1: *The main methods of contraception.*

Sexually transmitted infections (STIs)

Sexually transmitted infections (STIs) are transmitted mainly though sexual intercourse. However, some, like syphilis, can be transmitted in saliva whilst others, like candidiasis (thrush) can be transmitted by direct body contact. Using condoms during intercourse can prevent the transmission of many of these disease-causing micro-organisms. Perhaps the most significant of all the sexually transmitted infections is AIDS.

AIDS

AIDS (**A**cquired **I**mmune **D**eficiency **S**yndrome) is one of the most significant worldwide killers of the moment. It is caused by a virus called the human immunodeficiency virus or **HIV**. Figure 21.17 shows the structure of this virus.

HIV initially infects a type of white blood cell called a helper T-lymphocyte. These cells are necessary if other white blood cells (B-lymphocytes and other T-lymphocytes) are to become active and start fighting infections. The course of a typical HIV infection is described below:

1 The genetic material of HIV becomes incorporated in the DNA of the helper T-lymphocyte.

2 When the HIV DNA is activated, it 'instructs' the lymphocyte to make HIV proteins and genetic material.

3 Some of the HIV proteins and genetic material are assembled into new viruses.

4 Some of the HIV proteins end up as marker proteins (antigens) on the surface of the cell.

5 These HIV proteins are recognised as foreign proteins.

6 The lymphocyte is destroyed by the immune system.

7 The assembled virus particles escape to infect other lymphocytes.

8 This cycle repeats itself for as long as the body can replace the lymphocytes that have been destroyed.

9 Eventually the body will not be able to replace the lymphocytes at the same rate at which they are being destroyed.

10 The number of free viruses in the blood increases rapidly and HIV may infect other areas of the body, including the brain.

11 The immune system is severely damaged and other disease-causing micro-organisms infect the body.

12 Death is often as a result of 'opportunistic' infection by TB and pneumonia, due to the reduced capacity of the immune system that would normally destroy the organisms causing these diseases. Death can also be caused by rare cancers such as Kaposi's sarcoma (Figure 20.18).

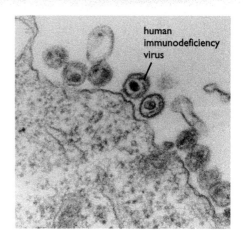

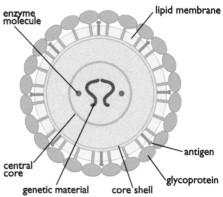

Figure 20.17 *The human immunodeficiency virus (HIV).*

Part of this immune response to infection by HIV involves the production of antibodies to destroy the virus. At this stage, the person is said to be **HIV positive**, because their blood gives a positive result when tested for HIV antibodies.

The period during which the body replaces the lymphocytes as fast as they are destroyed is called the **latency period**. It can last for up to 20 years. The person shows no symptoms of AIDS during this period, but will be highly infective to others.

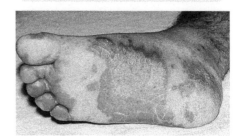

Figure 20.18 *A person suffering from Kaposi's sarcoma.*

Figure 20.19 shows how the levels of virus particles in the blood, HIV antibodies and helper T-lymphocytes change during the course of an AIDS infection.

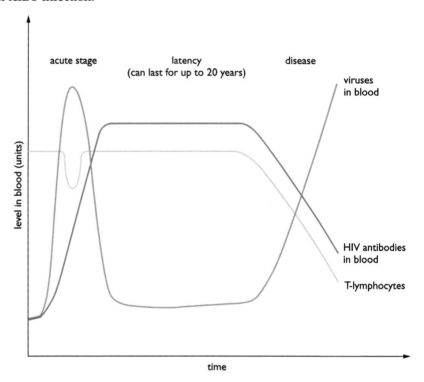

Figure 20.19 *The stages of an AIDS infection.*

The real significance of AIDS is that the cells that are being destroyed are the cells that are needed to help the other lymphocytes destroy the infected cells.

AIDS is primarily transmitted by unprotected sexual intercourse, (either homosexual or heterosexual) or by blood-to-blood contact (for example, when drug users share an infected needle, or if infected blood is given in a transfusion). AIDS can **not** be transmitted by kissing, sharing utensils (e.g. a cup or a glass), giving blood, skin-to-skin contact or sitting on the same toilet seat as an infected person.

As yet, there is no effective vaccine against HIV, although a number of anti-viral drugs can delay the onset of AIDS. The transmission of AIDS can be controlled, however, by a number of measures. These include:

- use of condoms – although transmission can still occur, the incidence of transmission is greatly reduced

- non-sharing and use of only sterile needles by drug users

- limiting the number of sexual partners – in particular, avoiding sex with high risk groups such as prostitutes.

The latest estimates (2013) show that, on average, about 1% of people in territories of the Caribbean are infected with HIV/AIDS. The figure varies from country to country, for example, it is about 1.5% in Jamaica and Barbados. These are some of the highest figures for any region of the world outside Africa. Apart from affecting the health and well-being of the people affected, infection with HIV/AIDS impacts on many other aspects of life, including the productivity of the countries, and their economic growth and development.

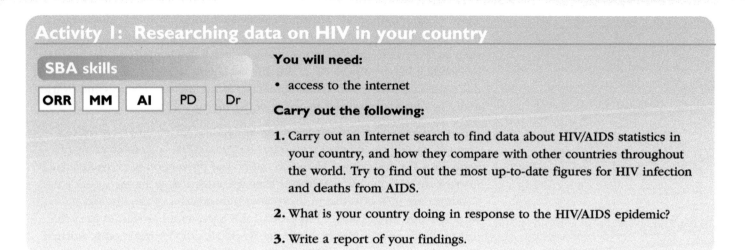

SBA skills

ORR	MM	AI	PD	Dr

You will need:

- access to the internet

Carry out the following:

1. Carry out an Internet search to find data about HIV/AIDS statistics in your country, and how they compare with other countries throughout the world. Try to find out the most up-to-date figures for HIV infection and deaths from AIDS.

2. What is your country doing in response to the HIV/AIDS epidemic?

3. Write a report of your findings.

Syphilis and gonorrhoea

There are a number of other sexually transmitted infections. Two of the most well known are **syphilis** and **gonorrhoea**. Unlike HIV, both these STIs can be successfully treated with antibiotics. A comparison of the two diseases is shown in Table 20.2.

Aspect of disease	Gonorrhoea	Syphilis
cause	• the bacterium *Neisseria*	• the bacterium *Treponema*
method of transmission	• sexual intercourse	• sexual intercourse • from mother to foetus in the uterus
first symptoms	• pain on passing urine • discharge from penis • headache and fever	• painless sore at site of infection
treatment	• injection of effective antibiotic (the bacterium is resistant to many antibiotics)	• early stages – a single dose of penicillin or other antibiotic • later stages – injections of antibiotics over a longer period
prevention	• trace contacts quickly for early treatment • educate about symptoms and dangers • avoid sexual intercourse, or use condoms	• trace contacts quickly for early treatment • educate about symptoms and dangers • avoid sexual intercourse, or use condoms

Table 20.2 *Comparison of gonorrhoea and syphilis.*

Both syphilis and gonorrhoea are caused by bacterial infections. Gonorrhoea is caused by the bacterium *Neisseria gonorrhoeae*, which is passed from person to person during sexual intercourse. Symptoms usually appear a few days after infection, although about half of all infected people, mainly women, show no symptoms at all. Most men show symptoms, including discharge of pus from the penis and pain on urinating. There are also general symptoms, such as fever or headaches. Some women have a discharge from the vagina, or bleeding between periods. If the disease is left untreated, it becomes much more serious, with the infection spreading to the uterus, oviducts and ovaries, and it can lead to a woman becoming infertile.

It is possible to cure gonorrhoea using antibiotics. Often a person with gonorrhoea will be infected with other bacterial STIs, so combinations of antibiotics are prescribed. The main method of prevention is to avoid sexual intercourse with people who might have the infection, or to use a condom. Condoms are 99% effective in preventing transmission of the disease. As with so many other bacterial infections, the bacterium has started to evolve resistance to some antibiotics, and the World Health Organization is worried that antibiotic resistance will soon make it difficult to treat the disease.

Syphilis is a much more serious disease than gonorrhoea and is often fatal if left untreated. It is caused by the bacterium *Treponema pallidum*, and transmitted via sexual intercourse or from mother to foetus during pregnancy. The first symptom is a painless sore on the penis or vagina, which develops 3 to 90 days after infection. If left untreated, syphilis spreads to many other tissues in the body, including the brain and nervous system, often resulting in death. Syphilis (and other STIs) is more common in people infected with HIV because of the lowered effectiveness of their immune system, and because syphilis increases the risk of transmission of HIV.

12. Name the hormone that:
    (a) causes a new lining to develop in the uterus each month and
    (b) causes the new uterine lining to become vascular.
13. What happens when ovulation occurs?
14. What is menstruation?
15. Explain how the safe period method of contraception works.
16. Why is the pill an effective method of contraception?
17. Explain why AIDS is such a serious disease.
18. Name two other sexually transmitted infections and describe the first symptoms of each infection.

Chapter summary

In this chapter you have learnt that:

- sexual reproduction differs from asexual reproduction in that it involves the production of sex cells, fertilisation and produces offspring that are genetically different

- asexual reproduction produces offspring that are genetically identical because they are all produced by mitosis; this has a survival advantage in a stable environment

- sex cells are produced by meiosis; they are haploid cells and show genetic variation

- fertilisation restores the normal diploid number of chromosomes in the resulting zygote

- mitosis ensures that all the cells of an organism formed from a zygote are genetically identical diploid cells

- mitosis and meiosis differ in that:

  - there is only one cell division in mitosis, but two in meiosis

  - mitosis forms two cells, whereas meiosis forms four cells

  - mitosis forms diploid cells, whereas meiosis forms haploid cells

  - mitosis produces genetically identical cells, whereas meiosis produces cells that vary genetically

- in the human male reproductive system, sperm are produced in the testes and move from the testes to the penis through the sperm duct

- the seminal vesicle adds a fluid to sperm to activate them and the penis transfers the sperm to the vagina during intercourse

- in the human female reproductive system, the ovaries release ova once a month from developing follicles; they also secrete the reproductive hormones oestrogen and progesterone

- the Fallopian tubes are the normal site of fertilisation and they also transfer ova to the uterus

- the uterus produces a new lining each month in which a developing embryo can implant

- the cervix dilates at the onset of birth and the baby leaves the body via the vagina

- during pregnancy, the placenta allows the exchange of materials between the mother's blood and the fetus's blood

- the three stages of birth are the breaking of the waters, delivery of the baby and delivery of the afterbirth

Chapter summary (continued)

- the menstrual cycle is a 28-day cycle during which time the levels of reproductive hormones secreted by the ovaries rise then fall, the uterine lining thickens, becomes more vascular and then breaks down and is lost in menstruation if there is no fertilisation resulting in a pregnancy
- contraception can be achieved by natural methods, barrier methods sterilisation, hormonal methods and by mechanical methods
- AIDS is caused by a virus (HIV), which infects cells of the body's immune system
- AIDS causes death by reducing the capacity of the immune system to fight other infections.

Questions

1. Sex cells are:

- A   diploid cells produced by meiosis
- B   diploid cells produced by mitosis
- C   haploid cells produced by meiosis
- D   haploid cells produced by mitosis

2. Fertilisation involves:

- A   fusion of two diploid cells
- B   fusion of two haploid cells
- C   splitting of a diploid cell
- D   splitting of a haploid cell

3. Which of the following is **not** a function of the ovaries?

- A   production of progesterone
- B   production of oestrogen
- C   production of ova
- D   acting as the site of fertilisation

4. The correct sequence of the stages of birth is:

- A   breaking of the waters → delivery of the afterbirth → delivery of the baby
- B   breaking of the waters → delivery of the baby → delivery of the afterbirth
- C   delivery of the baby → breaking of the waters → delivery of the afterbirth
- D   delivery of the baby → delivery of the afterbirth → breaking of the waters

5. Use of the condom falls into which of the following categories of birth control?

- A   Barrier methods
- B   Hormonal methods
- C   Mechanical methods
- D   Natural methods

6. The diagram shows a baby about to be born.

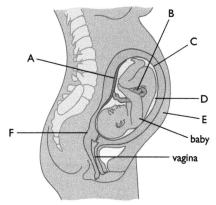

- *a)*   Name parts A to F on the diagram.   (6)
- *b)*   What is the function of A during pregnancy?   (2)
- *c)*   What must happen to D and E just before birth?   (2)
- *d)*   What must E and F do during birth?   (2)

7. The diagram shows *Hydra* (a small water animal) reproducing in two ways.

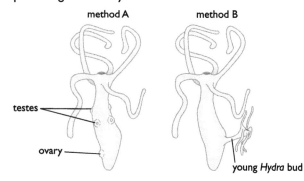

a) Which of the two methods shows asexual reproduction? Give a reason for your answer. (2)

b) Explain why organisms produced asexually are genetically identical to each other and to the organism that produced them. (3)

c) When the surroundings do not change for long periods, *Hydra* reproduces mainly asexually. When the conditions change, *Hydra* begins to reproduce sexually. How does this pattern of sexual and asexual reproduction help *Hydra* to survive? (4)

8. *a)* The diagram shows the female reproductive system.

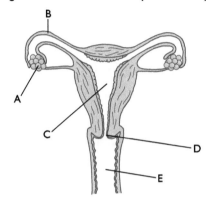

Which letter represents:

i) the site of production of oestrogen and progesterone

ii) the structure where fertilisation usually occurs

iii) the structure that must dilate when birth commences

iv) the structure that releases ova? (4)

b) The graph shows the changes in the thickness of the lining of a woman's uterus over 100 days.

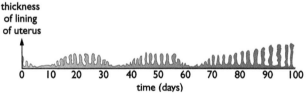

thickness of lining of uterus

time (days)

i) Name the hormone that causes the thickening of the uterine lining. (1)

ii) Use the graph to determine the duration of *this* woman's menstrual cycle. Explain how you arrived at your answer. (2)

iii) From the graph, deduce the approximate day on which fertilisation leading to pregnancy took place. Explain how you arrived at your answer. (2)

iv) Why must the uterus lining remain thickened throughout pregnancy? (2)

9. The diagram represents a typical menstrual cycle.

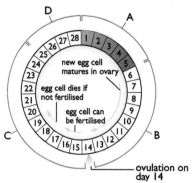

new egg cell matures in ovary

egg cell dies if not fertilised

egg cell can be fertilised

ovulation on day 14

a) During which of the stages A, B, C or D does:

i) the level of the hormone oestrogen increase in the blood

ii) the level of the hormone progesterone increase in the blood

iii) the uterine lining become more vascular

iv) the levels of oestrogen and progesterone in the blood fall

v) the uterine lining begin to break down? (5)

b) Explain how knowledge of the menstrual cycle can be used to avoid pregnancy. (3)

10. Gonorrhoea is a sexually transmitted infection caused by a bacterium. The graph shows changes in the incidence of gonorrhoea from 1925 to 1990.

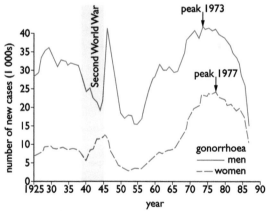

peak 1973

Second World War

peak 1977

number of new cases (1 000s)

gonorrhoea
— men
— — women

year

a) Name *two* other sexually transmitted infections. (2)

b) Describe the general trends in the numbers of cases of gonorrhoea from 1925 to 1990. (2)

c) Suggest reasons for the increases in the number of cases of gonorrhoea:

i) during the 1960s

ii) during the early 1970s. (2)

d) Suggest how the increase in AIDS cases in the 1980s may be linked to the decrease in the incidence of gonorrhoea. (1)

e) Explain why penicillin is not always effective in treating gonorrhoea. (1)

When you have completed this chapter, you will be able to:

- relate the parts of a flower to their functions
- compare the structure of an insect-pollinated flower with that of a wind-pollinated flower
- understand the difference between cross-pollination and self-pollination
- distinguish between the processes of pollination and fertilisation
- explain how the processes of fruit and seed formation occur
- describe fruit structure and adaptations for fruit and seed dispersal.

Flowers are the sexual organs of a flowering plant. They produce pollen and ovules, which contain the plant's gametes. Pollination is followed by fertilisation, after which the zygote develops into a new plant.

Haploid cells have only half the number of chromosomes of other body cells of that species.

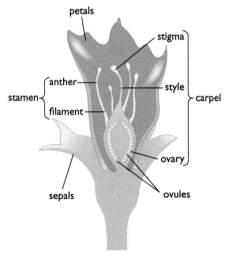

Figure 21.1 *The main structures in an insect-pollinated flower.*

Sexual reproduction in plants

Plants produce specialised, haploid sex cells in their flowers. The male sex cells are the **pollen grains** and the female sex cells are **ova**. Just as in animals, the male sex cells must be transferred to the female sex cells. This is called **pollination**. Pollination is normally carried out either by wind or insects. Following pollination, **fertilisation** takes place and the **zygote** formed develops into a **seed**, which, in turn, becomes enclosed in a **fruit**.

Production of sex cells and pollination

The sex cells are produced by meiosis in structures in the flowers. Pollen grains are produced in the **anthers** of the **stamens**. The ova are produced in **ovules** in the ovaries.

In pollination, pollen grains are transferred from the anthers of a flower to the stigma. If this occurs within the same flower, it is called **self-pollination**. If the pollen grains are transferred to a different flower, it is called **cross-pollination**. Pollination can occur by wind or by insect in either case.

Plants that are wind-pollinated produce flowers with a different structure to those of insect-pollinated flowers. These differences are related to the different methods of pollination of the flowers. Figure 21.1 shows the structure of typical insect-pollinated flower. Figure 21.2 on page 309 shows the structure of a typical wind-pollinated flower. Table 21.1 on page 309 summarises the main differences between insect-pollinated flowers and wind-pollinated flowers.

Feature	Insect-pollinated flower	Wind-pollinated flower
flower/ petals	large flowers with brightly coloured petals to attract insects	small, inconspicuous flowers with green petals
nectaries	present – sugary nectar is a 'reward' for insects	absent
scent	present – to attract insects	absent
anthers	enclosed within flower on a stiff filament to ensure insect makes contact with pollen	dangle loosely outside flower so wind can blow pollen away
stigma	enclosed within flower to ensure insect makes contact; sticky, so pollen grains attach to insect's body	exposed, outside the flower; feathery, to catch pollen blowing in the wind
pollen grains	small amounts produced; grains are sticky or spiky so attach to insect's body	large amounts of smooth, light pollen grains are produced that blow away in the wind

Table 21.1 *The main differences between insect-pollinated and wind-pollinated flowers.*

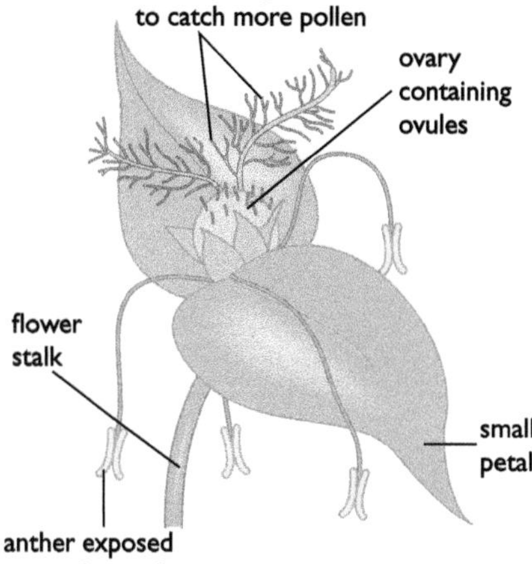

Figure 21.2 *A wind-pollinated flower.*

Activity 1: Observing insect-pollinated flowers and wind-pollinated flowers

SBA skills

| ORR | MM | AI | PD | Dr |

You will need:

- some flowers that are insect-pollinated and some that are wind-pollinated
- a hand lens

Carry out the following:

1. Observe each flower carefully.
2. Make a large, labelled drawing of each flower.
3. Decide whether the flower is insect-pollinated or wind-pollinated. Make a note of your decision beneath the drawing and give the reasons for your decision.

1. Name the male and female sex cells in plants.
2. What are the two main methods of pollination?
3. What develops from the zygote formed by fertilisation?
4. Describe three differences between insect-pollinated flowers and wind-pollinated flowers.

Fertilisation

Pollination transfers the pollen grain to the stigma. However, for fertilisation to take place, the nucleus of the pollen grain must fuse with the nucleus of the ovum, which is inside an ovule in the ovary. To transfer the nucleus to the ovum, the pollen grain grows a tube, which digests its way through the tissue of the style and into the ovary. Here it grows around to the opening in an ovule. The tip of the tube dissolves and allows the pollen grain nucleus to move out of the tube and into the ovule. Here it fertilises the ovum (egg cell) nucleus. These events are summarised in Figure 21.3 on page 310.

It is important to remember the difference between pollination and fertilisation. Pollination is the transfer of pollen from the anther to the stigma. Fertilisation is the fusion of the male pollen grain nucleus with the female nucleus.

(a)

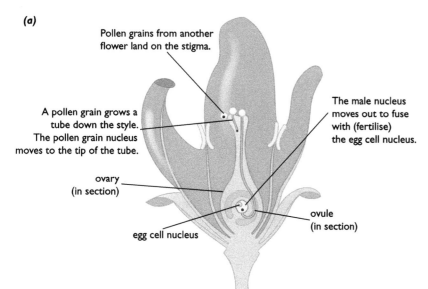

Pollen grains from another flower land on the stigma.

A pollen grain grows a tube down the style. The pollen grain nucleus moves to the tip of the tube.

The male nucleus moves out to fuse with (fertilise) the egg cell nucleus.

ovary (in section)

egg cell nucleus

ovule (in section)

Figure 21.3 *Pollination and fertilisation.*

(b) Details of fertilisation in the flower.

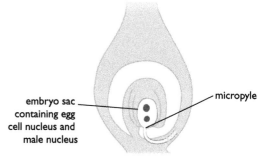

embryo sac containing egg cell nucleus and male nucleus

micropyle

The pollen tube enters the ovule through a hole called the micropyle. The structure in the middle of the ovule that contains the female nucleus is called the embryo sac. After fertilisation, the embryo sac will develop into the embryo plant and its food store.

Seed and fruit formation

Once fertilisation has occurred, a number of changes take place in the ovule and ovary that will lead to the fertilised ovule becoming a seed and the ovary in which it is found becoming a fruit. Different flowers produce different types of fruits, but in all cases the following changes take place:

- The zygote develops into an embryonic plant with a small root (**radicle**) and shoot (**plumule**).

- The other contents of the ovule develop into **cotyledons**, which will be a food store for the young plant when the seed **germinates**.

- The ovule wall becomes the seed coat or **testa**.

- The ovary wall becomes the fruit coat; this can take many forms depending on the type of fruit.

Any structure that contains seeds is a fruit. A pea pod is a fruit. The 'pod' is the fruit wall, formed from the ovary wall, and the peas are individual seeds. Each one was formed from a fertilised ovule.

Figure 21.4 summarises these changes as they occur in the plum flower. A coconut fruit forms in the same way, but the outer coat is fibrous rather than fleshy as it is in the plum.

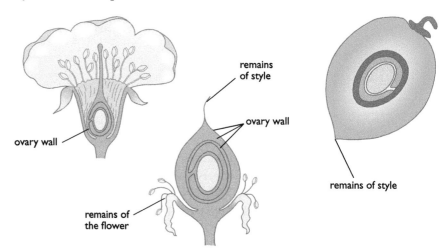

ovary wall

remains of style

ovary wall

remains of the flower

remains of style

Figure 21.4 *How a plum fruit forms.*

Dispersal of fruits and seeds

If all the seeds produced by a plant began to **germinate** (grow) in the same place, there would be too much competition for the available resources such as water, mineral ions and oxygen. To avoid this, plants **disperse** their seeds. Some are dispersed still inside the fruits; others are dispersed as seeds. Some seeds are dispersed by wind and others by animals. Figure 21.5 shows some wind-dispersed seeds and fruits. Figure 21.6 shows some seeds and fruits dispersed by animals.

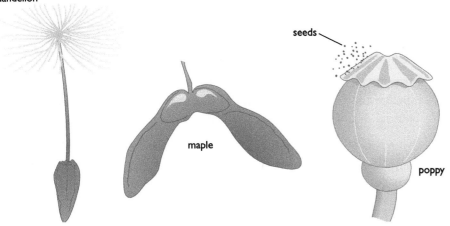

dandelion

seeds

maple

poppy

Figure 21.5 *These seeds are dispersed by wind. They are either very small and light, or their fruits have 'wings' or a 'parachute' to catch the wind.*

The fruits of coconut palms are dispersed by water. They can float many hundreds of miles before reaching a beach where they can germinate. Figure 21.7 shows a coconut palm beginning to grow from the fruit on a beach where it has landed.

Perhaps the simplest method of fruit and seed dispersal is mechanical dispersal. This is used by members of the pea and bean family. When the seeds are ready, the fruit pod dries up. As it does so, the inside of the pod dries faster than the outside. This makes the pod twist into a spiral shape and split suddenly, flicking out the seeds as it does so (Figure 21.8).

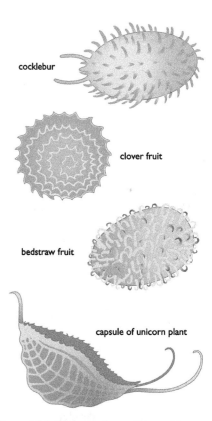

cocklebur

clover fruit

bedstraw fruit

capsule of unicorn plant

Figure 21.6 *These seeds are dispersed by animals. They have hairs or hooks that catch in the animal's fur. Seeds can also be dispersed by animals that eat the fruits and pass the undigested seeds out in their faeces.*

Figure 21.7 *A coconut palm growing from a fruit on a beach.*

Figure 21.8 *Pods of the pride of Barbados plant, a member of the pea family.*

Chapter summary

5. Explain how the pollen grain nucleus is transferred from the pollen grain on the stigma to the ovule in the ovary.
6. Following fertilisation, what is formed from the structures forming the ovule?
7. Explain why seeds must be dispersed.
8. State three ways in which fruits and seeds can be dispersed.

In this chapter you have learnt that:

- the male parts of the flower are the anthers, which produce pollen; the female parts are the ovaries, which contain ovules
- insect-pollinated flowers have large, brightly coloured petals, scent, and nectaries; their stigmas and stamens are enclosed and their pollen sticks to insects' bodies
- wind-pollinated flowers have small, green petals, no scent, and no nectaries; they have feathery stigmas and stamens that hang outside the flower, and light pollen
- the male nucleus is transferred to the female nucleus by means of a pollen tube that grows from the stigma through the style, to an opening in the ovule called the micropyle
- following fertilisation, the ovule contents become the seed and the ovary becomes the fruit
- seeds can be dispersed by animals, wind, water or self-dispersal.

Questions

1. Wind-pollinated flowers have:

   A large petals, a scent and 'sticky' pollen grains
   B large petals, no scent and 'light' pollen grains
   C small petals, a scent and 'sticky' pollen grains
   D small petals, no scent and 'light' pollen grains

2. Which is the correct sequence for the main stages of sexual reproduction?

   A gamete production → gamete transfer → fertilisation → development
   B gamete production → development → gamete transfer → fertilisation
   C development → fertilisation → gamete production → gamete transfer
   D fertilisation → development → gamete transfer → gamete production

3. During pollination, pollen grains are transferred from:

   A stigma to anther
   B style to anther
   C anther to stigma
   D anther to style

4. Following fertilisation, which of the following changes take place?

   A The ovule becomes the fruit, the ovary becomes the seed.
   B The stigma becomes the fruit, the style becomes the seed.
   C The style becomes the fruit, the stigma becomes the seed.
   D The ovary becomes the fruit, the ovule becomes the seed.

5. The diagram shows five different fruits.

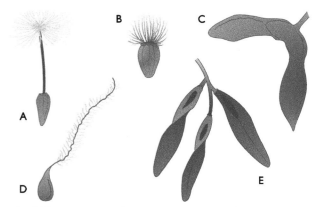

   a) i) How are these fruits dispersed from their parent plants? Give reasons for your answer. (3)

       ii) Describe *two* other ways in which fruits can be dispersed. Give an example for each method of dispersal. (4)

b) Explain the benefit of fruits being dispersed from their parent plant. *(3)*

c) Would the plants that grew from the fruits in the diagram be identical to the parent plants? Explain your answer. *(3)*

6. The drawing shows a wind-pollinated flower.

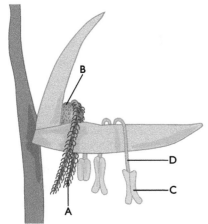

a) Name the structures labelled A, B, C and D. *(4)*

b) Give *three* pieces of evidence *visible in the diagram*, which show that this flower is wind-pollinated. *(3)*

c) Describe how fertilisation takes place once a flower has been pollinated. *(4)*

d) Describe *four* ways in which you would expect an insect-pollinated flower to be different from the flower shown. *(4)*

7. The diagram below shows a section through an insect-pollinated flower. Pollination happens when pollen grains land on part X.

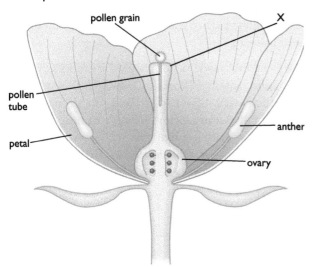

a) Name part X. *(1)*

b) Name two ways that insects are attracted to a flower like this. *(2)*

c) Copy the diagram and extend the pollen tube to show where it would go when fully grown. *(3)*

8. The drawing shows a strawberry plant reproducing in two ways.

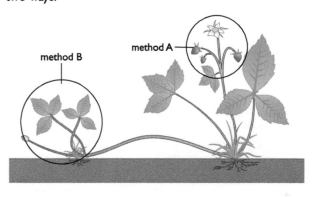

a) Which of the two methods of reproduction shown will result in offspring that show genetic variation? Explain your answer. *(2)*

b) Is the strawberry flower likely to be wind-pollinated or insect-pollinated? Give reasons for your answer. *(3)*

c) Many animals eat strawberry fruits. How does this help to disperse the strawberry seeds? *(2)*

Chapter 22: Growth

When you have completed this chapter, you will be able to:

- understand the different ways in which growth can be measured
- measure growth in a plant
- interpret patterns of growth in animals
- interpret patterns of growth in plants
- describe the structure of a dicotyledonous seed
- describe the processes that take place in a seed during germination.

In a multicellular organism, growth involves an increase in size, mass and number of cells. This chapter looks at the ways in which growth can be measured, and the different patterns of growth shown by animals and plants.

Measuring growth

There are various ways to measure the growth of an organism. When a baby is born, the midwife will weigh the baby and measure the length of its body. This is done in order to compare these measurements with charts, to see if the baby fits into the normal range of values. As we grow up, we still use measurements of weight (mass) and height as measurements of our growth.

Height and mass can be used to measure growth in a plant too, but there are other possibilities. For example, you could count the number of leaves, or record the total surface area of the leaves. These are important measurements in a plant, because the leaves carry out photosynthesis. Or, you may be interested in the growth of another part of the plant, such as the roots, and will want to measure these instead.

The mass of any organism is mainly made up of water, for example, a plant that is well watered might contain 90% water. However, this will vary considerably, depending on the availability of water, and in a drought the water content of the plant will fall. An organism's mass, including water, is called the **fresh mass** or **wet mass**. Often a biologist will want to know how much growth has taken place in terms of solid matter in particular – protein, and so on. This is especially important for measuring the growth of crops. For this, we can find the dry mass, which is the weight of matter after all the water has been removed by drying the plant in an oven.

Of course, drying the plant kills it, so you can only measure the dry mass of a particular plant once. To estimate the increase in dry mass of a crop, samples of plants are taken at intervals.

Activity 1: Measuring growth of a plant

SBA skills

ORR	MM	AI	PD	Dr

You will need:

- a healthy seedling
- a ruler
- graph paper

Carry out the following:

1. Construct a table like this to record your results.

Day	Height of plant (cm)	Number of leaves	Leaf area (cm²)
0			
1			
2			
3 (etc.)			

2. Measure the height of the seedling on the first day of measurement (day 0). See Figure 22.1.

3. Count the number of leaves on day 0.

4. Measure the total area of leaves on day 0. This is easiest to do by drawing round them on graph paper. Take care not to damage the leaves.

5. Take these measurements every day for a few weeks, until the plant is fully grown.

6. Plot line graphs of your results. Describe the shape of the graph curves.

Measure from top of soil to tip of plant.

Figure 22.1 *How to measure plant height.*

1. What is growth?
2. Explain the limitations of measuring plant growth as an increase in height.
3. What type of cell division is involved in growth?
4. Explain the difference between fresh mass and dry mass.

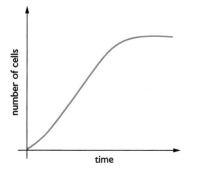

Figure 22.2 *How cell number increases with growth.*

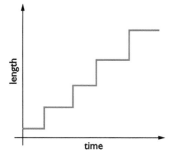

Figure 22.3 *Growth of a grasshopper measured by increase in length.*

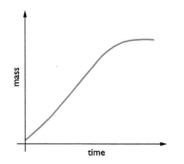

Figure 22.4 *Growth of a grasshopper measured by increase in mass.*

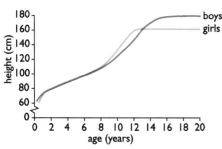

Figure 22.6 *The change in height of males and females from birth to adulthood.*

Growth in animals

All animals are multicellular – they are made of many cells. However, they all begin life as a single-celled zygote. Therefore to grow, they must make more cells. This is a continuous process for an animal. Until it reaches its full adult mass, the number of cells is constantly increasing. This is shown in Figure 22.2.

However, this constant increase in the number of cells is not always obvious. If you measure the length of a grasshopper from the time it hatches from an egg until it reaches full adult length, the increase is anything but constant. However, if we measure the increase in mass, we do get a constant increase. These different patterns are shown in Figures 22.3 and 22.4.

Having an exoskeleton explains the difference in the two patterns. As the grasshopper grows, its exoskeleton becomes too small and it must shed it. Underneath, it has already formed a new one, but this is still soft. Before it hardens, the grasshopper inflates its body to stretch the new exoskeleton. It then hardens in this inflated size, leaving room for the grasshopper to grow. Eventually, this will become too small and the process, called **ecdysis**, will be repeated.

Growth in humans, even when measured by height, shows a gradual increase in size until adulthood. However, the body does not grow at a constant rate and different parts of the body grow at different rates at different times. This is shown in Figures 22.5 and 22.6.

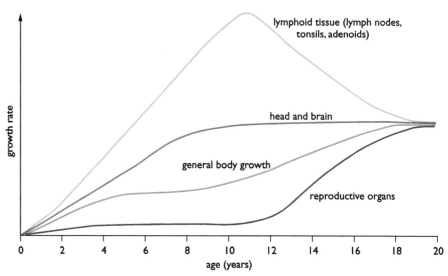

Figure 22.5 *Growth rates of different parts of the body.*

Boys and girls grow at more or less the same rate until they are in their teens. Then, girls begin an **adolescent growth spurt** before the boys and generally reach their adult body size at an earlier age. Boys, however, are usually bigger than girls when they do reach adult body size. An increased secretion of sex hormones stimulates the pituitary gland to secrete growth hormone, which causes the increased growth of the adolescent growth spurt. This increase lasts only a short time and from then onwards, growth rate slows down and, when we reach adulthood, growth stops. The changes in human growth rate are shown in Figure 22.7 on page 317.

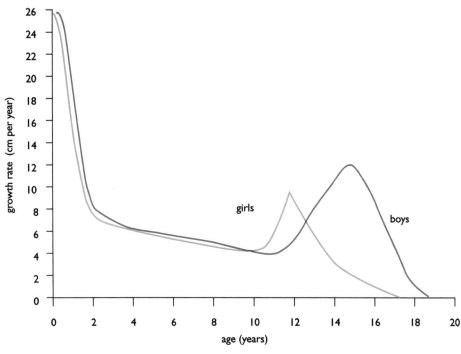

Figure 22.7 *The change in growth rate of males and females from birth to adulthood.*

Pituitary growth hormone causes cells to absorb amino acids more readily. Because of this, they are able to synthesise proteins more quickly from these amino acids. This leads to increased growth and division of cells, resulting in more cells and, therefore, growth.

Growth in plants

Like animals, all plants are multicellular, and growth involves an increase in the number of cells. In flowering plants, this growth occurs in two phases:

- An initial growth period following fertilisation, which results in the formation of an embryo plant and its food store within a seed; the seed may lie dormant for long periods (sometimes for many years) before growth starts again

- Germination of the seed, followed by growth of a new plant

Growth of the embryo in a seed

Broad-leaved species are called **dicotyledonous plants** (or dicots). This means that their embryos have two seed leaves, or **cotyledons**. (**Monocotyledons** are plants such as grasses and cereals, which have one seed leaf). The structure of a typical dicot seed is shown in Figure 22.8 on page 318.

The seed is surrounded by an outer seed coat or **testa**, which protects the seed and stops it from drying out. The **hilum** is a scar, showing the place where the seed was attached inside the fruit. The **micropyle** is a tiny hole in the seed coat, which allows water to enter the seed when it germinates (The micropyle is also the place where the pollen tube entered the ovule during fertilisation – see Chapter 21).

Inside the seed are the two cotyledons, which contain the food reserves for germination, such as starch. The embryo consists of an embryonic shoot called a **plumule**, and root called a **radicle**.

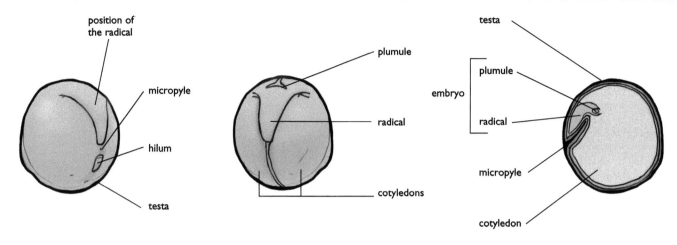

(a) The external appearance of a pea seed.   (b) The pea seed with the testa (seed coat) removed.   (c) A cross-section of the pea seed.

Figure 22.8 *Structure of a pea seed.*

Activity 2: Examining a seed of a dicotyledonous plant

SBA skills

ORR	MM	AI	PD	Dr

You will need:

- some dicotyledonous seeds, such as peas or beans, that have been soaked in water for 24 hours
- a hand lens
- a pair of forceps

Carry out the following:

1. Examine the outside of a seed. Using a hand lens, identify the testa, hilum, the position of the radicle underneath the testa, and the micropyle.

2. Make a drawing of the external features of the seed. Add a title and scale to your drawing.

3. Using a pair of forceps, carefully remove the testa. Identify the plumule and radicle of the embryo plant, and the two cotyledons.

4. Make another drawing of these features.

5. Separate the two cotyledons and examine the embryo plant in more detail. Make a third drawing to show the embryo attached to one of the two cotyledons.

Germination

For seeds to germinate, a number of conditions are needed, including:

- a supply of oxygen for the cells of the embryo to respire
- water to hydrate the tissues, enabling chemical reactions to proceed and food reserves to be moved around inside the seed
- a suitable temperature, that is, one that will activate the enzymes inside the seed.

SBA skills

ORR	MM	AI	PD	Dr

You will need:

- some bean sprout seeds
- four Petri dishes
- four pieces of filter paper
- a fridge (5°C), a freezer (–15°C) and two incubators (one at 20°C, one at 40°C)

Carry out the following:

1. Moisten each piece of filter paper and place one in each of the Petri dishes.

2. Count 20 bean sprout seeds into each Petri dish.

3. Place one Petri dish in the freezer, one in the fridge, and one in each of the two incubators.

4. Leave the Petri dishes for 48 hours.

5. Count the number of seeds that have germinated at each temperature and convert this to a percentage germination.

6. Write up your experiment and explain your results.

The cotyledons contain insoluble food reserves such as starch. When germination begins, enzymes digest these food reserves, forming soluble substances such as glucose. These products move into the embryo where they can be used for energy and growth. As this process continues, the mass of the cotyledons decreases, while the embryo grows in size and in number of cells.

The radicle grows, breaking through the testa and forming the root. Root hairs develop to absorb water and minerals. The plumule emerges next, growing upwards above the soil, and forming the first true leaves of the plant (Figure 22.9). In some species of plant (not in the pea), the seed leaves are pushed above the ground as the seed germinates.

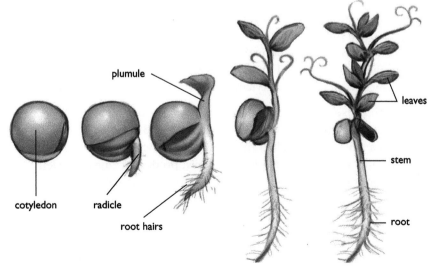

Figure 22.9 *Germination of a pea seed.*

SBA skills

ORR MM AI PD Dr

You will need:

- some dicotyledonous seeds, such as peas or beans, at different stages of germination

- a hand lens

Carry out the following:

1. Examine some seeds at different stages of germination, as shown in Figure 22.9, page 319.

2. Make labelled drawings of the different stages.

3. Annotate your drawings to explain the processes taking place.

4. Give the drawings a title and scale.

SBA skills

ORR MM AI PD Dr

You will need:

- some dicotyledonous seeds, such as peas or beans, at different stages of germination

- apparatus for carrying out tests for starch, reducing sugar, non-reducing sugar, protein and lipid (see Chapter 10).

Carry out the following:

1. Carry out food tests on cotyledons from seeds at different stages of germination.

2. Compare the food types present in the cotyledons of a seed before it has germinated to a seed (of the same species) after germination.

3. Explain your results.

Primary and secondary growth

Once the seed has germinated and grown into a young plant with a root and shoot, growth occurs mainly just behind the tips of the root and shoot. This growth occurs in tissues called **apical meristems**, in which cells divide by mitosis. Cell division is followed by the cells elongating and differentiating into new tissues, such as phloem and xylem. This is called **primary growth**.

Primary growth causes the shoot and root to elongate, but these structures also need to become thicker as they grow. Between the bundles of xylem and phloem is a ring of cells called the **cambium**, which is another meristem tissue. The cambium cells also divide by mitosis, producing rings of xylem vessels on the inside of the cambium, and rings of phloem cells on the outside. By this means, the stem and root increase in girth, which is known as **secondary growth** (Figure 22.10, page 321).

Another aspect of secondary growth takes place in woody plants (shrubs and trees). In these plants, there is another layer of cambium in the outer tissues of the stem and root, which divides to form the bark of the shrub or tree.

> Any region of a plant where cells are actively dividing by mitosis is called a meristem. The end of a root or a shoot is the apex.

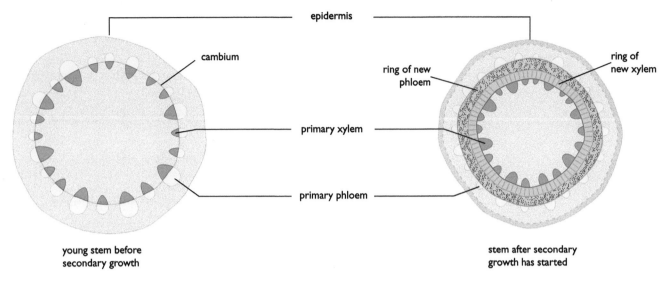

Figure 22.10 *Secondary growth in a stem.*

5. How is growth in animals and plants similar?

6. Why is the growth curve for change in length of a locust different from the growth curve for change in mass of the insect?

7. Explain what is meant by an adolescent growth spurt.

8. Name the parts of a plant embryo.

9. What happens to the contents of the cotyledons during germination?

10. What is:
- *(a)* an apical meristem
- *(b)* the cambium?

The extra rings of cells produced each year can be used to give an estimate of the age of a tree. Because a new ring of cells is produced each year, they are called annual rings. Figure 22.11 shows **annual rings**.

Figure 22.11 *Annual rings in the trunk of a larch tree.*

Chapter summary

In this chapter you have learnt that:

- growth is the increase in the amount of living tissue in an organism, accompanied by an increase in the number of cells

- growth can be estimated by measuring height or length, or wet or dry body mass, or by counting leaves or measuring leaf area; measurements are taken at intervals over a period of time

- growth of insects, as measured by an increase in length, appears to take place at certain times only; this is due to moulting of the skin

- in humans, different body tissues grow at different rates at different times

- growth in humans is most rapid at birth, it decreases until puberty, when it increases again, and then it then slows down until it ceases at adulthood

- seeds need a supply of oxygen and water, and a suitable temperature, in order to germinate

- as a seed germinates, food reserves in the cotyledons are digested and used by the embryo

- primary growth in plants takes place by mitosis at the apical meristems

- secondary growth in plants takes place by mitosis at the cambium.

Questions

1. Growth is best defined as:

   A   an increase in length
   B   an increase in size
   C   an increase in mass
   D   an increase in the amount of living material in an organism

2. In a multicellular organism, new cells produced during growth are produced by:

   A   meiosis
   B   mitosis
   C   fertilisation
   D   fission

3. Growing insects increase in length in a series of abrupt steps. This is because:

   A   they only grow at certain times of their life cycle
   B   they only get longer when the pressure of growth is enough to stretch the exoskeleton
   C   they shed their old exoskeleton periodically and stretch the new one
   D   the exoskeleton grows periodically

4. In order to germinate, seeds need:

   A   a supply of carbon dioxide, water and a suitable temperature
   B   a supply of carbon dioxide, water and a low temperature
   C   a supply of oxygen, water and a low temperature
   D   a supply of oxygen, water and a suitable temperature

5. Primary growth in plants takes place:

   A   throughout the stems and roots
   B   at the tips of the stems and roots
   C   in the centre of the stems and roots
   D   in the leaves

6. Some pupils measured the lengths of a group of caterpillars every 2 days from hatching for 30 days. They also measured the lengths of a group of tadpoles from hatching for 30 days. The table below shows their results.

   a)   Plot a graph of their results. Plot both sets of data on the same axes. *(4)*

   b)   Why is it good practice to use average values? *(2)*

   c)   Use your knowledge of growth in animals to explain the differences between the two lines. *(4)*

   d)   *i)*   Describe another way in which growth of these two animals could have been measured. *(2)*

   　　　ii)   Describe and explain any differences you would expect in the pattern of growth of these two animals over 30 days, measured by the method described in (i). *(3)*

Days from hatching	2	4	6	8	10	12	14	16	18	20	22	24	26	28	30
Average length of caterpillars (mm)	4	5	10	10	13	17	17	17	24	24	25	35	35	36	42
Average length of tadpoles (mm)	12	15	16	18	20	25	31	37	43	50	58	66	74	81	88

7. The graph shows the change in average height of males and females from birth to 20 years old.

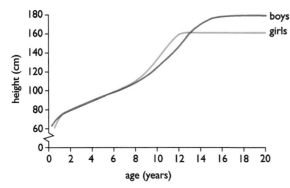

a) From the graph, what is the average height of:

   i) boys at age 8

   ii) boys at age 14

   iii) girls at age 8

   iv) girls at age 14? (4)

b) Use your answers from (a) to calculate:

   i) the average growth rate of boys between age 8 and age 14

   ii) the average growth rate of girls between age 8 and age 14. (4)

   Give your answers in cm per year.

c) What causes the increase in growth rate during adolescence? (3)

d) How else could growth in humans be measured? Give on advantage of this method over measuring height. (3)

8. The table gives information about the average change in height and growth rate of bean plants in the 50 days after sowing the bean seeds.

Days after sowing seed	Average height (cm)	Average growth rate (cm per day)
0	0	0.0
5	2	0.4
10	5	0.6
15	10	1.0
20	16	1.2
25	23	1.4
30	32	1.8
35	41	1.8
40	46	1.0
45	49	0.6
50	50	0.2

a) Plot graphs of change in height and growth rate over the 50-day period. (4)

b) Explain any differences between the two curves. (2)

c) In which regions of a plant does:

   i) primary growth take place

   ii) secondary growth take place? (2)

d) Explain what is meant by the term meristematic. (2)

9. The graph shows the changes in growth rate in humans from birth to adulthood.

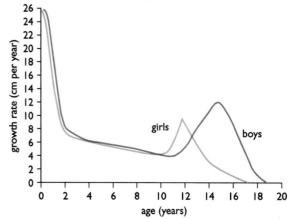

a) From the graph, when is the growth rate the greatest? (1)

b) Describe and explain the changes in growth that occur between the ages of 10 and 18 in boys and girls. (4)

c) Describe *two* ways in which growth in humans can be measured. Give an advantage and a disadvantage for each method you describe. (4)

d) Different regions of the body do not all grow at the same rate. Suggest reasons why:

   i) lymphoid tissue (which includes tissue responsible for our immune responses) grows quicker during childhood than many other tissues (2)

   ii) the head and brain are almost full size by the age of 10 (2)

   iii) reproductive organs do not develop fully until early (girls) or late (girls) teenage years. (2)

10. a) Draw a diagram to show the internal structure of a named seed. Label your diagram. (5)

b) Explain the function of:

   i) the testa

   ii) the micropyle

   iii) the cotyledons. (3)

c) Describe the changes that take place when a seed germinates. (5)

Chapter 23: Disease in Humans

When you have completed this chapter, you will be able to:

- distinguish between different types of disease
- understand how infectious diseases can be transmitted from one person to another
- discuss the role of the mosquito in the transmission of diseases
- identify the stages in the life cycle of a mosquito
- suggest appropriate methods of controlling mosquitoes
- discuss the treatment and control of different types of disease.

Some diseases are caused by an unhealthy lifestyle, for example, smoking-related diseases. Other diseases are genetic, caused by faulty genes. Many diseases are infectious – they are caused by micro-organisms. This chapter looks at different types of diseases and how they are treated and controlled.

Types of disease

It is difficult to define the term disease. One definition is 'a condition with a specific cause, in which part or all of the body functions abnormally and less efficiently'. The cause can be unhealthy activities, such as smoking or drinking alcohol; it can be genetic, such as the gene mutations responsible for certain types of cancer; or more commonly, the cause is a micro-organism.

Diseases can be classified into the four broad types described below:

- **Deficiency** diseases, which are caused by the lack of a dietary component – for example, a lack of vitamin D causes rickets, and a lack of food causes kwashiorkor. This type of disease is discussed in Chapter 12.

- **Hereditary (genetic)** diseases, which are caused by faulty genes – a change to a normal gene is called a gene **mutation**. We have known for many years that some diseases are hereditary, for example:

  – red–green colour blindness, which is more commonly found in males

  – albinism, a condition where the skin lacks the pigment melanin

- cystic fibrosis, a disease where the mutated gene causes the airways to the lungs to produces a thick, sticky mucus
- sickle-cell anaemia, where a person produces faulty haemoglobin in their red blood cells.

Over the last 20 years or so, scientists have come to realise that many other diseases have a genetic basis. For example, many cancers are the result of inherited gene mutations, such as the faulty genes BRCA1 and BRCA2, which cause one type of breast cancer. You will find out about mutations and hereditary diseases in Chapter 25.

- **Physiological** disease – a general term that refers to a problem with the physiology of the body, that is, how the organ systems work. For example, the disease diabetes results in the inability to control the level of glucose in the blood. As we shall see, there are different types of diabetes, and they have different causes, which can either be inherited or result from a poor diet and lifestyle. Similarly hypertension (where a person has high blood pressure) also has more than one factor contributing to a person developing the disease.

- **Pathogenic** diseases – are caused by a pathogen. There are some pathogens that are larger organisms, such as tapeworms and roundworms, but most pathogens are micro-organisms such as bacteria and viruses.

  Pathogenic diseases are **infectious** – they are transmitted by the pathogen being passed from one person to another. The other sorts of diseases are all non-infectious since they are not caused by a pathogen.

Deficiency diseases and hereditary diseases are covered elsewhere in this book. In this chapter, we will look at some physiological and pathological diseases.

> Hereditary diseases are also passed from one person to another, but in a different sense. They are passed from parents to their offspring through genes and not by a pathogen.

Diabetes

Diabetes mellitus (also known as **sugar diabetes** or just **diabetes)** is not one disease, but several. The common factor in these diseases is that the patient's blood contains a high concentration of blood glucose, either because their pancreas does not make enough insulin, or because cells of the target organs do not respond to the insulin produced. (See Chapter 18 for details of the normal role of insulin in controlling blood glucose levels.)

There are two main types of diabetes:

- **Type 1 diabetes** results from the pancreas being unable to make enough insulin. It most commonly begins in childhood, and onset happens suddenly. It is not associated with a poor diet or obesity. It can be treated only by injecting insulin.

- **Type 2 diabetes** results from receptor cells for insulin in the liver and muscle losing their ability to recognise the hormone. It may be accompanied by lower than normal production of insulin. Type 2 diabetes most commonly begins in adulthood, and onset happens gradually. It is associated with people who have a poor diet and are obese, and can usually be treated by improving their diet and taking medication.

> There is another disease called diabetes insipidus, which, despite having the same name, has nothing at all to do with sugar diabetes. Diabetes insipidus is a physiological disease caused by a lack of antidiuretic hormone (see Chapter 17). It results in a person losing large amounts of water in their urine.

Both types of diabetes produce similar symptoms. The person is unable to control their blood glucose levels, resulting in high concentrations of glucose in the blood. This reaches such a high level that glucose is excreted in the urine. This can be shown up by using coloured test strips (Figure 23.1).

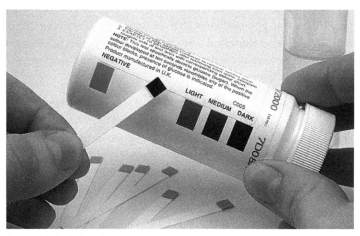

Figure 23.1 *Coloured test strips are used to detect glucose in urine.*

Another symptom of diabetes is a constant thirst. This is because the high blood glucose concentration stimulates receptors in the hypothalamus of the brain. Drinking more water dilutes the blood and switches off the 'thirst' receptors.

The causes and treatments of diabetes

Severe diabetes is very serious. If it is untreated, the sufferer loses weight and becomes weak. Eventually they lapse into a coma and die.

Type 2 diabetes can often be treated by controlling the patient's lifestyle and eating habits. Glucose in the blood is derived from carbohydrates such as starch, so the patient must eat less carbohydrate, and if they are obese, must diet to achieve a healthy body weight. The same conditions are effective in preventing a person getting type 2 diabetes in the first place – a healthy diet, exercising, and not being overweight will drastically reduce the chances of someone contracting the disease. However, although type 2 diabetes is mainly caused by a poor lifestyle and overeating, there is a genetic component – a person with type 2 diabetes in the family may inherit genes for the disease.

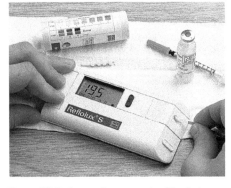

Figure 23.2 *Sensor for measuring blood glucose.*

With type 1 diabetes (or more severe cases of type 2), the only answer is for the patient to receive daily injections of insulin. People with diabetes can check their blood glucose level by using a special sensor. They prick their finger and place a drop of blood onto a test strip. The strip is then put into the sensor, which gives them an accurate reading of how much glucose is in their blood (Figure 23.2). They can then tell whether or not they need to inject insulin.

A more controlled way of 'delivering' the insulin is to use an insulin pump (Figure 23.3, page 327). This is a device that delivers insulin to the patient throughout the day and night, matching it to the body's needs. The pump is connected to a tiny tube called a cannula, which is inserted under the skin. The pump doesn't test the patient's blood glucose levels; instead the patient learns to adjust the pump throughout the day to vary their dose according to their diet, activity and blood glucose levels.

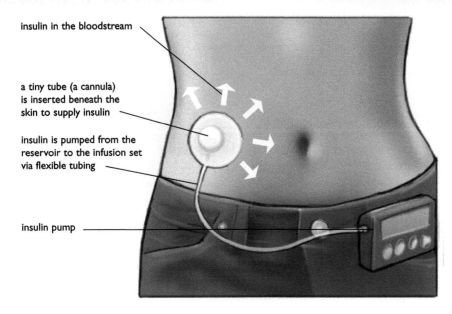

insulin in the bloodstream

a tiny tube (a cannula) is inserted beneath the skin to supply insulin

insulin is pumped from the reservoir to the infusion set via flexible tubing

insulin pump

Figure 23.3 *How an insulin pump delivers insulin to the body.*

Type 1 diabetes is also partly inherited, but the latest evidence is that the disease may have to be triggered by a viral infection. It is not brought on by poor diet or obesity. However it can cause many other complications, such as heart disease, kidney failure and damage to the retinas of the eyes. For this reason, it is important that treatment for diabetes doesn't just involve correcting the blood glucose levels. The patient also needs to keep their body in as healthy a condition as possible, by not smoking, maintaining a healthy body weight, exercising regularly and maintaining a normal blood pressure.

Hypertension

Having **hypertension** means having a high blood pressure. When a doctor takes your blood pressure, he or she records it as two numbers, for example, 120/80 (120 over 80). The units are old-fashioned ones used to measure pressure – millimetres of mercury. The first number is your systolic blood pressure (the highest pressure when the left ventricle contracts) and the second is the diastolic pressure (the lowest pressure when the left ventricle relaxes) – see Chapter 13.

A blood pressure of 120/80 is in the normal range for a healthy adult. If a person has a blood pressure that is consistently 140/90 or higher over a period of time, this is classed as hypertension. Hypertension normally shows no symptoms, so the only way to find out if a person has it is to measure their blood pressure. Many things affect a person's blood pressure during the day, so the doctor or nurse will need to take several readings to be sure. Having a high blood pressure is very dangerous. In particular, it increases the risk of a person having a **stroke**. A stroke is a disruption in the blood supply to the brain, either a blood clot or a bleed. Strokes starve the brain of oxygen and nutrients, and damage brain tissue; they are sometimes fatal.

We do not know exactly what causes hypertension, but there are several lifestyle factors that increase the risk of developing the disease. These are:

- eating too much salt

- not eating enough fruits and vegetables

- being overweight or obese

- drinking too much alcohol

- not getting enough exercise.

African-Caribbean people are at a particularly high risk of developing hypertension. We do not know exactly why this is, but it is most likely a combination of genetic causes along with the lifestyle factors given above. If you want to avoid developing high blood pressure, you will need to choose to follow a lifestyle that avoids these factors.

Pathogenic (infectious) diseases

There are four main groups of micro-organisms that cause disease (Table 23.1 and Figure 23.4).

Type of micro-organism	How the micro-organism causes disease	Examples of diseases caused
bacteria	Bacteria release toxins (poisons) as they multiply. The toxins affect cells in the region of the infection, and sometimes in other parts of the body as well.	typhoid, tuberculosis (TB), gonorrhoea, cholera, pneumonia
viruses	Viruses enter a living cell and disrupt the metabolic systems of that cell. The genetic material of the virus takes over the cell and instructs it to produce more virus.	influenza ('flu), poliomyelitis (polio), human immunodeficiency virus (HIV), measles, rubella, common cold
fungi	When fungi grow in or on the body, their fine threads (hyphae) secrete digestive enzymes onto the tissues, breaking them down. Growth of hyphae also physically damages body tissues. Some fungi secrete toxins. Others cause an allergic reaction.	thrush, athlete's foot, ringworm (a skin disease), 'farmer's lung'
protozoa	(there is no set pattern as to how protozoa cause disease)	malaria, trypanosomiasis (sleeping sickness)

Table 23.1: *The main types of pathogenic micro-organism*

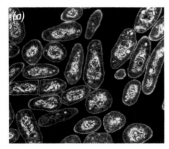 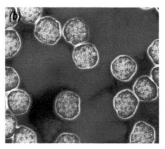 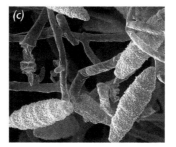 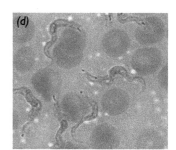

Figure 23.4 *Some pathogenic micro-organisms: (a) the bacterium that causes tuberculosis; (b) the measles virus; (c) the mould fungus responsible for ringworm; (d) the protozoan* Trypanosoma gambiense – *the cause of sleeping sickness.*

Transmission of diseases

Transmission of a disease is the means by which the pathogen is transferred from one host to another. Table 23.2 on page 329 summarises the main methods of transmission of infectious diseases.

Method of transmission	How the transmission route works	Examples of diseases
droplet infection	Many of these are respiratory diseases (diseases that affect the airways of the lungs). The organisms are carried in tiny droplets through the air when an infected person coughs or sneezes. They are inhaled by other people.	common cold, influenza, tuberculosis, pneumonia
drinking contaminated water	The micro-organisms transmitted in this way often infect regions of the gut. When a person drinks unclean water containing the organisms, they colonise a suitable area of the gut and reproduce. They are passed out with faeces and find their way back into the water.	cholera, typhoid fever, polio
eating contaminated food	Most food poisoning is bacterial, but some viruses are transmitted this way. The organisms initially infect a region of the gut.	typhoid fever, polio, salmonellosis, listeriosis, botulism
direct contact	Many skin infections, such as athlete's foot, are spread by direct contact with an infected person or contact with a surface carrying the organism.	athlete's foot, ringworm
sexual intercourse	Organisms infecting the sex organs can be passed from one sexual partner to another during intercourse. Some (such as the fungus thrush, which causes candidiasis) are transmitted by direct body contact. Others (such as the AIDS virus) are transmitted in semen or vaginal secretions. Some (such as syphilis) can be transmitted in saliva.	chlamydia, syphilis, AIDS, gonorrhoea
blood-to-blood contact	Many sexually transmitted diseases can also be transmitted in this way. Drug users sharing an infected needle can transmit HIV.	AIDS, hepatitis B
animal vectors	Many diseases are transmitted by insect bites. Mosquitoes spread malaria and tsetse flies spread sleeping sickness. In both cases, the disease-causing organism is transmitted when the insect bites humans in order to suck blood. Also, flies can carry micro-organisms from faeces onto food.	malaria, sleeping sickness, typhoid fever, salmonellosis

Table 23.2: *Methods of disease transmission.*

Insects such as the housefly are responsible for transmitting many diseases. Houseflies are attracted to animal or human faeces, and transmit many bacteria and viruses on their body or in their saliva. They do not bite humans, but will feed on human food if it is left uncovered (Figure 23.5). As the fly feeds, it releases its saliva onto the food. The pathogenic micro-organisms may then be transferred to humans when they eat the food. Many serious diseases, such as diphtheria, meningitis, typhoid, cholera and polio, are transmitted in this way.

The general course of a disease

After a person has been infected with a disease, there is an **incubation period**. This is the time between when a person is first infected with the pathogen and when they first show signs and symptoms of the disease. During the incubation period, the infected person may not feel sick, but could be infectious to others. Incubation periods vary greatly between diseases, from hours to months. Some diseases like leprosy can have an incubation period lasting years.

'Signs' and 'symptoms' of a disease have slightly different meanings.

- A **sign** of a disease is visible to other people. It can be seen, heard or measured. For example, a doctor might listen to a patient's chest with a stethoscope to hear signs of a chest infection, or measure their blood pressure to check for a heart problem.

Figure 23.5 *A fly feeding on human food. The fly releases saliva onto the food as it feeds. The saliva may contain disease-causing organisms.*

- A **symptom** is not usually visible to other people. It is what the patient is experiencing as a result of the disease, such as pain, chills, dizziness or nausea. The symptoms are the first thing that a patient notices that makes them go to the doctor.

Distribution of diseases

Some diseases are only found in certain parts of the world. For example, malaria is common in tropical and subtropical countries (Figure 23.6). If a disease is always present in a population of a particular geographical area, it is **endemic**.

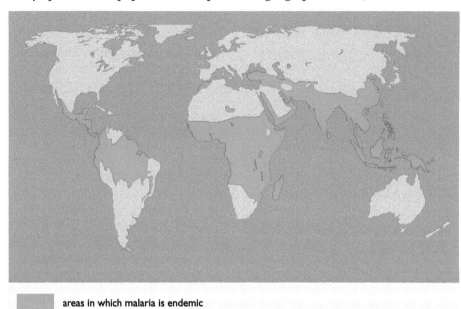

areas in which malaria is endemic

Figure 23.6 *Regions of the world where malaria is endemic.*

This is not to be confused with an **epidemic** – which is a widespread outbreak of an infectious disease, with many people becoming infected at the same time, spreading over a wide area. For a disease to be classified as an epidemic, there must be an increase in the area it affects, not just the numbers infected. If the disease spreads across the world, we call it a **pandemic**. There have been 'flu pandemics in 1918 (Spanish 'flu), 1957 (Asian 'flu), 1968 (Hong Kong 'flu) and 2009 (swine 'flu).

The expansion in worldwide travel over the past 30 years has exposed many people to diseases that are not normally endemic in their own country. Potentially, this can increase the spread of diseases from an area where they are currently endemic to other new areas.

Mosquito-borne diseases

Mosquitoes are the most dangerous of all blood-sucking insects. In tropical and sub-tropical regions of the Earth, different species of mosquitoes transmit a wide range of diseases, including malaria, yellow fever, dengue fever – a type of elephantiasis – encephalitis, and many more viral diseases, many not yet named. It is estimated that every year mosquitoes transmit disease to more than 700 million people.

It has been estimated that over 50 million people died from the Spanish 'flu pandemic in 1918–19 – more than were killed in the First World War of 1914–18.

Yellow fever

Yellow fever is a viral disease, transmitted by a species of mosquito called *Aedes aegypti*. It is found in Africa, South America and parts of the Caribbean. For the first few days after infection, yellow fever just causes fever, chills, loss of weight, nausea, muscle pain and headache. Most people recover after this stage, but in some cases, toxins (poisons) are produced. This toxic phase results in liver damage, jaundice, and internal bleeding (haemorrhaging) in the mouth, eyes and intestine. About 20% of people who enter the toxic phase eventually die. Every year there are about 300 000 infections with yellow fever, and 30 000 deaths from the disease.

However there is an effective vaccine against yellow fever, and the disease can be combatted by killing the mosquito vector (as with malaria – see below).

Dengue fever

Dengue fever is another infectious disease. It is caused by the dengue virus, which is also transmitted by the *Aedes* mosquito. Most people that are infected show no symptoms, or only mild ones such as a fever. However, in a small proportion of cases, the disease develops into dengue haemorrhagic fever, resulting in internal bleeding, a drop in blood pressure and sometimes death. Dengue is a major problem around the world, and is endemic in 110 countries in South and Central America, the Caribbean, Africa and tropical Asia. It is estimated to infect tens of millions of people every year, and to cause about 25 000 deaths. As yet, there is no vaccine against dengue fever. All that can be done is to manage the symptoms and hope that the person recovers.

Malaria

Malaria is the most widely known mosquito-borne disease. It also affects the most people. The World health Organization estimated that in 2010 there were 219 million cases of malaria, and between 660 000 and 1.2 million people died of the disease.

Malaria is caused by five different species of a protozoan pathogen called *Plasmodium* (Figure 23.7), which is passed to humans via an insect host – the mosquito.

The female *Anopheles* mosquito transmits *Plasmodium* in her saliva when she sucks blood from humans (Figure 23.8).

Figure 23.9 on page 332 shows the life cycle of the *Anopheles* mosquito and Figure 23.10 shows the life cycle of the malarial parasite.

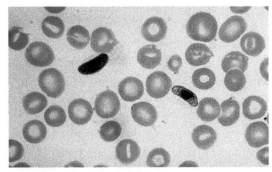

Figure 23.7 Plasmodium *parasites in human blood.*

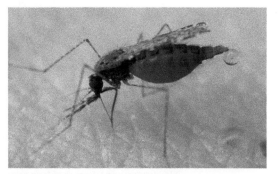

Figure 23.8 *A female* Anopheles *mosquito sucking human blood.*

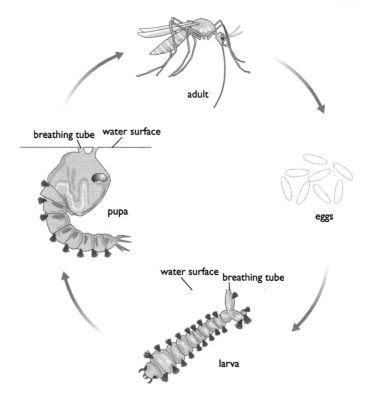

breathing tube   water surface

pupa

adult

eggs

water surface   breathing tube

larva

Figure 23.9 *Life cycle of the* Anopheles *mosquito.*

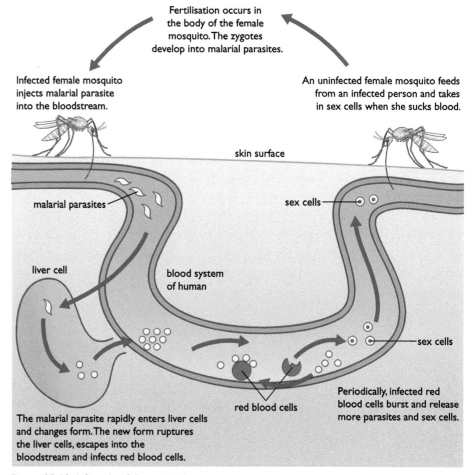

Fertilisation occurs in the body of the female mosquito. The zygotes develop into malarial parasites.

Infected female mosquito injects malarial parasite into the bloodstream.

An uninfected female mosquito feeds from an infected person and takes in sex cells when she sucks blood.

skin surface

malarial parasites

sex cells

liver cell

blood system of human

sex cells

red blood cells

Periodically, infected red blood cells burst and release more parasites and sex cells.

The malarial parasite rapidly enters liver cells and changes form. The new form ruptures the liver cells, escapes into the bloodstream and infects red blood cells.

Figure 23.10 *Life cycle of the malarial parasite.*

Malaria is widespread in tropical and sub-tropical parts of the world, and is estimated to cause the death of several million people every year, as well as making hundreds of millions so ill that they cannot work – contributing to poverty in the developing countries where it is endemic.

The malaria parasite spends anywhere between 2 weeks and several months in the person's liver before the next stage in the life cycle infects red blood cells. The well-known symptoms of malaria then appear – alternating cold sweats and fever, vomiting, joint pains and anaemia. Severe malaria causes coma and death, especially in young children.

If the life cycle of either the malarial parasite or the mosquito host can be broken at any point, then the transmission of malaria can be controlled. Controlling numbers of the mosquito means there will be fewer insects to transmit the protozoan pathogen. Controlling numbers of the *Plasmodium* parasite means there will be fewer opportunities for mosquitoes to take in the parasite and transmit it to other humans.

Control measures that have been tried include:

- the use of insecticides to kill the adult mosquitoes

- draining swamps that form the natural habitat of the larvae of the mosquitoes

- the use of drugs to target the various stages of the protozoan's life cycle

- stocking ponds with a fish called *Tilapia*, which feeds on the larvae of mosquitoes

- the use of insect repellents, wearing long-sleeved shirts and sleeping under mosquito nets to prevent bites from the adult mosquitoes.

Research is currently underway to develop a vaccine against *Plasmodium*.

> Malaria is a difficult disease for our immune system to attack, because the protozoan parasite spends much of its time 'hidden' inside red blood cells, and changes form several times inside the body.

Activity 1: Researching data on mosquito-borne diseases

SBA skills

| ORR | MM | AI | PD | Dr |

You will need:

- access to the Internet

Carry out the following:

1. Carry out an Internet search to find some data on the incidence of malaria, dengue fever and yellow fever in your country and throughout the Caribbean.

2. Summarise your findings in writing, illustrating the account using maps, tables and charts.

SBA skills

ORR	MM	AI	PD	Dr

You will need:

- specimens of adult mosquitoes, and mosquito eggs, larvae and pupae
- a hand lens

Carry out the following:

1. Observe specimens of the different stages in the life cycle of a mosquito. Use a hand lens. Compare the specimens with the diagrams in Figure 23.9.

2. Make labelled drawings of the stages. Add a title and magnification to your drawing.

The economic and social implications of disease

Animals other than humans suffer from disease. Pathogenic bacteria and viruses can enter their bodies in the same way that they enter ours. Mammals such as farm animals contract diseases very similar to those of humans, such as pneumonia, influenza and hepatitis; they also get diseases that only affect particular species of mammal.

Diseases of farm animals such as cattle, sheep, pigs and poultry result in lowered productivity and loss of income for the farmer if the animal dies.

Plants too are susceptible to diseases. There are bacterial, viral and fungal infections of plants (Figure 23.11), and these will also result in lowered productivity and reduced profit for the farmer.

Figure 23.11 *A bean plant infected with a fungal disease called 'rust'.*

Diseases don't just cause suffering and damage to the person with the disease, or to their animals and plants. Diseases cause a loss of production and a loss of income while the person recovers from the disease.

1. What is meant by the term disease?
2. List the four main types of disease and give an example of each type.
3. List four ways in which infectious diseases can be transmitted, and give an example of each method of transmission.
4. Name two insects that act as vectors for disease. In each case, explain how the insect is responsible for the spread of the disease.
5. Describe three methods that can be used to control the spread of malaria. Explain how each method works.

Chapter summary

In this chapter you have learnt that:

- diseases can be classified into four groups:
  - deficiency disease – caused by a lack of a nutrient or nutrients
  - hereditary disease – caused by a genetic mutation
  - physiological disease – caused by a problem with an organ system
  - pathogenic disease – caused by an infection with a pathogen

- there are two types of diabetes – type 1 and type 2

- type 1 diabetes results when a person cannot make enough insulin; it is not associated with a poor diet or obesity, and can be treated only by injecting insulin

- type 2 diabetes results when receptor organs don't respond to insulin; it is associated with a poor diet and obesity, and may be treated by a change of lifestyle

- hypertension means having a high blood pressure; there are several factors that increase the likelihood of a person having hypertension

- there are four main types of pathogenic micro-organism – bacteria, viruses, protozoa and fungi

- infectious diseases can be transmitted by droplet infection, drinking contaminated water, eating contaminated food, direct contact, sexual intercourse, blood-to-blood contact and animal vectors

- yellow fever, dengue fever and malaria are all transmitted by mosquitoes

- knowledge of the life cycle of the mosquito is needed to suggest appropriate methods of control

- diseases in humans, as well as in their animals and crop plants, can result in lowered productivity and loss of income.

Questions

1. A pathogen is an organism that:

   A causes disease
   B is a bacterium
   C damages the liver
   D can transmit a disease

2. Diseases that result from incorrect working of body organs are called:
   A genetic
   B deficiency
   C physiological
   D infectious

3. Diabetes can be caused by:
   A overproduction of insulin
   B lack of response to insulin in the liver
   C lack of response to insulin in the pancreas
   D injection of insulin

4. Hypertension can be improved by:
   A taking salt tablets
   B getting regular exercise
   C drinking alcohol
   D eating fewer vegetables

5. Malaria is caused by:
   A mosquitoes
   B bacteria
   C a virus
   D a protozoan

6. The table below shows four types of disease, with examples. Copy and complete the table, giving three examples of each type of disease. *(8)*

Type of disease	Examples
1. deficiency	a) rickets
	b)
	c)
2.	a) cystic fibrosis
	b) red–green colour blindness
	c)
3. physiological	a) type 1 diabetes
	b) type 2 diabetes
	c)
4.	a) malaria
	b)
	c)

7. Copy and complete the following account about diabetes.

   Diabetes is a disease where there is an unusually high concentration of glucose in the patient's blood. One symptom of diabetes is a constant _____, another is _____ in the urine. The cause of type 1 diabetes is that the _____ does not make enough insulin. Insulin is a hormone that acts at the _____ and muscles, reducing the concentration of glucose in the blood. In type 2 diabetes, cells of the target organs do not respond to insulin. This can be treated by changing the patient's _____. *(5)*

8. *a)* Name three diseases that are transmitted by drinking contaminated water or eating contaminated food. *(3)*

   b) Explain the meaning of the term droplet infection. *(2)*

   c) Explain the meaning of the following terms:

   i) endemic

   ii) epidemic

   iii) pandemic *(3)*

9. *a)* The mosquito is the vector for malaria. What does the term vector mean? *(1)*

   b) Name two other diseases besides malaria that are spread by mosquitoes. *(2)*

   c) Describe the life cycle of the malaria mosquito. *(4)*

   d) Name three control methods that can be used against malaria mosquitoes and explain how each method works. *(6)*

Chapter 24: Cell Division

When you have completed this chapter, you will be able to:

- understand mitosis and its role in organisms
- understand why asexual reproduction produces genetically identical offspring and explain the value of this in a stable environment
- describe the process of meiosis and appreciate its importance in the human life cycle
- describe the differences between mitosis and meiosis.

The chemical that is the basis of inheritance in nearly all organisms is **DNA**. DNA is usually found in the nucleus of a cell, in the **chromosomes** (Figure 24.1). A small section of DNA that determines a particular feature is called a gene. Genes determine features by instructing cells to produce particular proteins, which then lead to the development of the feature. So a gene can also be described as a section of DNA that codes for a particular protein.

DNA can replicate (make an exact copy of) itself. When a cell divides by mitosis, each new cell receives exactly the same type and amount of DNA. The cells formed are **genetically identical**.

DNA is short for deoxyribonucleic acid. It gets its 'deoxyribo' name from the sugar in the DNA molecule. This is deoxyribose – a sugar containing five carbon atoms.

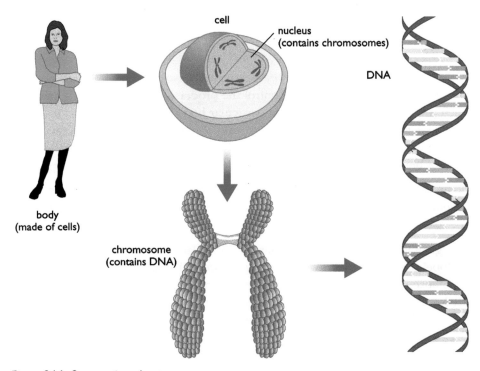

cell

nucleus (contains chromosomes)

DNA

body (made of cells)

chromosome (contains DNA)

Figure 24.1 *Our genetic make-up.*

The structure of chromosomes

Each chromosome contains one double-stranded DNA molecule. The DNA is folded and coiled so that it can be packed into a small space. The DNA is coiled around proteins called **histones** (Figure 24.2).

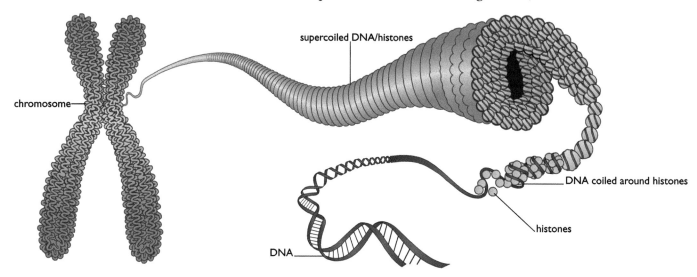

Figure 24.2 *The structure of a chromosome.*

Because a chromosome contains a particular DNA molecule, it will also contain the genes that make up that DNA molecule. Another chromosome will contain a different DNA molecule, and so will contain different genes.

Human body cells have 46 chromosomes in 23 pairs called **homologous pairs**. Chromosomes in an homologous pair carry genes for the same features in the same sequence (Figure 24.3). They do not necessarily have the same *alleles* of every gene. These body cells are diploid cells – they have *two* copies of each chromosome. The sex cells, with 23 chromosomes, (only one copy of each chromosome) are haploid cells.

Genes and alleles

Genes are sections of DNA that control the production of proteins in a cell. Each protein contributes towards a particular body feature. Sometimes the feature is visible, such as eye colour or skin pigmentation. Sometimes the feature is not visible, such as the type of haemoglobin in red blood cells or the type of blood group antigen on the red blood cells.

Some genes have more than one form. For example, the genes controlling several facial features have alternate forms, which result in alternate forms of the feature (Figure 24.4, page 339).

The gene for earlobe attachment has the forms 'attached earlobe' and 'free earlobe'. These different forms of the gene are called **alleles**. Homologous chromosomes carry genes for the same features in the same sequence, but the alleles of the genes may not be the same (Figure 24.5, page 339). The DNA in the two chromosomes is not quite identical.

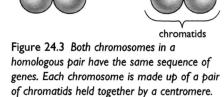

Figure 24.3 *Both chromosomes in a homologous pair have the same sequence of genes. Each chromosome is made up of a pair of chromatids held together by a centromere.*

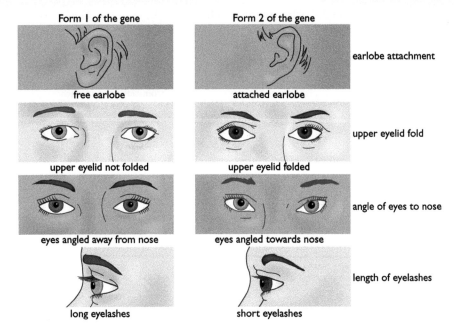

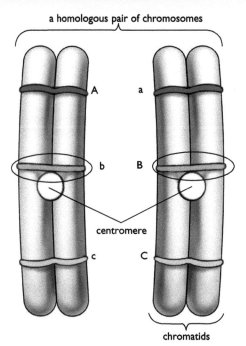

Figure 24.4 *The alternate forms of four facial features.*

Cell division

In most parts of the body, cells need to divide so that organisms can grow and replace worn out or damaged cells. The cells that are produced in this type of cell division should be exactly the same as the cells they are replacing. This is the most common form of cell division.

Only in the sex organs is cell division different. Here, some cells divide to produce gametes (sex cells), which contain only half the original number of chromosomes.

There are two kind of cell division: **mitosis** and **meiosis**. When cells divide by mitosis, two cells are formed. These have the same number and type of chromosomes as the original cell. Mitosis forms all the cells in our bodies except the sex cells.

When cells divide by meiosis, four cells are formed. These have only half the number of chromosomes of the original cell. Meiosis forms sex cells.

Mitosis

When a **parent cell** divides, it produces **daughter cells**. Mitosis produces two daughter cells that are genetically identical to the parent cell – both daughter cells have the same number and type of chromosomes as the parent cell. To achieve this, the dividing cell must do two things:

- It must copy each chromosome before it divides. Each daughter cell will then be able to receive a copy of each chromosome when the cell divides.

- It must divide in such a way that each daughter cell receives one copy of every chromosome. If it does not do this, both daughter cells will not contain all the genes.

These two processes are shown in Figure 24.6.

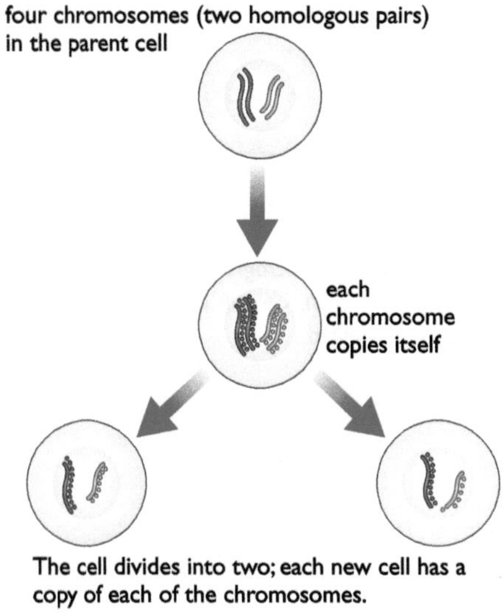

Figure 24.5 *A and a, B and b, C and c are different alleles of the same gene. They control the same feature but code for different expressions of that feature.*

> Growth and reproduction are two characteristics of living things. Both involve cell division, which is the subject of this chapter.

> Meiosis is sometimes called **reduction division**. This is because it produces cells with only half the number of chromosomes of the original cell.

four chromosomes (two homologous pairs) in the parent cell

each chromosome copies itself

The cell divides into two; each new cell has a copy of each of the chromosomes.

Figure 24.6 *A summary of mitosis.*

A number of distinct stages occur when a cell divides by mitosis. These are shown in Figure 24.7. Figure 24.8 is a photomicrograph of some cells from the root tip of an onion dividing by mitosis.

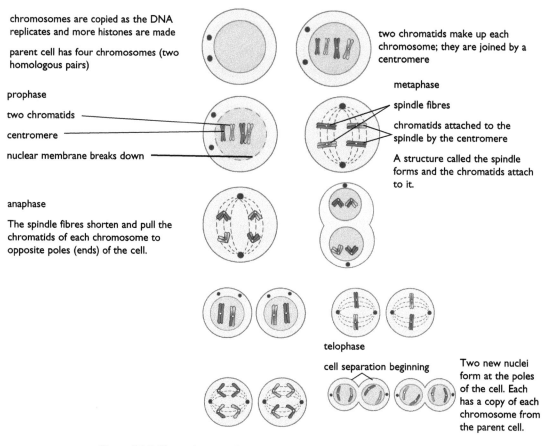

chromosomes are copied as the DNA replicates and more histones are made

parent cell has four chromosomes (two homologous pairs)

prophase

two chromatids

centromere

nuclear membrane breaks down

anaphase

The spindle fibres shorten and pull the chromatids of each chromosome to opposite poles (ends) of the cell.

two chromatids make up each chromosome; they are joined by a centromere

metaphase

spindle fibres

chromatids attached to the spindle by the centromere

A structure called the spindle forms and the chromatids attach to it.

telophase

cell separation beginning

Two new nuclei form at the poles of the cell. Each has a copy of each chromosome from the parent cell.

Figure 24.7 *The main stages in mitosis.*

Each daughter cell formed by mitosis receives a copy of every chromosome, and therefore every gene, in the parent cell. Each daughter cell is genetically identical to the others. All the cells in our body (except the sex cells) are formed by mitosis from the zygote (single cell formed at fertilisation). They all, therefore, contain copies of all the chromosomes and genes of that zygote. They are all genetically identical.

The role of mitosis in the body

Mitosis plays an important role in the normal functioning of the body. This division process enables the following processes to take place –

Growth

To increase the number of cells in an organism, existing cells must divide. The two new daughter cells that are formed have the same number of chromosomes in their nuclei as the parent nucleus; therefore, they are genetically identical. This facilitates the growth of multicellular organisms.

Repair and replacement of cells

Whenever cells need to be replaced in our bodies, cells divide by mitosis to make them. This happens more frequently in some regions than in others.

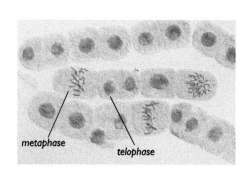

metaphase

telophase

Figure 24.8 *Photomicrograph of cells in the root tip of an onion. Cells in the main stages of mitosis are labelled.*

- The skin loses thousands of cells every time we touch something. This adds up to millions every day that need replacing. A layer of cells beneath the surface is constantly dividing to produce replacements.

- Cells are scraped off the lining of the gut as food passes along it. A layer of cells beneath the gut lining constantly divides to produce replacement cells.

- Cells in our spleen destroy worn out red blood cells at the rate of 100 000 000 000 per day! These are replaced by cells in the bone marrow dividing by mitosis. In addition, the bone marrow forms all our new white blood cells and platelets (Figure 24.9).

- Cancer cells divide by mitosis. The cells formed are exact copies of the parent cell, including the mutation in the genes that makes the cells divide uncontrollably.

Asexual reproduction

When organisms reproduce asexually, there is no fusion of gametes. A part of the organism grows and somehow breaks away from the parent organism. The cells the part contains were formed by mitosis, so they contain exactly the same genes as the parent. Asexual reproduction produces offspring that are genetically identical to the parent, and genetically identical to one another.

Asexual reproduction is common in plants. For example, the leaf of life and *Bryophyllum crenatodaigremontianum* produce young plants on their leaves, which eventually fall off.

Even though mitosis is the basis for asexual reproduction, it can take many forms namely, binary fission, budding, fragmentation and vegetative propagation.

- **Binary fission** is common in unicellular organisms, for example, Amoeba. First the nucleus of the organism divides by mitosis, and then the cytoplasm separates in two. This creates two new daughter cells (Figure 24.11).

- **Budding** occurs when cells in the wall of the parent divide to form a small outgrowth called a bud. The bud may break off the parent and develop on its own or remain on the parent (Figure 24.12).

- **Fragmentation** occurs when a piece of the body of an organism breaks off and undergoes mitosis until a completely new organism is generated (Figure 24.13). Echinoderms, such as starfish, reproduce in such a manner.

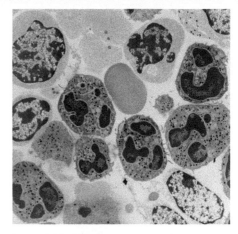

Figure 24.9 *Cells in bone marrow dividing to produce blood cells.*

Figure 24.10 *The leaf of life plant: plantlets grow in the notches of the leaves.*

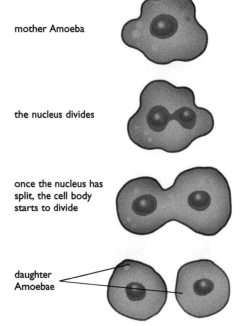

mother Amoeba

the nucleus divides

once the nucleus has split, the cell body starts to divide

daughter Amoebae

Figure 24.11 *Binary fission in an* Amoeba.

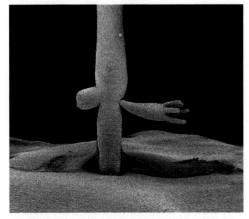

Figure 24.12 *A Hydra reproducing asexually by budding.*

Figure 24.13 *Starfish reproducing asexually by fragmentation.*

Vegetative propagation

There are many different methods of **vegetative propagation** in plants. Most methods involve some part of the plant growing and then breaking away from the parent plant, before growing into a new plant (Figures 24.14–24.16). The vegetative propagation methods of runners, buds and bulbs were discussed in Chapter 16.

Figure 24.14 *Ginger: the rhizome is a modified underground stem.*

Figure 24.15 *Corms are swollen underground stems.*

Figure 24.16 *Banana plant: a sucker grows from the base of an old plant to form a new plant.*

Cuttings

A portion of the parent plant is cut off. In many cases, this will be a shoot with a few leaves. The cut end is treated with a substance to encourage the formation of roots, after which it is inserted into soil to grow into a new plant (Figure 24.17).

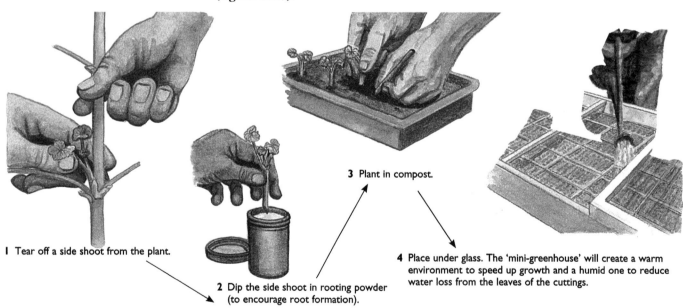

1 Tear off a side shoot from the plant.

2 Dip the side shoot in rooting powder (to encourage root formation).

3 Plant in compost.

4 Place under glass. The 'mini-greenhouse' will create a warm environment to speed up growth and a humid one to reduce water loss from the leaves of the cuttings.

Figure 24.17 *Taking stem cuttings.*

Grafting

Two plants are combined to form one new plant. An angled cut is first made into the trunk of a rooted plant (called the stock). A cutting (the scion) is then inserted into the cut on the stock and held in position with grafting tape (Figure 24.18, page 343).

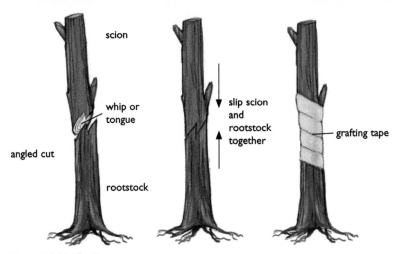

Figure 24.18 *Grafting.*

Tissue culture

Tissue culture is a process that produces several new plants, which are all clones. It involves taking tissue from plants (that may generate new individuals) and providing them with relevant nutrients and hormones to promote and support growth. Table 24.1 describes the main stages in tissue culture or micropropagtion.

Stages	Illustrations	
The tips of the stems and side shoots are removed from the plant to be cloned. These parts are called **explants**. The explants are trimmed to a size of about 0.5–1 mm. They are then placed in an agar medium that contains nutrients and plant hormones to encourage growth. More explants can be taken from the new shoots that form on the original ones. This can be repeated until there are enough to supply the demand.		Figure 24.19 *Explants growing in a culture medium.*
The explants with shoots are transferred to another culture medium containing a different balance of plant hormones to induce root formation.		Figure 24.20 *Explants forming roots.*
When the explants have grown roots, they are transferred to greenhouses and transplanted into compost. They are then gradually acclimatised to normal growing conditions. The atmosphere in the greenhouse is kept very moist to reduce water loss from the young plants. Because of the amount of water vapour in the air, they are often called 'fogging greenhouses'.		Figure 24.21 *Young plants being grown in compost in a greenhouse.*

Table 24.1: *The main stages of tissue culture.*

There are many advantages to propagating plants in this way.:

- Large numbers of genetically identical plants can be produced rapidly.

- Species that are difficult to grow from seed or from cuttings can be propagated by this method.

- Plants can be produced at any time of the year.

- Large numbers of plants can be stored easily (many can be kept in cold storage at the early stages of production and then developed as required).

- Genetic modifications can be introduced into thousands of plants quickly, after modifying only a few plants.

Cloning animals

We have been able to clone plants by taking cuttings for thousands of years. It is now possible to make genetically identical copies of animals. The first, and best-known example of this is the famous cloned sheep, Dolly.

Dolly was produced by persuading one of her mother's ova (egg cells) to develop into a new individual without being fertilised by a sperm. The nucleus of the ovum was removed and 'replaced' with a cell taken from the udder of another sheep. The cell that was formed had the same genetic information as all the cells in the donor and so developed into an exact genetic copy. The stages in the procedure are shown in Figure 24.22. Figure 24.23 shows how an udder cell is inserted into an egg cell that has had its nucleus removed.

The nucleus of an ovum is haploid. It cannot develop into a new individual because it only has half the chromosomes of normal body cells. An ordinary diploid body cell, even though it has all the chromosomes, is too specialised. Transferring a diploid nucleus into an egg cell that has had its nucleus removed creates a cell that is capable of developing into a new individual. In practice, it is easier to transfer a small whole cell rather than attempt to transfer just the nucleus, as the nucleus alone could too easily be damaged.

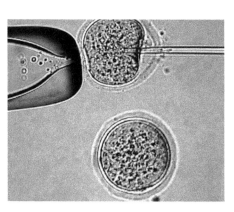

Figure 24.23 *Inserting an udder cell into an egg cell that has had its nucleus removed.*

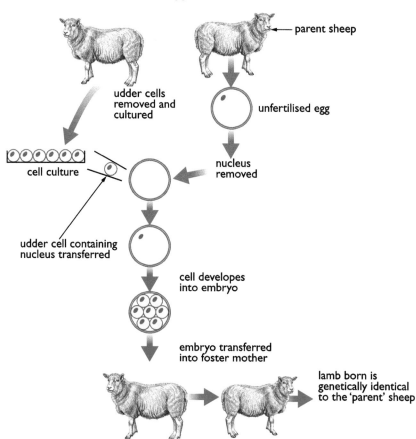

Figure 24.22 *How 'Dolly' was produced.*

Dolly was produced only after many unsuccessful attempts. Since then, the procedure has been repeated using other sheep as well as rats, mice and pigs. Some of the animals produced are born deformed. Some do not survive to birth. Biologists believe that these problems occur because the genes that are transferred to the egg are 'old genes'. These genes came from an animal that had already lived for several years and from cells specialised to do things other than produce sex cells. It will take much more research to make the technique reliable.

Asexual reproduction produces identical offspring

All the offspring produced when *Hydra* buds are genetically identical – they have exactly the same genes. This is because all the cells of the new individual are produced by mitosis from just one cell in the body of the adult. When cells divide by mitosis, the new cells that are produced are exact copies of the original cell (see Chapter 22. As a result, all the cells of an organism that is produced asexually have the same genes as the cell that produced them – the original adult cell. So *all* asexually produced offspring from one adult will have the same genes as the cells of the adult. They will *all* be genetic copies of that adult and so will be identical to each other.

Asexual reproduction is useful to a species when the environment in which it lives is relatively stable. If an organism is well adapted to this stable environment, asexual reproduction will produce offspring that are also well adapted. However, if the environment changes significantly, then *all* the individuals will be affected equally by the change. It may be such a dramatic change that none of the individuals are adapted well enough to survive. The species will die out in that area.

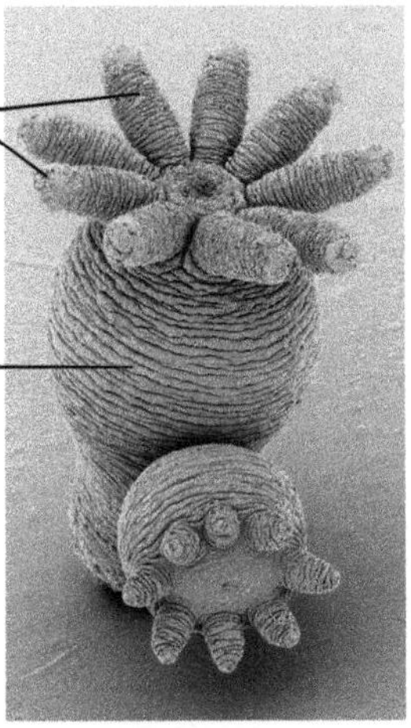

Meiosis – production of sex cells

Sperm are produced in the male sex organs – the testes. Ova are produced in the female sex organs – the ovaries. Both are produced when cells inside these organs divide. These cells do not divide by mitosis but by meiosis. Meiosis produces cells that are not genetically identical and have only half the number of chromosomes as the original cell. Figure 24.25 shows how a cell with just four chromosomes divides by meiosis. These four chromosomes are in two pairs called homologous pairs. Homologous pairs of chromosomes carry the same genes in the same sequence.

In meiosis, the cell divides twice, rather than just once as in mitosis. Also, because each of the sex cells formed only receives one chromosome from each original pair, they only have *half* the original number of chromosomes. They are haploid cells. This is important because it ensures that when the male and female gametes fuse together (fertilisation), the resulting cell (zygote) will have the full complement of chromosomes and can then divide and develop into a new individual.

The gametes formed by meiosis don't all have the same combinations of alleles – there is **genetic variation** in the cells. During the two cell divisions of meiosis, the chromosomes are divided between the two daughter cells independently of each other. Three things occur:

- After the homologous pairs are formed, chromatids cross over, break and rejoin onto the adjoining homologous chromatid. This exchanges genetic material between the chromatids, creating chromosomes with different genes.

- During the first division of meiosis, one chromosome from each homologous pair goes into each daughter cell.

A gene is a section of DNA that determines a particular characteristic or feature. Genes are found in the nucleus of a cell on the chromosomes.

Individuals produced asexually from the same adult organism are called clones.

buds – young *Hydra* growing from the parent – they will eventually break off and become independent

parent *Hydra*

Figure 24.24 A Hydra *reproducing asexually by budding.*

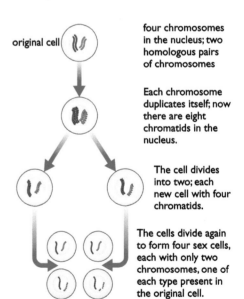

original cell

four chromosomes in the nucleus; two homologous pairs of chromosomes

Each chromosome duplicates itself; now there are eight chromatids in the nucleus.

The cell divides into two; each new cell with four chromatids.

The cells divide again to form four sex cells, each with only two chromosomes, one of each type present in the original cell.

Figure 24.25 A *summary of meiosis.*

Cells that have the full number of chromosomes in homologous pairs are called diploid cells. Cells that only have half the normal number of chromosomes are called haploid cells.

- During the second division of meiosis, the chromosomes separate into two parts. One part goes into each daughter cell. This allows for a lot of variation in the daughter cells.

More variation is created when a zygote is created by random fertilisation. In humans, any one of the billions of sperm formed by a male during his lifetime could, potentially, fertilise any one of the thousands of ova formed by a female. Therefore, if there are changes in the environment, only some of the individuals will be unable to adapt to the changes. As different genes are present in the population, some individuals will be successful in adapting to the change and will survive.

Table 24.2 compares mitosis and meiosis.

Feature	Mitosis	Meiosis
number of cell divisions	1	2
number of cells formed	2	2
number of chromosomes in cells formed	same as original cell (diploid)	half the number of original cell (haploid)
type of cells formed	body cells	sex cells
genetic variation in cells formed	none	variation

Table 24.2: *Mitosis and meiosis compared.*

Chapter summary

In this chapter you have learnt that:

- asexual reproduction produces offspring that are genetically identical because they are all produced by mitosis; this has a survival advantage in a stable environment

- methods of asexual reproduction include binary fission, budding, fragmentation and vegetative propagation

- sex cells are produced by meiosis; they are haploid cells and show genetic variation

- fertilisation restores the normal diploid number of chromosomes in the resulting zygote

- mitosis ensures that all the cells of an organism formed from a zygote are genetically identical diploid cells

- plants can reproduce asexually by producing runners, tubers, bulbs and corms, as well as by other methods

- mitosis and meiosis differ in that:
  - there is only one cell division in mitosis, but two in mitosis
  - mitosis forms two cells, whereas meiosis forms four cells
  - mitosis forms diploid cells, whereas meiosis forms haploid cells
  - mitosis produces genetically identical cells, whereas meiosis produces cells that vary genetically.

Questions

1. *Amoeba* reproduces by binary fission. This is an example of asexual reproduction because:

   A the chromosome number is halved
   B the nucleus is not divided in the cell division
   C one parent produces two offspring
   D the nucleus divides by meiosis

2. Asexual reproduction is involved in which of the following processes?

   A Artificial insemination
   B Vegetative propagation
   C Insect pollination
   D Wind pollination

3. Each daughter cell created by mitosis has:

   A the same number of chromosomes as the parent
   B a variable number of chromosomes
   C twice the number of chromosomes as the parent
   D half the number of chromosomes as the parent

4. In which of the following does meiosis occur?

   A A developing embryo
   B The skin of a mammal
   C The ovaries of a mammal
   D The tip of the root of a plant

5. Cells can divide by mitosis or by meiosis. Human cells contain 46 chromosomes. The graphs show the changes in the number of chromosomes per cell as two different human cells undergo cell division.

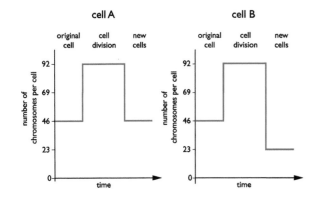

a) Which of the two cells, A or B, is dividing by meiosis? Explain how you arrived at your answer. (3)

b) Explain the importance of meiosis, mitosis and fertilisation in maintaining the human chromosome number constantly at 46 chromosomes per cell, generation after generation. (6)

c) Give *three* differences between mitosis and meiosis. (3)

6. The diagram shows a potato plant producing new tubers (potatoes). Buds on the parent plant grow into stems that grow downwards, called stolons. The ends of each stolon develop into a new tuber.

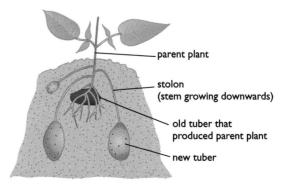

a) Give *two* pieces of evidence that show that this is an asexual method of reproduction. (2)

b) Explain why all the new tubers will be genetically identical. (3)

c) Even though the tubers are genetically identical, the plants that grow from them may not be the same height. Explain why. (2)

d) Why do wild plants need to reproduce sexually as well as asexually? (3)

7. All banana plants are reproduced asexually. Biologists are concerned for their future, as a new strain of fungus has appeared, which is killing all the banana plants in some plantations.

a) Explain why the fungus is able to kill *all* the banana plants in some plantations. (3)

b) Explain why this would be less likely to happen if banana plants reproduced sexually. (3)

c) Describe the benefits of reproducing banana plants asexually. (4)

8. The diagrams A to F show an animal cell during cell division.

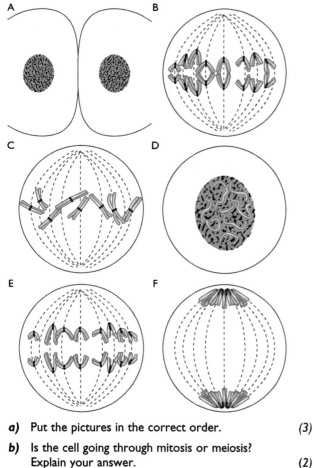

a) Put the pictures in the correct order. *(3)*

b) Is the cell going through mitosis or meiosis? Explain your answer. *(2)*

c) This cell has eight chromosomes, which is its diploid number. How many chromosomes would a diploid human cell have? *(1)*

d) Describe *two* differences between mitosis and meiosis. *(2)*

9. Some cells divide by mitosis, others divide by meiosis. For each of the following examples, say whether mitosis or meiosis is involved. In each case give a reason for your answers.

a) Cells in the testes dividing to form sperm. *(2)*

b) Cells in the lining of the small intestine dividing to replace cells that have been lost. *(2)*

c) Cells in the bone marrow dividing to form red blood cells and white blood cells. *(2)*

d) Cells in an anther of a flower dividing to form pollen grains. *(2)*

e) A zygote dividing to form an embryo. *(2)*

10. In an investigation into mitosis, the distance between a chromosome and the pole (end) of a cell was measured. The graph shows the result of the investigation.

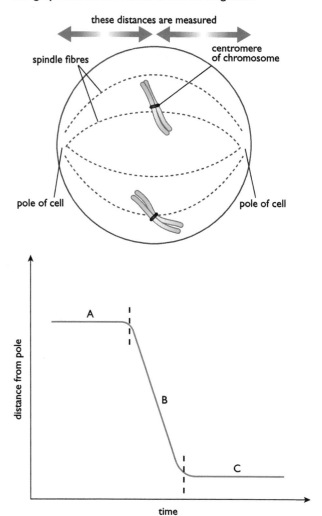

a) Describe the **two** events that occur during stage A. *(4)*

b) Explain what is happening during stage B. *(2)*

c) Describe **two** events that occur during stage C. *(2)*

Section C: Continuity and Variation

Chapter 25: Genetics

When you have completed this chapter, you will be able to:

- appreciate how the basic laws of inheritance were discovered
- understand how genes are passed from one generation to the next
- explain the patterns of inheritance shown by features that are determined by:
  - dominant and recessive alleles
  - codominant alleles
- understand, define and use the terms dominant, recessive, codominant, genotype, phenotype, homozygous, heterozygous
- use genetic diagrams to describe patterns of inheritance
- interpret genetic diagrams and pedigrees
- appreciate that certain human conditions are determined genetically and that their inheritance follows a similar pattern to many other features
- explain how gender is determined.

Genes are sections of DNA that determine a particular feature by instructing cells to produce particular proteins. As the DNA is part of a chromosome, we can also define a gene as 'part of a chromosome that determines a particular feature'.

Each cell with two copies of a chromosome also has two copies of the genes on those chromosomes. Suppose that, for the gene controlling earlobe attachment, a person has one allele for attached earlobes and one for free earlobes. What happens? Is one ear free and the other attached? Are they both partly attached? Neither. In this case, both earlobes are free. The 'free' allele is **dominant** and 'switches off' the 'attached' allele, which is **recessive**.

Gregor Mendel

The ground-breaking research that uncovered the basic rules of how features are inherited was carried out by Gregor Mendel and published in 1865.

Gregor Mendel was a monk and lived in a monastery in Brno in what is now the Czech Republic. He became interested in inheritance and his first attempts at controlled breeding experiments were with mice. This was not well received in the monastery and he was advised to use pea plants instead. As a result of the experiments with pea plants, he was able to formulate the basic laws of inheritance.

Mendel established that for each feature he studied:

- a 'heritable unit' (we now call it a gene) is passed from one generation to the next

- the heritable unit (gene) can have alternate forms (we now call these different forms alleles)

- each individual must have two alternate forms (alleles) per feature

- the sex cells have only one of the alternate forms (allele) per feature

- one allele can be dominant over the other.

Mendel was able to use his ideas to predict outcomes from breeding certain types of pea plant and then test his predictions by experiment. Mendel published his results and ideas in 1865, but very few people took any notice.

At that time, biologists had little knowledge of chromosomes and cell division, so Mendel's ideas had no physical basis. Also, biology then was very much a descriptive science and biologists of the day were not interested in the mathematical treatment of results. Mendel's work went against the ideas of the time that inheritance resulted from some kind of blending of features. The idea of a distinct 'heritable unit' just did not fit in.

It was not until 1900 that Hugo DeVries and other biologists working on inheritance rediscovered Mendel's work and recognised its importance. In 1903, Walter Sutton pointed out the connection between Mendel's suggested behaviour of genes and the behaviour of chromosomes in meiosis. The science of genetics was well and truly born.

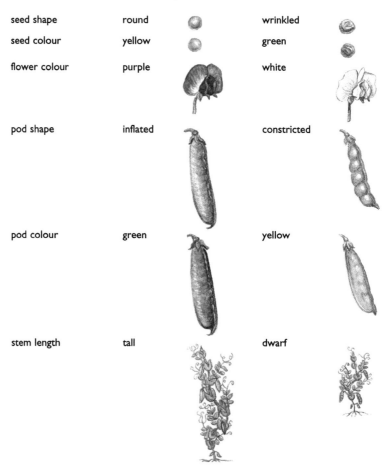

seed shape	round		wrinkled	
seed colour	yellow		green	
flower colour	purple		white	
pod shape	inflated		constricted	
pod colour	green		yellow	
stem length	tall		dwarf	

Figure 25.1 *Some features of pea plants used by Mendel in his breeding experiments.*

Mendel's experiments on inheritance

Mendel noticed that many of the features of pea plants had just two alternate forms. For example, plants were either tall or dwarf, they either had purple or white flowers, they produced yellow seeds or green seeds. There were no intermediate forms, no pale purple flowers or green/yellow seeds or intermediate height plants. Figure 25.1 on page 350 shows some of the contrasting features of pea plants that Mendel used in his breeding experiments.

Mendel decided to investigate systematically the results of cross breeding plants that had contrasting features. These were the 'parent plants', referred to as 'P' in genetic diagrams. He transferred pollen from one experimental plant to another. He also made sure that the plants could not be self-fertilised.

He collected all the seeds formed, grew them and noted the features that each plant developed. These plants were the first generation of offspring, or the 'F$_1$' generation. He did not cross-pollinate these plants, but allowed them to self-fertilise. Again, he collected the seeds, grew them and noted the features that each plant developed. These plants formed the second generation of offspring or the 'F$_2$' generation. When Mendel used pure-breeding tall and pure-breeding dwarf plants as his parents, he obtained the results shown in Figure 25.2.

> In his breeding experiments, Mendel initially used only plants that had 'bred true' for several generations. For example, any tall pea plants he used had come from generations of pea plants that had all been tall.

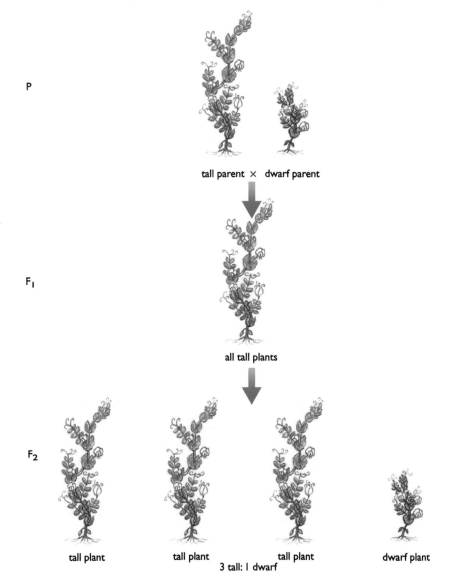

P

tall parent × dwarf parent

F$_1$

all tall plants

F$_2$

tall plant   tall plant   tall plant   dwarf plant

3 tall: 1 dwarf

Figure 25.2 *A summary of Mendel's results from breeding tall pea plants with dwarf pea plants.*

Mendel obtained very similar results when he carried out breeding experiments using plants with different pairs of contrasting characters (Figure 25.3). He noticed two things in particular:

- All the plants of the F_1 generation were always of just one type. This type was not a blend of the two parental features, but one or the other. Every time he repeated the experiment with the same feature, it was always the same type that appeared in the F_1 generation. For example, when tall and dwarf parents were cross-bred, the F_1 plants were always all tall.

- There was always a 3:1 ratio of types in the F_2 generation. Three-quarters of the plants in the F_2 generation were of the type that appeared in the F_1 generation. One-quarter showed the other parental feature. For example, when tall and dwarf parents were cross-bred, three-quarters of the F_2 plants were always tall and one-quarter were dwarf.

Mendel was able to use these patterns in his results to work out how features were inherited, without any knowledge of genes and chromosomes.

Whenever you have to work out a genetic cross, you should choose suitable symbols to represent the dominant and recessive alleles, give a key to explain which symbol is which and write out the cross exactly like the one shown.

Genotype describes the alleles each cell has for a certain feature, e.g. TT. **Phenotype** is the observable feature that results from the genotype.

phenotype of parents	tall	dwarf	Both parents are pure breeding. The tall parent has two alleles for tallness in each cell. The dwarf parent has two alleles for dwarfness in each cell. Because each has two copies of just one allele, we say that they are **homozygous** for the height gene.
genotype of parents	TT ×	tt	

gametes (sex cells)   (T)   (t)   The sex cells are formed by meiosis. As a result, they only have one allele each.

genotype of F_1 Tt   The F_1 plants have one tall allele and one dwarf allele. We say that they are **heterozygous** for the height gene.

phenotype of F_1 all tall   The plants are tall because the tall allele is dominant.

The F_1 plants are allowed to self-fertilise.

gametes from the F_1 plants   male gametes   female gametes   The sex cells are formed by meiosis and so only have one allele. Because the F_1 plants are heterozygous, half of the gametes carry the T allele and half carry the t allele.

(T) or (t)   (T) or (t)

genotypes of F_2 female gametes   The diagram opposite is called a **Punnett square**. It allows you to work out the results from a genetic cross. Write the genotypes of one set of sex cells across the top of the square and those of the other sex cells down the side. Then combine the alleles in the two sets of gametes; the squares represent the possible fertilisations.

male gametes	(T)	(t)
(T)	TT	Tt
(t)	Tt	tt

1 TT : 2 Tt : 1 tt   You can now work out the *ratio* of the different genotypes.

phenotypes of F_2 3 tall : 1 dwarf   Because T is dominant, all the genotypes containing at least one T will be tall.

Figure 25.3 *Results of crosses using true-breeding tall and dwarf pea plants.*

Explaining Mendel's results

We can now explain Mendel's results using the ideas of chromosomes, genes, mitosis and meiosis:

- Each feature is controlled by a gene, which is found on a chromosome.

- There are two copies of each chromosome and each gene in all body cells, except the sex cells.

- The sex cells have only one copy of each chromosome and each gene

- There are two alleles (forms) of each gene.

- One allele is dominant over the other allele, which is recessive.

- When two different alleles (one dominant and one recessive) are in the same cell, only the dominant allele is expressed (is allowed to 'work').

- An individual can have two dominant alleles, two recessive alleles or a dominant allele and a recessive allele in each cell.

We can use the cross between tall and dwarf pea plants as an example (Figure 25.3 on page 352). In pea plants, there are tall and dwarf alleles of the gene for height. We will use the symbol T for the tall allele and t for the dwarf allele. The term genotype describes the alleles each cell has for a certain feature (e.g. TT). The phenotype is the observable feature that results from the genotype (e.g. a tall plant).

Working out genotypes – the test cross

You cannot tell just by looking at it whether a tall pea plant is homozygous (TT) or heterozygous (Tt). Both these genotypes would appear equally tall because the tall allele is dominant. It would help if you knew the genotypes of its parents. You could then write out a genetic cross and perhaps work out the genotype of your tall plant. If you don't know the genotypes of the parents, the only way you can find out is by carrying out a breeding experiment called a **test cross**.

In a test cross, the factor under investigation is the unknown genotype of an organism showing the dominant feature. A tall pea plant could have the genotype TT or Tt. You must control every other possible variable *including the genotype of the plant you breed it with*. The only genotype you can be *certain* of is the genotype of plants showing the *recessive* feature (in this case dwarf plants). They *must* have the genotype tt.

> In a test cross, you breed an organism showing the dominant feature with one showing the recessive feature.

In this example, you must breed the 'unknown' tall pea plant (TT or Tt) with a dwarf pea plant (tt). You can write out a genetic cross for both possibilities (TT × tt and Tt × tt) and *predict* the outcome for each (Figure 25.4 on page 354). You can then compare the results of the breeding experiment with the predicted outcomes to see which one matches most closely.

> The genotype of an organism is represented by two letters, each letter representing one allele of the gene that controls the feature. Normally, we use the initial letter of the dominant feature to represent the gene. Writing it as a capital letter indicates the dominant allele, the lower case letter represents the recessive allele. For the feature height in pea plants, plants can be either tall or dwarf. Using the initial letter of the word 'tall', the alleles are shown as T and t. TT means that a plant has two alleles for tallness.
>
> A plant with two alleles for dwarfness would be represented by tt. Although we use pea plants to illustrate Mendel's ideas, the same principles apply to other organisms also.

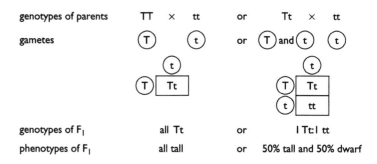

| genotypes of parents | TT × tt | or | Tt × tt |
| gametes | (T) (t) | or | (T) and (t) (t) |

genotypes of F₁ ... all Tt ... or ... I Tt:I tt

phenotypes of F₁ ... all tall ... or ... 50% tall and 50% dwarf

Figure 25.4 *A test cross.*

From our crosses we would expect:

- *all* the offspring to be tall if the tall parent was homozygous (TT)

- *half* the offspring to be tall and *half* to be dwarf if the tall parent was heterozygous (Tt).

> **1.** Describe three of the features investigated by Mendel.
> **2.** What are pure lines?
> **3.** Describe the pattern of inheritance Mendel found over two generations when he crossed two pure bred lines.
> **4.** List the seven points that we use to explain Mendel's results.
> **5.** Explain how a test-cross can be used to decide whether an organism showing the dominant feature is homozygous or heterozygous for that feature.

Ways of presenting genetic information

Writing out a genetic cross is a useful way of showing how genes are passed through one or two generations, starting with just two parents. To show a proper family history of a genetic condition requires more than this. We use a diagram called a **pedigree**. Polydactyly is an inherited condition in which a person develops an extra digit (finger or toe) on the hands and feet. It is determined by a dominant allele. The recessive allele causes the normal number of digits to develop.

If we use the symbol D for the polydactyly allele and d for the normal-number allele, the possible genotypes and phenotypes are:

- DD – person has polydactyly (has two dominant polydactyly alleles)

- Dd – person has polydactyly (has a dominant polydactyly allele and a recessive normal allele)

- dd – person has the normal number of digits (has two recessive, normal-number alleles).

We don't use P and p to represent the alleles as you would expect, because P and p look very similar and could easily be confused. A pedigree for polydactyly is shown in Figure 25.5 on page 355.

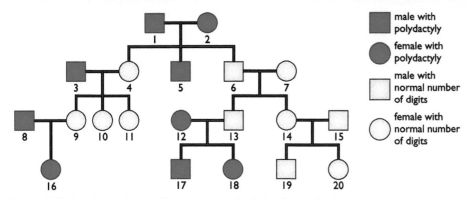

Figure 25.5 *A pedigree showing the inheritance of polydactyly in a family.*

We can extract a lot of information from a pedigree. In this case:

- there are four generations shown (individuals are arranged in four horizontal lines)

- individuals 4, 5 and 6 are children of individuals 1 and 2 (a family line connects each one directly to 1 and 2)

- individual 4 is the first-born child of 1 and 2 (the first-born child is shown to the left, then second born to the right of this, then third born, and so on)

- individuals 3 and 7 are not children of 1 and 2 (no family line connects them directly to 1 and 2)

- 3 and 4 are father and mother of the same children – as are 1 and 2, 6 and 7, 8 and 9, 12 and 13, 14 and 15 (a horizontal line joins them).

It is usually possible to work out which allele is dominant from pedigrees. Look for a situation in which two parents show the same feature and at least one child shows the contrasting feature. In this pedigree, 1 and 2 both have polydactyly, but children 4 and 6 do not. We can explain this in only one way:

- The normal alleles in 4 and 6 can only have come from their parents – 1 and 2, so 1 and 2 have normal alleles.

- 1 and 2 show polydactyly, so they *must* have polydactyly alleles as well.

- If they have both polydactyly alleles *and* normal alleles but show polydactyly, the polydactyly allele must be the dominant allele.

Now that we know which allele is dominant, we can work out most of the genotypes in the pedigree. All the people with the normal number of digits *must* have the genotype dd (if they had even one D allele, they would show polydactyly). All the people with polydactyly must have *at least one* polydactyly allele (they must be either DD or Dd).

From here, we can begin to work out the genotypes of the people with polydactyly. To do this, we need to bear in mind that people with the normal number of digits must inherit one 'normal-number' allele from each parent, and also that people with the normal number of digits will pass on one 'normal-number' allele to each of their children.

From this we can say that any person with polydactyly who has children with the normal number of digits must be heterozygous (the child must have inherited one of their two 'normal-number' alleles from this parent), and also that any person with polydactyly who has one parent with the normal number of digits must also be heterozygous (the normal parent can only have passed on a 'normal-number' allele). Individuals 1, 2, 3, 16, 17 and 18 fall into one or both of these categories and must be heterozygous.

We can now add this genetic information to the pedigree. This is shown in Figure 25.6.

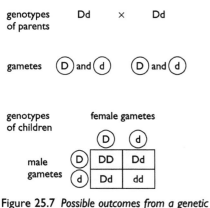

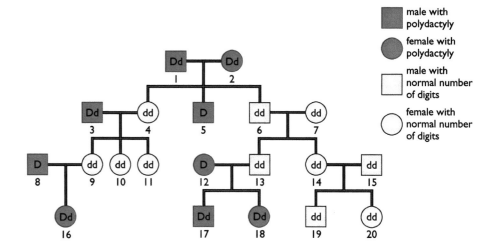

Figure 25.7 *Possible outcomes from a genetic cross between two parents, both heterozygous for polydactyly.*

Figure 25.6 *A pedigree showing the inheritance of polydactyly in a family, with details of genotypes added.*

We are still left uncertain about individuals 5, 8 and 12. They could be homozygous or heterozygous. For example, individuals 1 and 2 are both heterozygous. Figure 25.7 shows the possible outcomes from a genetic cross between them. Individual 5 could be any of the outcomes indicated by the shading. It is impossible to distinguish between DD and Dd.

Look at the pedigree of PTC tasting in a family (Figure 25.8). PTC has a very bitter taste, which some people cannot detect.

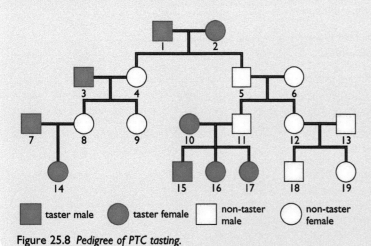

Figure 25.8 *Pedigree of PTC tasting.*

PTC tasting is determined by a dominant allele, T. Non-tasting is determined by a recessive allele, t.

6. How many generations are represented in the pedigree?
7. What relation are (a) individuals 4 and 5, and (b) individuals 5 and 6?
8. What are the genotypes of individuals 4, 8, 13 and 18? Give evidence from the pedigree to justify your answers.
9. What are the genotypes of individuals 1 and 2? Give evidence from the pedigree to justify your answers.
10. What is the chance that, if individuals 10 and 11 have another child, he or she will be a non-taster? Give evidence from the pedigree to justify your answer.

Codominant alleles

The alleles that determine a particular feature are not always completely dominant or recessive. They can sometimes interact with each other so that the resulting heterozygote does not have the same phenotype as one that is homozygous for the dominant allele, but shows a completely new characteristic.

Flower colour in four o'clock

The two alleles are R (produces red flowers) and r (produces white flowers).

The possible genotypes and phenotypes are:

- RR – plants with this genotype produce red flowers

- rr – plants with this genotype produce white flowers

- Rr – these plants are heterozygous, but because neither allele is completely dominant, both alleles express themselves and interact and the flowers are pink.

We can use a standard genetic diagram to predict the genotypes and phenotypes of the offspring that result when two pink–flowered four o'clock are crossed (Figure 25.9).

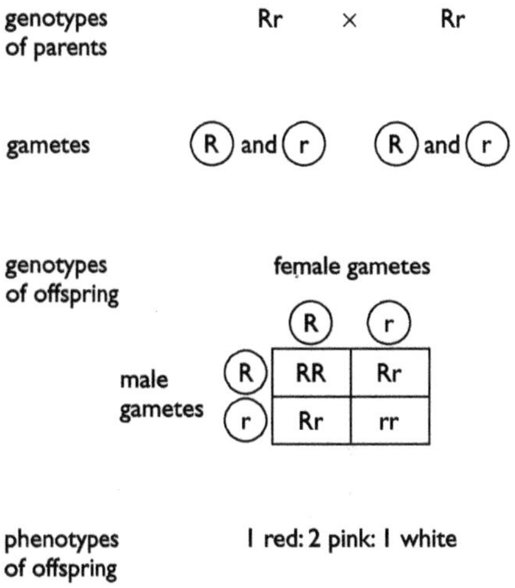

Notice that we do not get a 3:1 ratio of phenotypes, but a 1:2:1 ratio instead. This is because the heterozygotes do not develop the same colour as the homozygous dominant types as a result of codominance.

Human blood groups

The inheritance of human blood groups also shows codominance. The pattern of inheritance is more complex than flower colour in four o'clock as three different alleles are involved.

The blood group of a person is the result of the presence or absence of two antigens, the A antigen and the B antigen, on the surface of the red blood cells. There are three alleles involved in the inheritance of these antigens:

- I^A – determines the production of the A antigen

- I^B – determines the production of the B antigen

- I^O – determines that neither antigen is produced.

The alleles I^A and I^B are codominant, but I^O is recessive to both. Any one person can only inherit two alleles. The possible genotypes and phenotypes are shown in Table 19.1.

Figure 25.9 *Possible outcomes from a genetic cross between two pink-flowered four o'clock.*

Genotype	Antigen produced	Blood group
$I^A I^A$, $I^A I^O$	A	A
$I^B I^B$, $I^B I^O$	B	B
$I^A I^B$	A and B	AB
$I^O I^O$	neither	O

Table 19.1: *Genotypes in blood groups.*

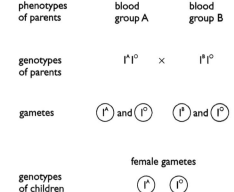

		female gametes	
genotypes of children		I^A	I^O
male gametes	I^B	$I^A I^B$	$I^B I^O$
	I^O	$I^A I^O$	$I^O I^O$

phenotypes of children   IAB : IA : IB : IO

Figure 25.10 *Possible outcome from two parents heterozygous for blood groups A and B.*

When a recessive allele determines a genetic disease, people who are heterozygous for the condition appear outwardly normal. They can, however, pass on the allele to their children. Because of this, they are often called **carriers**.

Parents who are heterozygous for blood group A and blood group B could produce four children, each with a different blood group (Figure 25.10).

You can interpret pedigrees of blood groups in the same way as other pedigrees. For example, in Figure 25.11, what are the blood groups of individuals 5 and 8?

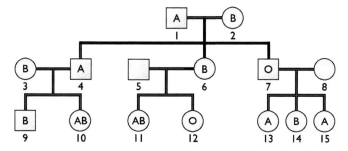

Figure 25.11 *Pedigree of human blood groups.*

Individual 12 has two I^O alleles (as she is blood group O) and so must inherit one of them from individual 5. Individual 11 must inherit her I^A allele from individual 5 as her other parent is blood group B. *Individual 5 therefore has the genotype $I^A I^O$ and so must be blood group A.*

Individuals 7 and 8 produce children with blood group A and blood group B. Individual 7 is blood group O and so both the I^A and the I^B alleles must come from individual 8. *Individual 8 is blood group AB.*

Inherited disorders

Our genes determine all our features. In some cases, they can determine conditions that we recognise as 'genetic diseases'. Two such conditions are cystic fibrosis and sickle-cell anaemia.

Cystic fibrosis

This condition is determined by a recessive allele of the gene that controls the production of mucus by cells in glands throughout the body. The gene has a high mutation rate, and if mutations occur in the sex cells (or in the cells that form the sex cells), the mutant allele could be inherited.

The normal, dominant allele results in normal mucus being secreted. The mutated, recessive allele causes the production and secretion of viscous (very thick) mucus. This has several adverse effects:

- It blocks the pancreatic duct so that pancreatic enzymes cannot reach the small intestine. This affects the digestion of carbohydrates, lipids and proteins (see Chapter 12).

- It cannot be easily moved out of the lungs by the cilia as can normal mucus. Gas exchange suffers as a result.

Because of these effects, people suffering from cystic fibrosis often die young. However, treatment is now much better and can extend their life-span. Gene therapy may offer a cure in the future.

To be affected, a person must inherit two recessive alleles (one from each parent), so each parent must carry at least one recessive cystic fibrosis allele. The disease is most commonly inherited when both parents are heterozygous for the condition (Figure 25.12). There is a 1 in 4 (25%) chance of a child from such a relationship developing cystic fibrosis.

Sickle-cell anaemia

Like cystic fibrosis, this condition is determined by a mutant, recessive allele. It is an allele of the gene that codes for the production of haemoglobin. The haemoglobin produced in sufferers is abnormal. It causes red blood cells to distort when the oxygen concentration of the surroundings is low. Instead of being the usual disc shape, they become sickle shaped (Figure 25.13). This can happen in any active tissues that use up oxygen very rapidly as they respire.

There are two main consequences as a result of the red blood cells becoming sickle shaped:

- The sickle-cells tend to form blood clots that block capillaries, leading to a 'sickle-cell crisis'. This is a painful condition in which one or more organs are damaged due to a lack of blood. If the blood supply to the brain is affected, a stroke may result.

- The sickle-cells are more fragile than normal cells and easily burst. They are also destroyed at a higher rate than normal red blood cells by the spleen. The drastic reduction in the numbers of red blood cells causes the anaemia that gives the disease its name.

To be fully affected, a person must be homozygous for the sickle-cell allele. In other words they must inherit a 'sickle-cell' allele from each parent. The condition develops soon after birth and infant mortality is very high, usually due to lack of blood to an organ during a sickle-cell crisis. Sufferers who survive childhood have an increased risk of infection by bacteria that cause a certain type of pneumonia. They also have an increased risk of developing gallstones.

H = normal allele resulting in production of normal haemoglobin

h = allele resulting in the production of abnormal (sickling) haemoglobin

heterozygous carrier = Hh

| genotypes of parents | Hh | × | Hh |

| gametes | (H) and (h) | (H) and (h) |

female gametes

genotypes of children		(H)	(h)
male gametes	(h)	HH	Hh
	(h)	Hh	hh

this child would develop sickle cell anaemia

Figure 25.14 *Inheritance of sickle-cell anaemia.*

N = dominant normal allele resulting in production of normal mucus

n = recessive cystic fibrosis resulting in the production of viscous mucus

heterozygous non-sufferer = Nn

| genotypes of parents | Nn | × | Nn |

| gametes | (N) and (n) | (N) and (N) |

female gametes

genotypes of children		(N)	(n)
male gametes	(N)	NN	Nn
	(n)	Nn	nn

this child would develop cystic fibrosis

Figure 25.12 *Inheritance of cystic fibrosis.*

A sickle is a crescent-shaped agricultural tool, used for cutting down vegetation.

Anaemia describes any condition in which the concentration of haemoglobin in the blood falls significantly below normal.

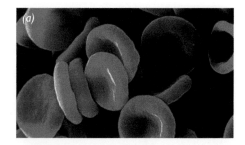
(a)

(b)

Figure 25.13 *(a) Normal red blood cells and (b) distorted red blood cells from a person suffering from sickle-cell anaemia.*

The inheritance of sickle-cell anaemia is not quite as straightforward as the inheritance of some other characteristics. This is because the normal allele is not actually dominant over the sickle-cell allele. The red blood cells of individuals who are heterozygous contain half normal and half abnormal haemoglobin. So *both* the alleles are actually having an effect. This is another example of codominance.

Heterozygotes do not suffer sickle-cell crises because there is sufficient normal haemoglobin in their red blood cells for them to maintain their normal shape. However, under very low oxygen concentrations, some of the cells will deform into the typical sickle-shaped cells. This can be seen under a microscope.

The disease is most commonly inherited when both parents are heterozygous carriers of the condition (Figure 25.14). There is a 1 in 4 (25%) chance of a child from such a relationship developing sickle-cell anaemia.

Sickle-cell anaemia can be treated by a blood transfusion following a crisis and can be cured by a bone marrow transplant. Bone marrow is the 'production line' for red blood cells. If bone marrow from a non-sufferer is transplanted successfully, it will produce normal red blood cells. The treatment is only available in specialist centres, but the success rate is around 85%. In the future, gene therapy may be used to introduce non-sickling alleles into the bone marrow of sufferers, resulting in the production of normal red blood cells.

Inheriting the sickle-cell allele need not be a bad thing. Heterozygous carriers of the allele usually show no symptoms of the disease at all. In fact, they can actually benefit from their condition – being a carrier for sickle-cell anaemia gives increased resistance to malaria.

Albinism

Albinism is another inherited condition, although this is not normally referred to as a genetic disease. In this condition, the pigment in the skin, hair and irises of the eyes fails to develop. Albinism is determined by a recessive allele (a) with the allele for normal pigmentation (A) being dominant. A person must inherit one recessive allele from each parent to inherit the condition. Possible genotypes and phenotypes are:

- AA, Aa – normal pigmentation
- aa – albinism (pigment fails to develop).

Two normally pigmented people could produce an albino child if both were heterozygous (Figure 25.15).

11. Describe, briefly, how sickle-cell anaemia is inherited.
12. Explain one possible benefit of inheriting the sickle-cell allele.

Questions 13, 14 and 15 refer to the pedigree of albinism in a family shown in Figure 25.16.

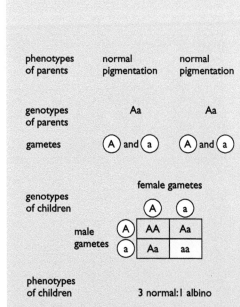

Figure 25.15 *Possible outcome of a genetic cross between two parents who are both heterozygous for albinism.*

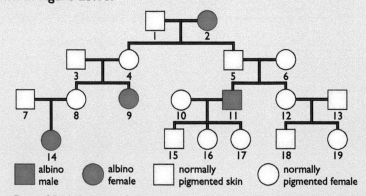

Figure 25.16 *Pedigree of albinism.*

13. What is albinism?
14. What are the genotypes of individuals 10 and 11? Give evidence from the pedigree to justify your answers.
15. If individuals 10 and 11 were to have another child, what is the chance that the child would be (a) female, (b) an albino and (c) both? Give evidence from the pedigree to justify your answers.

Sex determination

Nearly all human cells contain 46 chromosomes. The photographs in Figure 25.17 show the 46 chromosomes from the body cells of a human male and female.

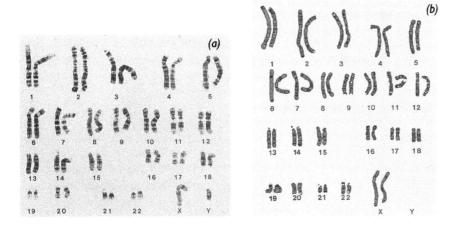

Figure 25.17 *Chromosomes of a human male (a) and female (b). A picture of all the chromosomes in a cell is called a* **karyotype**.

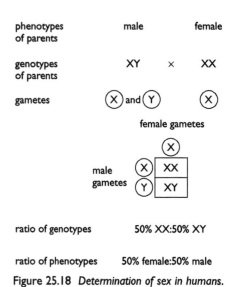

phenotypes of parents	male		female
genotypes of parents	XY	×	XX
gametes	Ⓧ and Ⓨ		Ⓧ

female gametes

		Ⓧ
male gametes	Ⓧ	XX
	Ⓨ	XY

ratio of genotypes	50% XX:50% XY
ratio of phenotypes	50% female:50% male

Figure 25.18 *Determination of sex in humans.*

The chromosomes are not arranged like this in the cell. The original photograph has been cut up and chromosomes of the same size and shape 'paired up'. The cell from the male has 22 pairs of chromosomes and two that do not form a pair – the X and Y chromosomes.

Our sex – whether we are male or female – is not under the control of a single gene. It is determined by the X and Y chromosomes – the sex chromosomes. As well as the 44 non-sex chromosomes, there are two X chromosomes in all cells of females (except the egg cells), and one X and one Y chromosome in all cells of males (except the sperm). Our sex is effectively determined by the presence or absence of the Y chromosome.

> The X and the Y chromosomes are the sex chromosomes. They determine whether a person is male or female.

The inheritance of sex follows the pattern shown in Figure 25.17. In any one family, however, this ratio may well not be met. Predicted genetic ratios are usually only met when large numbers are involved. The overall ratio of male and female births in all countries is 1:1.

Sex-linked genes

Females have two X chromosomes, so they have two copies of each gene. By contrast, males have only one X chromosome, so they have only one copy of specific genes. This creates a difference between the possible genotypes males and females possess on these sex chromosomes. Genes that are found on the X chromosome and not on the Y chromosome are called **sex-linked genes**.

So, in the case of the female, she may have either of the following genotypes for a sex-linked gene:

- B = normal allele resulting in normal condition
- b = allele resulting in sex-linked condition

- X^BX^B – homozygous dominant: normal

- X^BX^b – heterozygous: carrier

- X^bX^b – homozygous recessive: sex-linked condition is present.

Note that the Xs and Ys must be included when writing the genotypes involving sex-linked genes.

Because males have only one X chromosomes, there are only two possible genotypes for sex-linked genes:

- X^BY – normal

- X^bY – sex-linked condition is present.

Haemophilia

Haemophilia is an inherited disease of the blood where if the person gets grazed, bruised or cut, they bleed for a longer period than normal as the blood cannot clot properly. In a normal person, platelets activate clotting enzymes called factors, which are part of a step-by-step process required for the proper clotting of blood, which stops bleeding. If one clotting factor is missing, the process is inefficient.

People with haemophilia lack clotting factor 8 or 9. The genes for these clotting factors are found on the X chromosome, making them sex-linked genes. There are two alleles of the factor 8 gene. The allele H causes normal factor 8 to be formed. The allele h does not. H is dominant to h, so if a person has at least one H allele in their cells, they will make normal factor 8.

Genetic diagrams may be used to show the inheritance of sex-linked genes. The following genetic cross shows the probable genotypes of children born to a couple where the female is a carrier for haemophilia and the father is normal (Figure 25.19).

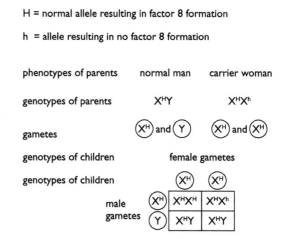

Figure 25.19 *Inheritance of haemophilia.*

Every time this couple has a child, there is a 25% chance that they will have a normal boy, a 25% chance that they will have a normal girl, a 25% chance that they will have a carrier girl, and a 25% chance that they will have a boy who is a haemophiliac.

Colour-blindness

Red–green colour-blindness is another inherited sex-linked condition. Men stand a higher chance of suffering from it than women. The normal allele causes a protein to be produced that forms one of the pigments in the cone cells in the retina of the eye. The recessive allele does not cause this protein to be formed. People without the normal allele cannot distinguish red from green; see Figure 25.20.

If a colour-blind man and a normal woman have children, all of the children will have normal vision (Figure 25.21).

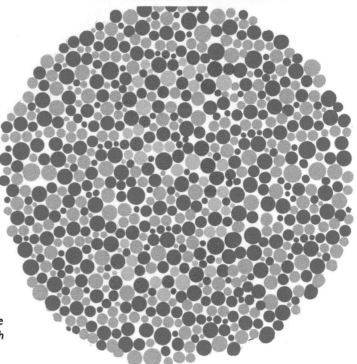

Figure 25.20 *People with normal vision can see the number 5 formed by green dots. People with red–green colour blindness see only dots.*

G = normal allele resulting in protein production for one of the pigments in cones

g = allele resulting in no protein production for one of the pigments in cones

phenotypes of parents   colour blind man   ×   normal female

genotypes of parents X^gY × X^GX^G

gametes X^gY X^GX^G

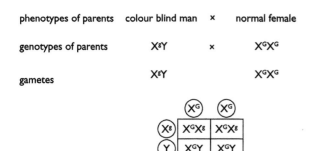

F₁ genotypes

The females are carriers and the males are normal. However, if one of the females marries a normal man, she passes on the recessive allele to her son.

phenotypes of parents   normal man   ×   carrier woman

genotypes of parents X^GY × X^GX^g

gametes X^GY X^GX^g

	X^G	X^g
X^G	X^GX^G	X^GX^g
Y	X^GY	X^gY

F₂ genotypes

There is a probability that 50% of the boys are normal or colour blind. There is another probability that 50% of the females have normal colour sight or are carriers.

Figure 25.21 *Inheritance of colour blindness.*

Chapter summary

In this chapter you have learnt that:

- the rules governing inheritance were first discovered by Gregor Mendel as a result of his breeding experiments with pea plants

- features of organisms are determined by genes, which are carried on chromosomes

- in most cells, there are two copies of each gene; one on each of the two chromosomes that make up a homologous pair

- sex cells only have one chromosome from the pair and so only have one copy of a gene

- alleles are different forms of the same gene

- a dominant allele of a gene expresses itself in the heterozygote

- a recessive allele only expresses itself in the homozygote

- it is impossible to tell whether an organism showing the feature determined by the dominant allele is homozygous or heterozygous without performing a test cross

- the alleles of some genes show codominance; in this condition, neither allele is completely dominant and the two interact in the heterozygote to produce a feature different to either determined by the two alleles individually

- pedigrees are a convenient way of representing patterns of inheritance of a feature across several generations in one family

- some conditions arise as a result of genes that cause parts of the body to not function normally; these are called genetic disorders

- the X and the Y chromosomes determine gender

- females have the genotype XX and males have the genotype XY

- haemophilia and colour-blindness are sex-linked conditions.

Questions

1. Each of our features is determined by:

   A the entire nucleus
   B a gene, which is part of a chromosome
   C a chromosome, which is part of a gene
   D a molecule of DNA

2. Allele A is dominant over allele a. Two heterozygous (Aa) individuals are crossed. Which phenotype ratio will the offspring show?

   A 1:1
   B 2:1
   C 3:1
   D 4:1

3. Which of the following statements is **not** correct?

   A Sex cells have only one chromosome from each homologous pair.
   B Sex cells have only one allele from each pair of alleles.
   C Sex cells have only one gene.
   D Sex cells have only one sex chromosome.

4. Alleles B and b are codominant. Two heterozygous individuals are crossed. Which phenotype ratio will the offspring show?

   A 1:2:1
   B 1:3:1
   C 1:1:1:1
   D 1:1

5. Tallness in pea plants is determined by a dominant allele. A tall pea plant could be homozygous or heterozygous for the tall allele. To find out which, you must:

   A cross it with another tall plant
   B compare its height with that of a known homozygous tall plant
   C compare its height with that of a known heterozygous tall plant
   D cross it with a dwarf (short) pea plant

6. Predict the *ratios* of offspring from the following crosses between tall/dwarf pea plants.

   a) TT × TT

   b) TT × Tt

   c) TT × tt

   d) Tt × Tt

   e) Tt × tt

   f) tt × tt

7. In cattle, a pair of alleles controls coat colour. The allele for black coat colour is dominant over the allele for red coat colour. The genetic diagram represents a cross between a pure-breeding black bull and a pure-breeding red cow.
B = dominant allele for black coat colour;
b = recessive allele for red coat colour.

   a) i) What term describes the genotypes of the pure-breeding parents? (1)

       ii) Explain the terms dominant and recessive. (2)

   b) i) What are the genotypes of the sex cells of each parent? (2)

       ii) What is the genotype of the offspring? (1)

   c) Cows with the same genotype as the offspring were bred with bulls with the same genotype.

       i) What genetic term describes this genotype? (1)

       ii) Draw a genetic diagram to work out the ratios of:
           • the genotypes of the offspring
           • the phenotypes of the offspring. (4)

8. In nasturtiums, a single pair of alleles controls flower colour.

The allele for red flower colour is dominant over the allele for yellow flower colour. The diagram represents the results of a cross between a pure-breeding red-flowered nasturtium and a pure-breeding yellow-flowered nasturtium.
R = dominant allele for red flower colour;
r = recessive allele for yellow flower colour

   a) Copy and complete the genetic diagram. (3)

   b) What are the colours of the flowers of A–D? (4)

9. Cystic fibrosis is an inherited condition. The diagram shows the incidence of cystic fibrosis in a family over four generations.

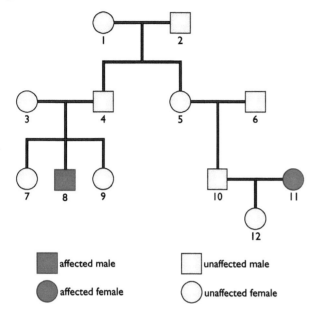

■ affected male	□ unaffected male
● affected female	○ unaffected female

a) What evidence in the pedigree suggests that cystic fibrosis is determined by a recessive allele? (3)

b) What are the genotypes of individuals 3, 4 and 11? Explain your answers. (3)

c) Draw genetic diagrams to work out the probability that the next child born to individuals 10 and 11 will:

i) be male

ii) suffer from cystic fibrosis. (4)

10. In guinea pigs, the allele for short hair is dominant to that for long hair.

a) Two short-haired guinea pigs were bred and their offspring included some long-haired guinea pigs. Explain these results. (3)

b) How could you find out if a short-haired guinea pig was homozygous or heterozygous for hair length? (3)

11. Sickle-cell anaemia is determined by a single mutant allele. Sufferers are homozygous for this allele. Heterozygotes show virtually no signs of the condition, but have an increased resistance to malaria.

a) What is a sickle-cell crisis? (3)

b) Why does a bone marrow transplant often cure sickle-cell anaemia? (3)

c) A sickle-cell sufferer survived to have children and married a person heterozygous for the condition. What proportion of their children would develop sickle-cell anaemia? Use a genetic diagram to explain your answer. (4)

d) Why do people heterozygous for the condition have an increased resistance to malaria? (2)

Chapter 26: Genetic Variation

When you have completed this chapter, you will be able to:

- understand that variation can be genetic or environmental
- define and describe variation
- distinguish between continuous and discontinuous variation
- explain the causes of genetic variation
- explain the importance of genetic variation.

Several differences exist between members of the same species. This is called variation. Variation is evident everywhere around you in the environment. Plants are different heights and people have different weights; other variations are less obvious, such as blood group and heart rate.

If a feature varies in such a way that it falls into two or more categories, we call this **discontinuous variation**, for example, eye colour. If the variation shows a range of values between two extremes, we call this **continuous variation**, for example, human height.

Genes and environment both produce variation

Pea plants are either tall or short because of the genes they inherit. There are no 'intermediate height' pea plants. However, all the tall pea plants are not *exactly* the same height and neither are all the short pea plants *exactly* the same height. Figure 20.1 illustrates the different types of variation in pea plants.

Examples of discontinuous variation include blood group, eye colour, fingerprints, tongue rolling, and horned or polled cows.

Examples of continuous variation include leaf size, height, weight, foot size, finger length and heart rate.

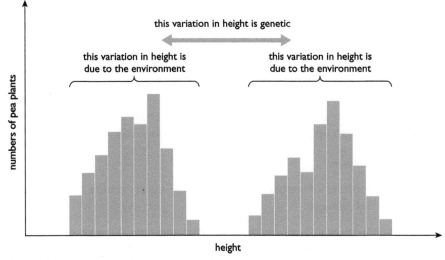

Figure 26.1 *Bar chart showing variation in height of pea plants.*

- Several **environmental factors** can influence the height of pea plants:

- They may not all receive the same amount of light and so some will not photosynthesise as well as others.

- They may not all receive the same amount of water and mineral ions from the soil: this could affect the manufacture of a range of substances in the plant.

- They may not all receive the same amount of carbon dioxide: again, some will not photosynthesise as well as others.

Similar principles apply in humans. Identical twins have the same genes, and often grow up to look very alike (although not quite identical). Also, they often develop similar talents. However, identical twins never look *exactly* the same. This is especially true if, for some reason, they grow up apart. The different environments affect their physical, social and intellectual development in different ways.

Activity 1: Investigating variation in humans

SBA skills

| ORR | MM | AI | PD | Dr |

You will need:

- a metre rule
- a 30-cm rule
- weighing scales

Carry out the following:

1. Record the height, maximum hand span of the right hand and body mass of all the students in your class and of as many other students in the same year group as possible.

2. Also note whether they are left-handed, right-handed or ambidextrous, their eye colour, and whether or not they can roll their tongue.

3. For the continuously variable features (height, hand span and body mass) divide the ranges into class intervals and record the numbers in each class interval.

4. Plot graphs showing the variation for all six features.

---

1. Why are tall and dwarf pea plants different heights?
2. Why aren't all tall pea plants the same height?
3. Explain why identical twins are often so similar.
4. Explain why identical twins reared apart are often more different than those reared together.

Genetic variation is caused by sexual reproduction and mutation

Sexual reproduction

The two main processes of sexual reproduction are the formation of gametes by meiosis and fertilisation:

- In meiosis, chromosomes are randomly sorted and crossing over occurs between homologous chromosomes (see Chapter 24). This creates gametes with various genotypes.

- In fertilisation, any sperm could fertilise any ovum. The possible combination of chromosomes (and genes) in the zygote is a very large number. This means that every individual is likely to be genetically unique.

The only exceptions are **identical twins** (and identical triplets and quadruplets). Identical twins are formed from the *same* zygote – they are sometimes called **monozygotic twins**. When the zygote divides by mitosis, the two *genetically identical* cells formed do not 'stay together'. Instead, they separate and each cell behaves as though it were an individual zygote, dividing and developing into an embryo (Figure 26.2). Because they have developed from genetically identical cells (and, originally, from the same zygote), the embryos (and, later, the children and the adults they become) will be genetically identical.

Non-identical twins or **fraternal twins** develop from different zygotes and so are not genetically identical.

Mutation

A mutation is a change in the DNA of a cell. It can happen in individual genes or in whole chromosomes. Sometimes, when DNA is replicating, errors are made that result in a **gene mutation**, which can alter the type of gene formed. This can lead to the gene coding for the wrong protein.

Mutations that occur in body cells, such as those in the heart, intestines or skin, will only affect that particular cell. If they are very harmful, the cell will die and the mutation will be lost. If they do not affect the functioning of the cell in a major way, the cell may not die. If the cell then divides, a group of cells containing the mutant gene is formed. When the organism dies, however, the mutation is lost with it; it is not passed to the offspring. Only mutations in the sex cells or in the cells that divide to form sex cells can be passed on to the next generation. This is how genetic diseases begin.

Sometimes a gene mutation can be advantageous to an individual. For example, as a result of random mutations, bacteria can become resistant to antibiotics. Resistant bacteria obviously have an advantage over non-resistant types if an antibiotic is being used. They will survive the antibiotic treatment and reproduce. All their offspring will be resistant and so the proportion of resistant types in the population of bacteria will increase as this happens in each generation. This is an example of **natural selection** (see Chapter 27). Pests can become resistant to pesticides in a similar way.

A certain amount of genetic variation within a population is brought about by genes reshuffling during meiosis. If the environment changes, the chance that some individuals will survive is increased because there is a variety of traits present in the population. The surviving offspring will be able to pass down their genes from generation to generation.

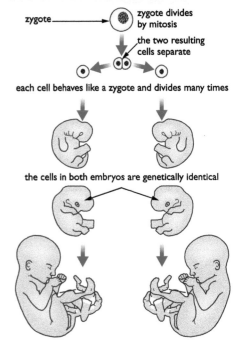

Figure 26.2 *How identical twins are formed.*

Chapter summary

In this chapter you have learnt that:

- variation that is genetic can be passed from one generation to the next, but environmental variation cannot be inherited
- the effects of genes can be modified by the environment
- meiosis, fertilisation and mutation create genetic variation
- genetic variation produces a variety of genes within a population. This increases the chance that some individuals within the population will survive if the environment changes.

Questions

1. Variation in which there are only a few categories is called:

   A continuous variation
   B discontinuous variation
   C proportionate variation
   D disproportionate variation

2. Which one of the following statements is the **least** accurate?

   A Discontinuous variation results entirely from genetic differences.
   B Continuous variation can result from genetic differences.
   C Discontinuous variation cannot be altered by environmental effects.
   D Continuous variation results from environmental effects.

3. Which types of variation can be inherited?

	Variation caused by genes	Variation caused by the environment
A	yes	yes
B	yes	no
C	no	yes
D	no	no

4. a) Do environmental or genetic factors cause the differences we see in plants with an identical genetic make-up?

   b) List some of the advantages of breeding plants that have the same genetic make-up.

   c) Explain why breeding a population of identical organisms can be disadvantageous. *(5)*

5. Complete the table by classifying each of the following variations based on what causes them:

   obesity, eye colour, tallness, singing ability, blood group, natural hair colour; sickle-cell anaemia, agility

Genetic effects only	A combination of genetic and environmental effects

   (5)

6. For each variation in question 5, give two examples in human populations of:

   a) continuous variation

   b) discontinuous variation. *(2)*

7. a) Give three examples of types of competition that occur between members of an animal species in the same population.

   b) In each case, suggest a variation that might help an individual to compete more effectively. *(6)*

8. The histogram shows the range and frequency of particular blood pressures (systolic) in a group of women in the 30–39 age group.

   a) Based on this evidence, could you say that blood pressure is a discontinuous variable?

   b) Justify your answer.

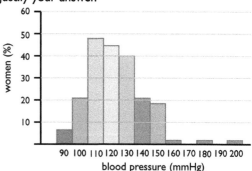

Variation in blood pressure in women aged 30–39.

Chapter 27: Natural Selection and the Formation of New Species

When you have completed this chapter, you will be able to:

- appreciate the importance of the work of Charles Darwin in producing the theory of natural selection
- understand the basic principles of natural selection
- quote specific examples of natural selection in operation
- understand how natural selection can lead to the formation of new species
- explain what is meant by artificial selection
- appreciate the difference between artificial selection and natural selection
- quote specific examples of artificial selection
- understand the basic principles of genetic engineering
- quote specific examples of the use of genetic engineering.

Humans have been asking the question 'Where did we come from?' for thousands of years. The theory of evolution, occurring by natural selection, is the most widely accepted scientific explanation of the answer to this question. The two terms are quite distinct:

- **Evolution** is a gradual change in the range of organisms on the Earth. New species continually arise from species that already exist, and other species become extinct.

- **Natural selection** is the *mechanism* by which new species arise. Natural selection 'allows' different forms of a species to survive in different areas. Over time, these different forms become increasingly different and may eventually become different species. If the environment of a species changes and that species is no longer adapted to survive in the new conditions, it may become extinct.

The inability of a species to survive changes in the environment will lead to that species becoming extinct. Factors that may lead to extinction of a species include diseases, habitat damage and overhunting.

Allopatric speciation – geographical isolation

New species are created when individuals within a population become isolated. Isolation may occur because of emigration of a group from the main population or because of a natural barrier, such as a mountain or river, separating members of a population. The barrier or distance prevents the separated groups from interbreeding. The two groups created may have differences between them that already existed because of selection pressure; however, with the isolation, more variations develop.

Variation arises in each group because of sexual reproduction and mutation (see Chapter 26). The changes that occur in one population may not occur in the other because of isolation. The two populations therefore evolve independently. With time, both populations become reproductively isolated. Even if they are reunited, the differences that exist would prevent them from interbreeding. Since reproductive isolation exists, the two populations have developed new species.

In the Galapagos Islands, Charles Darwin identified a number of species of finches. He found evidence to suggest that they had all evolved from one ancestral type that had colonised the islands from South America. They had been isolated from one another by the ocean. The main differences between the finches were in their beaks. Each species of finch had developed a unique beak that was adapted to the kinds of food it ate. Figure 27.1 shows some of the different beak types that evolved, and the beak of the likely ancestral finch.

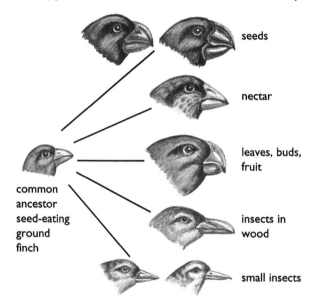

Figure 27.1 *Beak types of different species of finch that evolved and that of the likely common ancestor.*

Parapatric speciation – ecological and behavioural differences

A species may be dispersed over a large geographic area with no physical barrier to separate the individuals of the species. This allows any member of the species to interbreed. Even with the lack of a physical barrier, individuals tend to mate only with others within their own geographic region. The differences in the habitats across the region will lead to the development of distinct behaviours and characteristics within the organisms. This leads to reproductive isolation and the formation of a new species that has adapted to its environment in order to occupy different niches.

The Caribbean *Anolis* lizard has evolved into different species that live in a large geographic area with no physical barriers. The different species have evolved features to adapt to different areas of their environment. This is evident in characteristics such as their limb length and toe pad size. Ground-dwelling *Anolis* species forage through twigs, so they have slender bodies, short legs and small toe pads that allow them to move along surfaces that have a small diameter. Another *Anolis* species lives on tree trunks. These lizards are stocky and have long legs – characteristics that enhance running and jumping. Some species found high up in the trees have large toe pads, which are essential for clinging.

Figure 27.2 *The brown anole lizard (Anolis sagrei) is adapted to living on the ground.*

Figure 27.3 *The green anole lizard (Anolis carolinensis) is adapted to living in treetops.*

The work of Charles Darwin

At the age of 22, Charles Darwin became the ship's naturalist on *HMS Beagle*, which left England for a five-year voyage in 1831.

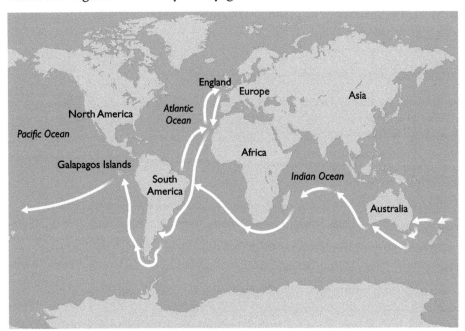

Figure 27.5 *The five-year journey of HMS Beagle.*

Figure 27.4 *Charles Darwin (1809–1882).*

During the voyage, Darwin collected hundreds of specimens and made many observations about the variety of organisms and the ways in which they were adapted to their environments. He gained much information, in particular, from the variety of life forms in South America and the Galapagos Islands.

On his return to England, Darwin began to evaluate his data and wrote several essays, introducing the ideas of natural selection. He arrived at his theory of natural selection from observations made during his voyage on the Beagle and from deductions made from those observations. Darwin's observations were that:

- organisms tend to produce more offspring than are needed to replace them – a single female salmon can release 5 million eggs per year; a giant puffball fungus produces 40 millions spores

The phrase 'survival of the fittest', does not refer to physical fitness, but to how well adapted an organism is to its environment. However, Darwin never used this phrase.

Darwin was not aware of genes and how they determine characteristics when he put forward his theory of natural selection. Gregor Mendel had yet to publish his work on inheritance.

Figure 27.6 *These moths are different forms of the same species – Biston betularia, the peppered moth. Both forms are food for several species of birds. Both are found in areas with clean air and in smoke-polluted areas. Which form is best adapted to which area?*

- despite the over-reproduction, stable, established populations of organisms tend to remain the same size – the seas are not overflowing with salmon, and you are not surrounded by piles of giant puffball fungi

- members of the same species are not identical – living things vary.

He made two important deductions from these observations:

- From the first two observations, he deduced that there is a 'struggle for existence'. Many offspring are produced, yet the population stays the same size. There must be competition for resources and many must die.

- From the third observation, he deduced that if some offspring survive whilst others die, those organisms best equipped or best suited to their environment will survive to reproduce. Those less suited will die. This gave rise to the phrase 'survival of the fittest'.

Notice a key phrase in the second deduction – the best-suited organisms survive *to reproduce*. This means that those characteristics that give the organism a better chance of surviving will be passed on to the next generation. Those organisms that are less suited to the environment, survive to reproduce in smaller numbers. The next generation will have more of the type that is adapted and fewer of the less well-adapted type. This will be repeated in each generation.

Evidence for natural selection

The theory of natural selection proposes that some factor in the environment 'selects' which forms of a species will survive to reproduce under those conditions. Forms that are not well adapted will not survive. Any evidence for natural selection must show that:

- there is variation within the species

- changing conditions in the environment (a **selection pressure**) favours one particular form of the species (which has a **selective advantage**)

- the frequency of the favoured form increases (it is selected *for*) under these conditions (survival of the fittest)

- the frequency of the less well-adapted form decreases under these conditions (it is selected *against*)

- the changes are not due to any other factor.

The peppered moth

The peppered moth is found throughout the United Kingdom. It has two forms, one is greyish-white with dark markings ('peppered') and one is much darker (Figure 27.6).

Initially, nearly all the moths were of the peppered type. The first record of a dark moth in Manchester (a city in the north of England) was in 1848. Around that time, Manchester was heavily industrialised. By 1895, 98% of the moths in Manchester were of the dark form. What had caused this change?

The following pieces of information will give you some clues:

- Peppered moths are food for birds.

- Increasing industrialisation during the 19th century killed off many of the lichens growing on tree trunks and also covered the trunks with soot.

- The distribution of the two forms of peppered moth was linked to the degree of industrialisation (Figure 27.7).

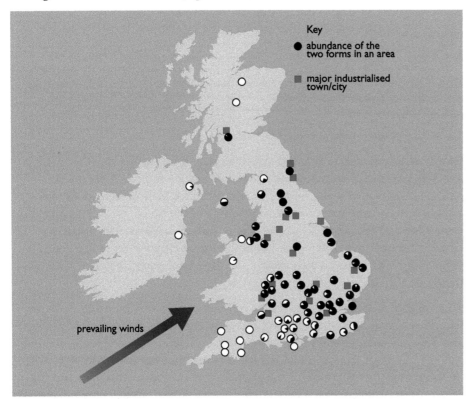

Figure 27.7 *The distribution of the two forms of the peppered moth in the United Kingdom in the 1950s.*

Natural selection explains the peppered moth distribution in the following way:

- In any area, there is an over production of offspring and so there is a struggle for existence. Because of this, the form most suited to its environment will survive.

- In the country, tree trunks are still relatively unpolluted; in the cities, the tree trunks and buildings are covered with soot.

- In the country, the peppered form is camouflaged; in the cities, the dark form is camouflaged.

- Camouflaged moths are less likely to be eaten by birds.

- In the country, more of the peppered form survive to reproduce; in the cities, more of the dark form survive to reproduce.

- Over many generations, the numbers of the dark form increase in the cities, while the numbers of the peppered form remain high in the countryside. Table 27.1 on page 376 illustrates this.

Feature of natural selection	Effect on population of peppered moths in:	
	Countryside	**City**
selection pressure	predation by birds on moths on clean tree trunks	predation by birds on moths on soot-covered tree trunks and other surfaces
natural variation in the species	some moths are 'peppered', others are dark	
type with selective advantage	peppered form (camouflaged on clean, lichen-covered tree trunks)	dark form (camouflaged on dark, soot-covered surfaces)
type selected for	peppered form	dark form
type selected against	dark form	peppered form
result of natural selection over many generations	percentage of peppered form increases or remains high, percentage of dark form decreases or remains low	percentage of dark form increases or remains high, percentage of peppered form decreases or remains low

Table 27.1: *Peppered moths as evidence for natural selection.*

Shortly after the data in the map was obtained, a law called the 'Clean Air Act' was passed. This limited the amount of smoke and other pollutants emitted from factories and houses in certain areas. 'Smoke free zones' were established. As time went by, the regulations of the Clean Air Act became more rigid and less and less smoke was permitted in cities. The surfaces of trees became less polluted and buildings were cleaned. More and more peppered forms survived to reproduce as the cleaner surfaces once again gave them camouflage.

This appears to be excellent evidence for natural selection in action. It shows natural selection operating in one direction as the cities became more polluted and then reversing as the cities became cleaner again. The independent variable was the nature of the surface of the tree trunks, the dependent variable was the percentage of each form of the peppered moth. Industrialisation and the Clean Air Act changed the independent variable in the cities. Throughout it all, the countryside acted as a kind of unchanging control experiment. This showed that over the same period, when the independent variable was *not* changed, the percentage of each form of the peppered moth was unaltered. Other factors were not causing the change – it *must* have been the nature of the surfaces offering camouflage to different forms of the moth.

Sickle-cell anaemia and malaria

Sickle-cell anaemia is caused by a mutant allele (see Chapter 25). It affects the formation of haemoglobin in red blood cells. The abnormal haemoglobin causes the red blood cells to become distorted (sickle shaped) when the oxygen concentration of the surroundings is low. The condition can be fatal in individuals homozygous for the allele.

Heterozygous 'carriers' of the allele usually show no symptoms of the disease at all, although 50% of the haemoglobin in their red blood cells is abnormal. They do have an important benefit, however. They are more resistant to malaria than people with 100% normal haemoglobin (homozygous for the normal allele).

The red blood cells of carriers look normal, but are slightly more fragile than normal red blood cells (because of the 50% abnormal haemoglobin).

Recently, some biologists have questioned the methods by which the peppered moth data was obtained. They do not, however, think that the conclusion about the way in which natural selection is thought to operate is necessarily wrong. They think that it presents a picture that is too clear cut and further data, obtained in a more rigorous manner, is needed to support the conclusion.

Homozygous means having two alleles of a gene that are the same (e.g. two alleles for sickle-cell or two normal alleles). Heterozygous means having two different alleles of a gene (e.g. one sickle-cell allele and one normal allele). People heterozygous for sickle-cell anaemia are called 'carriers'.

The malarial parasite is transmitted by the female *Anopheles* mosquito and spends part of its life cycle inside red blood cells (see Chapter 23). When these parasites enter the fragile red blood cells of carriers, the cells often burst before the parasite has time to develop and the parasite dies. The life cycle is broken. Table 27.2 shows how natural selection affects the incidence of sickle-cell anaemia in an area where malaria is common.

Feature of natural selection	Effect on incidence of sickle-cell anaemia
selection pressure	infection by the malarial parasite
natural variation in the species	carriers (people who are heterozygous for the sickle-cell allele) and people homozygous for the normal allele; people homozygous for the sickle-cell allele often die at an early age
type with selective advantage	carriers – malarial parasite cannot complete life cycle
type selected for	carriers
type selected against	people with 100% normal haemoglobin (homozygous for normal allele)
result of natural selection over many generations	numbers of heterozygotes in the population are higher than in areas where malaria is absent; numbers of people suffering from the disease are also higher than in other areas

Table 27.2: *Sickle-cell anaemia and natural selection.*

The carriers have a selective advantage over those homozygous for the normal allele in areas where malaria is common. However, if two carriers marry, they can produce children who are homozygous for the sickle-cell allele. As a result, sickle-cell anaemia is more common in these areas also (Figure 27.8).

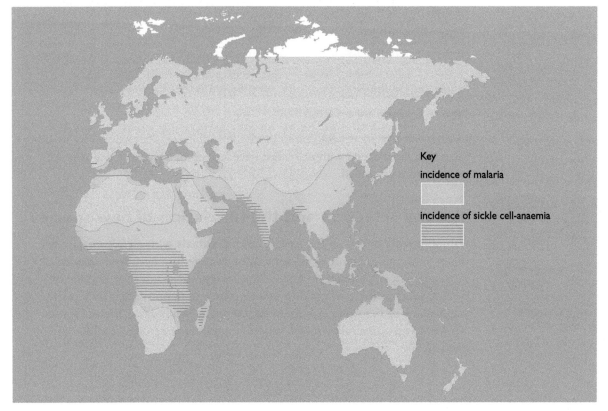

Key

incidence of malaria

incidence of sickle cell-anaemia

Figure 27.8 *A map showing areas of the world where sickle-cell anaemia and malaria are common.*

Natural selection in antibiotic-resistant bacteria

Alexander Fleming discovered penicillin, the first antibiotic, in 1929. Since then, other natural antibiotics have been discovered and many more have been synthesised in laboratories. The use of antibiotics has increased dramatically, particularly over the last 20 years. We now almost expect to be given an antibiotic for even the most trivial of ailments. This can be dangerous, as it leads to the development of bacterial resistance to an antibiotic (Figure 27.9).

Mutations happen all the time in all living organisms. In bacteria, a chance mutation could give a bacterium resistance to an antibiotic. In a situation where antibiotics are widely used, this new resistant bacterium has an advantage over non-resistant bacteria of the same type.

The resistant bacterium will survive and multiply in greater numbers than the non-resistant types. The generation time of a bacterium can be as short as 20 minutes. This means that there could be 72 generations in a single day – the equivalent of about 1 500 years of human generation time. The numbers of resistant types would increase with each generation. Very soon a population of bacteria could become almost entirely made up of resistant types. Table 27.3 shows how natural selection can introduce resistance to an antibiotic in a population of bacteria.

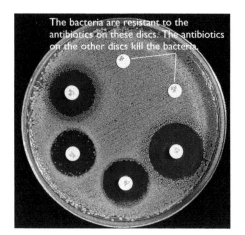

The bacteria are resistant to the antibiotics on these discs. The antibiotics on the other discs kill the bacteria.

Figure 27.9 *Bacterial resistance to an antibiotic.*

Feature of natural selection	Effect on population of non-resistant bacteria
selection pressure	repeated use of antibiotics
natural variation in the species	some are resistant (due to a chance mutation), others are not
type with selective advantage	resistant type – will survive antibiotic treatment
type selected for	resistant type
type selected against	non-resistant type
result of natural selection over many generations	percentage of resistant types in the population increases

Table 27.3: *Bacteria and natural selection.*

Doctors are now more reluctant to prescribe antibiotics. They know that by using them less, the bacteria with resistance have less of an advantage and will not become as widespread.

Parapatric speciation also occurs:
- when a population becomes separated because groups breed at different intervals
- when populations behave differently, for example, a group of birds may sing a particular song, which is different from the song of another group of birds; the songs will not be identified by either group of birds.

Natural selection and the formation of new species

Natural selection favours the survival of individuals with an advantage over others in the population. Consequently, over time, the least well-adapted members of a population do not survive to reproduce, and the population becomes increasingly adapted to its environment. Suppose that there were two populations of the same species in different environments. Different forms would have an advantage in the different environments (Figure 27.10, page 379).

1 A population of plants lives in a fairly normal type of soil, with normal rainfall.

2 Some of these plants colonise a different area, where water is found much deeper in the soil and the rainfall is considerably less. In this new environment, natural selection favours those plants with longer roots (able to reach the soil water) and smaller leaves with fewer stomata (to minimise water loss).

3 The two populations of plants are isolated from each other and cannot interbreed.

4 There is natural variation in these features in both populations as a result of sexual reproduction and gene mutation.

5 In the original population, longer roots and smaller leaves give no advantage and natural selection maintains the original form for as long as the environment remains stable.

6 In the new population, longer roots and smaller leaves give an advantage, as plants with these features gain more water and lose less than those without them. In each generation, more plants with these features survive than those without them.

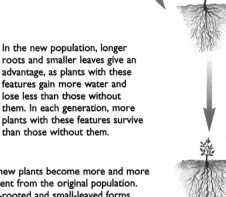

7 The new plants become more and more different from the original population. Long-rooted and small-leaved forms survive best and the population eventually consists almost entirely of this type.

8 Eventually, the two populations are so different that they cannot interbreed. At this point, we consider them to be separate species.

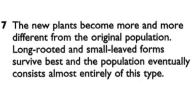
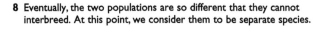
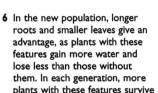

Figure 27.10 *How natural selection can lead to the formation of a new species.*

1. State Charles Darwin's main observations.
2. Describe the two main deductions that Darwin made from these observations.
3. What is meant by the survival of the fittest?
4. How does the changing distribution of the two forms of the peppered moth provide evidence for natural selection?
5. Explain how natural selection could lead to the formation of new species.

Artificial selection

The cultivation of the first wheat and barley, and the domestication of the first stock animals took place in the Middle East 12 000 years ago. Since then, humans have tried to obtain bigger yields from them. They cross-bred different maize plants (and barley plants) to obtain strains that produced more grain. They bred sheep and goats to give more milk and meat – selective breeding had begun. Today, animals and plants are bred for much more than food. They are bred to produce a range of medicines, and for research into spare-part surgery and the action of drugs.

Selective breeding is best described as the breeding of only those individuals with desirable features. It is sometimes called '**artificial selection**', as human choice, rather than environmental factors, is providing the **selection pressure**.

The methods used today for selective breeding are vastly different from those used only 50 years ago. Modern gene technology makes it possible to create a new strain of plant within weeks, rather than years. These new techniques raise serious moral and ethical questions, which will be discussed later.

Traditional selective breeding

Plants

Traditionally, farmers have bred crop plants of all kinds to obtain increased yields. Probably the earliest example of selective breeding was the cross-breeding of strains of wild wheat. The aim was to produce wheat with a much increased yield of grain (Figure 27.11). This wheat was used to make bread.

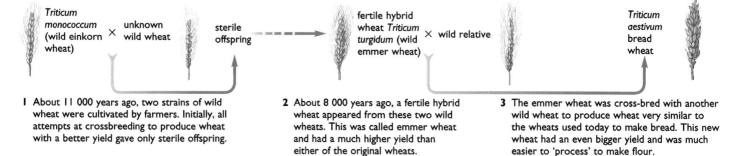

1 About 11 000 years ago, two strains of wild wheat were cultivated by farmers. Initially, all attempts at crossbreeding to produce wheat with a better yield gave only sterile offspring.

2 About 8 000 years ago, a fertile hybrid wheat appeared from these two wild wheats. This was called emmer wheat and had a much higher yield than either of the original wheats.

3 The emmer wheat was cross-bred with another wild wheat to produce wheat very similar to the wheats used today to make bread. This new wheat had an even bigger yield and was much easier to 'process' to make flour.

Figure 27.11 *Modern wheat is the result of selective breeding by early farmers.*

The production of modern bread wheats by selective breeding is probably one of the earliest examples of producing genetically modified food. Each original wild wheat species had 14 chromosomes per cell. The wild emmer hybrid had 28 chromosomes per cell. Modern bread wheat has 42 chromosomes per cell. Selective breeding has modified the genetic make-up of wheat.

Other plants have been selectively bred for certain characteristics. *Brassica* is a genus of cabbage-like plants. One species of wild brassica (*Brassica olera*) was selectively bred to give several strains, each with specific features. Some of the strains had large leaves, others had large flower heads, and others produced large buds.

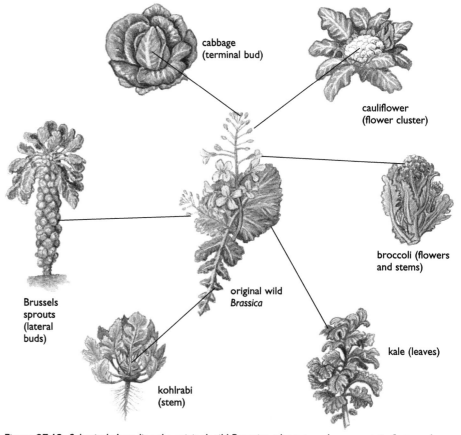

cabbage (terminal bud)

cauliflower (flower cluster)

broccoli (flowers and stems)

Brussels sprouts (lateral buds)

original wild *Brassica*

kohlrabi (stem)

kale (leaves)

Figure 27.12 *Selectively breeding the original wild Brassica plants to enhance certain features has produced several familiar vegetables.*

Selective breeding has produced many familiar vegetables. Besides the ones produced from *Brassica*, selective breeding of wild *Solanum* plants has produced the many strains of potatoes that are eaten today. Carrots and parsnips are also the result of selective breeding programmes.

Crop plants are bred to produce strains that:

- give higher yields
- are resistant to certain diseases (the diseases would reduce the yields)
- are resistant to certain insect pest damage (the damage would reduce the yield)
- are hardier (so that they survive in harsher climates or are productive for longer periods of the year)
- have a better balance of nutrients in the crop (for example, plants that contain more of the types of amino acids needed by humans).

Animals

Farmers have bred stock animals for similar reasons to the breeding of crops. They have selected for animals that:

- produce more meat, milk or eggs
- produce more fur or better quality fur
- produce more offspring
- show increased resistance to diseases and parasites.

For many thousands of years, the only way to improve livestock was to mate a male and a female with the features that were desired in the offspring. In cattle, milk yield is an important factor and so high yielding cows would be bred with bulls from other high-yielding cows. Examples of this are the Jamaica Hope and the Buffalypso of Trinidad and Tobago (Figure 27.13).

The **Jamaica Hope** was developed by Dr Thomas Lecky because the cattle in Jamaica, which were from temperate regions, could not tolerate the heat and were susceptible to pests and diseases. Dr Lecky crossed three different breeds of cattle to create a breed that is large in size, has a high milk yield, tolerates high temperatures, and is resistant to tropical pests and diseases.

Four breeds of water buffalo were introduced to Trinidad to work in the cane fields. Dr Steve Bennett started the selective breeding of the water buffalo – and the **buffalypso** was created. The breed was selectively bred for its strength, resistance to disease, efficient digestive system and high milk and meat quality.

Figure 27.13 *Jamaica Hope cattle.*

Figure 27.14 *The many different breeds of dog all originate from a common ancestor – the wolf.*

Since the Second World War, the technique of **artificial insemination (AI)** has become widely available. Bulls with many desirable features ('superior bulls') are kept in special centres. Semen obtained from these bulls is diluted, frozen and stored. Farmers can purchase quantities of this semen to inseminate their cows. AI makes it possible for the semen from one bull to be used to inseminate many thousands of cows.

Modern sheep are domesticated wild sheep, and pigs have been derived from wild boars. Just think of all the varieties of dogs that now exist. All these have been derived from one ancestral type. This original 'dog' was a domesticated wolf (Figure 20.14). In domesticating the wolf, humans gained an animal that was capable of herding stock animals. The sheepdog has all the same instincts as the wolf except the instinct to kill. This has been selectively 'bred out'.

6. In what way is artificial selection similar to natural selection?
7. How is artificial selection different to natural selection?
8. Describe one example of selective breeding in plants.
9. Describe one example of selective breeding in animals.
10. Give some of the aims of selective breeding in animals and plants.

Genetic engineering

DNA – the stuff of genes

A gene is a section of one strand of a DNA molecule that codes for the production of a protein. Each sequence of three bases (a triplet) in the DNA strand codes for one amino acid. Different genes produce different proteins because each has a unique sequence of bases that codes for a unique sequence of amino acids – that results in a unique protein (Figure 27.15).

one gene copied   travels out of nucleus into cytoplasm; at ribosome, tRNA brings amino acids in correct order

Figure 27.15 *The role of DNA in protein synthesis.*

The protein that is produced could be:

- an enzyme that controls a particular reaction inside a cell or in the digestive system

- a structural protein like the keratin in hair, collagen in skin or one of the many proteins found in the membranes of cells

- a hormone

- a protein with a specific function, such as haemoglobin or an antibody.

Recombinant DNA

Producing recombinant DNA is the basis of gene technology or genetic engineering. A section of DNA – a gene – is snipped out of the DNA of one species and inserted into the DNA of another. This new DNA is called **recombinant** DNA, as the DNA from two different organisms has been 'recombined'. The organism that receives the new gene is a **transgenic organism.**

The organism receiving the new gene now has an added capability. It will manufacture the protein its new gene codes for. For example, a bacterium receiving the gene from a human that codes for insulin production will make human insulin. If these transgenic bacteria are cultured by the billion in a fermenter, they become a human insulin factory.

Producing genetically modified (transgenic) bacteria

The breakthrough in being able to transfer DNA from cell to cell came when it was found that bacteria have two sorts of DNA – the DNA found in their bacterial 'chromosome', and much smaller circular pieces of DNA called **plasmids** (Figure 27.16).

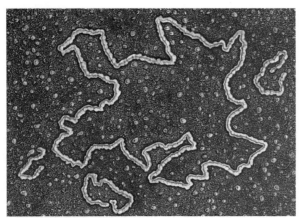

Figure 27.16 *Bacterial DNA.*

Bacteria naturally 'swap' plasmids, and biologists found ways of transferring plasmids from one bacterium to another. The next stage was to find molecular 'scissors' and molecular 'glue' that could snip out genes from one molecule of DNA and then stick them back into another. Further research found enzymes that were able to do this:

- **Restriction endonucleases** are enzymes that cut DNA molecules at specific points. Different restriction enzymes cut DNA at different places. They can be used to cut out specific genes from a molecule of DNA.

- **DNA ligases** are enzymes that join cut ends of DNA molecules.

Biologists now had a method of transferring a gene from any cell into a bacterium. They could insert the gene into a plasmid and then transfer the plasmid into a bacterium. The plasmid is called a **vector** because it is the means of transferring the gene. The main processes involved in producing a transgenic bacterium are shown in Figure 27.17 on page 384.

A transgenic organism is one that contains a gene or several genes from another species. For example, some bacteria have had human genes transferred to them that allow them to make human insulin. Some sheep secrete AAT in their milk because they have the human gene that directs the manufacture of this substance. Because they contain 'foreign' genes, they are no longer quite the same organisms. They are transgenic.

A bacterial 'chromosome' is not like a human chromosome. It is a continuous loop of DNA rather than a strand. Also, the DNA is 'naked' – there are no proteins in it – only DNA.

There is a lot more to producing recombinant DNA and transgenic bacteria than is shown in Figure 27.17. You could carry out an Internet search or search appropriate CD-encyclopaedias to find out more.

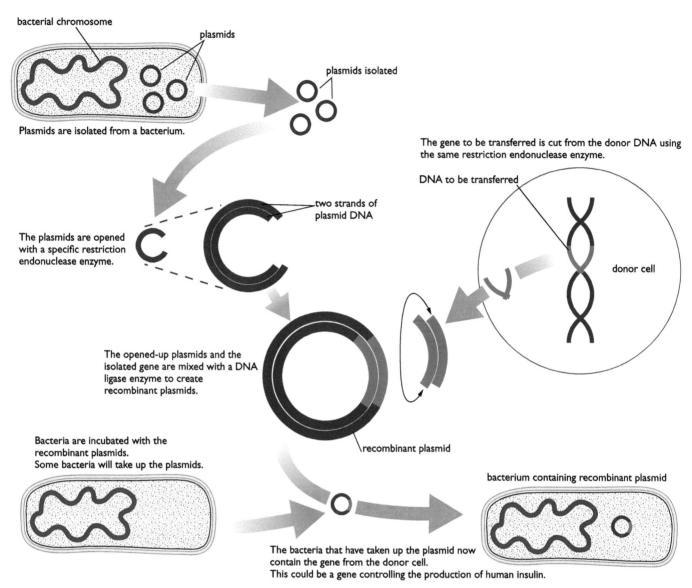

bacterial chromosome

plasmids

plasmids isolated

Plasmids are isolated from a bacterium.

The gene to be transferred is cut from the donor DNA using the same restriction endonuclease enzyme.

DNA to be transferred

two strands of plasmid DNA

The plasmids are opened with a specific restriction endonuclease enzyme.

donor cell

The opened-up plasmids and the isolated gene are mixed with a DNA ligase enzyme to create recombinant plasmids.

recombinant plasmid

Bacteria are incubated with the recombinant plasmids.
Some bacteria will take up the plasmids.

bacterium containing recombinant plasmid

The bacteria that have taken up the plasmid now contain the gene from the donor cell.
This could be a gene controlling the production of human insulin.

Figure 27.17 *Stages in producing a transgenic bacterium.*

Making use of genetically modified bacteria

Different bacteria have been genetically modified to manufacture a range of products. Once they have been genetically modified, they are cultured in fermenters to produce large amounts of the product. Some examples are described here:

- **Human insulin** – People suffering from diabetes need a reliable source of insulin. Before the advent of genetic engineering, the only insulin available came from other animals. This is not quite the same as human insulin and does not give quite the same control of blood glucose levels.

More insulin is required every year because the number of diabetics increases world-wide each year and diabetics now have longer life spans.

- **Enzymes for washing powders** – Many stains on clothing are biological. Blood stains are largely proteins, grease marks are largely lipids. Enzymes can digest these large, insoluble molecules into smaller, soluble ones. These then dissolve in the water. Amylases digest starch, proteases digest proteins and lipases digest lipids. Bacteria have been genetically engineered to produce enzymes that work at higher temperatures, allowing even faster and more effective action.

- **Enzymes in the food industry** – One bacterial enzyme used in the food industry is **glucose isomerase**. This enzyme turns glucose into a similar sugar called fructose. Fructose is much sweeter than glucose and so less is needed to sweeten foods. This has two advantages – it saves money (less is used) and it means that the food contains less sugar and is healthier.

- **Human growth hormone** – The pituitary gland of some children does not produce sufficient quantities of this hormone and their growth is retarded. Injections of growth hormone from genetically modified bacteria restore normal growth patterns.

- **Bovine somatotrophin (BST)** (a growth hormone in cattle) – This hormone increases the milk yield of cows and increases the muscle (meat) production of bulls. Giving injections of BST to dairy cattle can increase the milk yield by up to 10 kg per day. To do this, they need more food, but this increased cost is more than offset by the increased income from the increased milk yield (Table 27.4).

	Feed (kg day $^{-1}$)	Milk output (kg day $^{-1}$)	Milk to feed ratio
without BST	34.1	27.9	0.82
with BST	37.8	37.3	0.99

Table 27.4: *Effects of BST on milk yield.*

- **Human vaccines** – Bacteria have been genetically modified to produce the antigens of the Hepatitis B virus. These are used in the vaccine against Hepatitis B. The body makes antibodies against the antigens but there is no risk of contracting the actual disease from the vaccination.

Since the basic technique of transferring genes was worked out, many unicellular organisms have been genetically modified to produce useful products. Also, other techniques for transferring genes into larger organisms have been developed.

Some moral, ethical and practical concerns about genetic engineering

Morality is our personal sense of what is right, or acceptable, and what is wrong. It is not necessarily linked to legality. **Ethics** have a sense of right and wrong also, but they are the not one individual's opinions. Ethics represent the 'code' adopted by a particular group to govern its way of life.

Before human growth hormone from genetically modified bacteria was available, the only source of the hormone was from human corpses. This was a rather gruesome procedure and had health risks. A number of children treated in this way developed Creutzfeld–Jacob disease (the human form of 'mad cow' disease). When this became apparent, the treatment was withdrawn.

Issues	Concerns about genetic engineering
Genes transferred from one species to another species could not normally have got there.	This is a moral issue. Many people feel that a species should not be altered in any way.
We do not know enough about the long-term ecological effects of introducing genetically modified organisms into fields. They may out-compete wild plants and take over an area. They may be toxic to wild animals. Because we do not know, we should not, therefore, introduce such plants into fields. Some people believe that we should not even carry out trials to find out.	Again, this is a moral view. We could never know this with any new breed of plant. The early farmers who crossbred wild wheats had no idea of the impact their new plants would have. Does this make it wrong?
When genetically modified plants are planted, the 'new' genes may 'jump' into other species of plants. For instance, genes for herbicide (weedkiller) resistance may transfer into the weeds.	This is not a moral issue, but a purely practical one. Is it likely that resistance genes could be transferred into weeds in sufficient numbers to make a real difference? We can only find this out by conducting field trials.
Using genetic fingerprinting to combat crime will only really be useful if there is a genetic database – a file of the genetic fingerprints of everyone in the country. Once the human genome project finally identifies all the genes, who will have access to this information?	There are concerns that a genetic database would be subject to misuse and that evidence could be manufactured. Also, if insurance companies had access to the genetic database, they may refuse car insurance, or charge higher premiums, to a person with an increased risk of heart disease, for example. Employers could (secretly) refuse employment to a person because their 'genetic profile' did not meet the requirements.
Is it acceptable to genetically modify pigs so that they can be bred to provide organs for humans?	Again, this is a moral issue. While some people find this unacceptable, their opinion does not necessarily make it wrong. Other people find this just as acceptable as breeding pigs to produce more or leaner bacon for human consumption. This is especially true of many people on transplant waiting lists.
Gene technology may give doctors the ability to create designer babies. They may be able to obtain a newly fertilised human egg, determine its genotype and ask the parents which genes they would like to modify. They might start only by replacing genes that actually cause disease, but they may then be led into replacing other genes.	Most doctors would find this morally *and* ethically unacceptable. They may consider replacing genes that cause disease, but not replacing genes just to improve a child's image in the eyes of its parents. However, if and when such practices become possible, who will define what is ethically acceptable for doctors? What will be the dividing line between 'cosmetic gene therapy' and 'medical gene therapy'?

Table 27.5: *Some concerns about genetic engineering.*

11. What is recombinant DNA?
12. What is a transgenic organism?
13. Describe the roles of restriction enzymes and ligase enzymes in the genetic engineering of bacteria.
14. Describe two uses of genetic engineering.
15. Describe some of the concerns that people have about the use of genetic engineering.

Chapter summary

In this chapter you have learnt that:

- new species are formed due to geographical or behavioural isolation of organisms. This leads to reproductive isolation
- new species are specially adapted to their environment
- Charles Darwin's theory of natural selection describes the process by which species change over many generations
- the theory of natural selection proposes that:
  - because of over-reproduction that takes place, there is a *struggle for survival* between members of a species
  - because of the variation in a species, those most suited to their environment will survive to reproduce – the *survival of the fittest*
- the different proportions of the melanic and light forms of the peppered moth in cities and country environments provides evidence of natural selection in operation
- natural selection operating on two populations of the same species in different environments will result in the populations becoming more and more different over time; they may eventually become two different species
- artificial selection also allows certain types to survive whilst others do not, but humans choose the types that will be most advantageous to them and use selective breeding to ensure the survival of those types
- genetic engineering often involves the transfer of genes from one species into another; the new organisms formed are called transgenic organisms
- bacteria have been genetically engineered to produce many useful products, such as human insulin and bovine somatotrophin
- many scientists and other people are concerned about the long-term effects that genetic engineering may have on the environment, as well as being concerned about our right to carry out genetic engineering at all.

Questions

1. Natural selection suggests that the best adapted individuals in a population survive because:

   A more offspring are born than can survive
   B there is variation among individuals
   C neither of these
   D both of these

2. A gene is:

   A a feature within an organism
   B a molecule of DNA
   C a section of a molecule of DNA that codes for a feature
   D a structure that holds chromosomes together

3. Dominant alleles:

   A will only express themselves if there is a pair of alleles in an organism
   B will express themselves even if they are with the recessive allele
   C determine the most favourable of a pair of alternate features
   D will always be inherited in preference to recessive alleles

4. Tallness in pea plants is dominant. If two heterozygous tall pea plants are crossed, the ratio of tall to dwarf plants in the offspring will be:

   A 2:1
   B 1:1
   C 4:0
   D 3:1

5. Selective breeding is sometimes called 'artificial selection'.

   a) How is selective breeding similar to natural selection? *(4)*

   b) How is selective breeding different from natural selection? *(3)*

6. The changes in the distribution of the peppered moth as areas became industrialised is often used as evidence of natural selection. There are two forms of the peppered moth, the peppered form and the dark form.

   a) Explain why the dark form of the moth became much more common in industrialised cities. *(4)*

   b) Explain why the dark form of the moth remained extremely rare in the countryside during the same period. *(3)*

   c) Suggest why the two types of the peppered moth have not become separate species. *(3)*

7. Warfarin is a pesticide that was developed to kill rats. When it was first used in 1950, it was very effective. Some rats, however, had a mutant allele that made them resistant to warfarin. Nowadays the pesticide is much less effective.

   a) Use the ideas of natural selection to explain why warfarin is much less effective than it used to be. *(4)*

   b) Suggest what might happen to the number of rats carrying the warfarin resistance allele if warfarin were no longer used. Explain your answer. *(4)*

8. The diagram shows the results of a breeding programme to improve the yield of maize (sweetcorn).

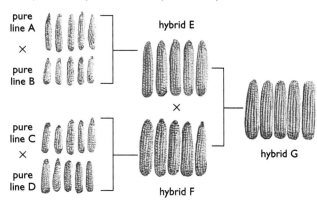

   a) Describe the breeding procedure used to produce hybrid G. *(3)*

   b) Describe *three* differences between the corn cobs of hybrid G and those of hybrid C. *(3)*

   c) How could you show that the differences between hybrid G and hybrid C are genetic? *(4)*

9. Write an essay about the benefits and concerns of selective breeding of animals. You should produce about one side of A4 word-processed work. Use books and the Internet to find out more information. *(10)*

10. A propagation progamme that selects the characteristics of a population of organisms is called artificial selection.

    a) Describe how this process is different from natural selection.

    b) Describe how this process is similar to natural selection. *(4)*

11. **a)** Describe the difference between artificial selection and genetic engineering.

    b) Genetic engineering has been used to change the characteristics of organisms. What are some of the concerns people have with this process? *(6)*

12. For natural selection to operate, some factor has to exert a 'selection pressure'. In each of the following situations, identify both the selection pressure and the likely result of this selection pressure.

    a) Near old copper mines, the soil becomes polluted with copper ions that are toxic to most plants. *(2)*

    b) In the Serengeti of Africa, wildebeest are hunted by lions. *(2)*

    c) A farmer uses a pesticide to try to eliminate pests of a potato crop. *(2)*

13. The theory of natural selection was proposed by Charles Darwin to explain the variation in living organisms.

    a) Explain how natural selection is thought to operate using a named example to illustrate your answer. *(4)*

    b) Explain how natural selection operating on two populations of a species in different environments could eventually lead to the formation of two distinct species. *(4)*

    c) The graph shows the yield of oysters in a bay in Canada over a 25-year period. A new disease was introduced into the area in 1915.

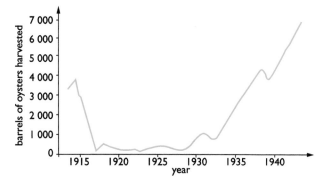

    Explain how natural selection could account for:

    i) the initial decrease in yield from 1915–1917

    ii) the increase from 1930 onwards. *(4)*

Appendix 1: School-Based Assessment Guide

The School-Based Assessment, or SBA, is an important part of the marks a student will achieve in CXC Biology. In the syllabus the teacher can find comprehensive advice on conducting the SBA. Here we provide sample SBA practicals with suggested mark schemes. This is for the benefit of both the student and the teacher. These practicals test the full range of SBA skills, namely:

- ORR   Observation, Recording, Reporting

- MM   Manipulation, Measurement

- AI   Analysis and Interpretation

- PD   Planning and Design

- Dr   Drawing

It is often possible to assess more than one of the above skills within the same practical, as will be shown. The SBA component of the syllabus is examined over a two-year period. These practical activities are meant *'to assist the students in acquiring certain knowledge, skills, and attitudes that are critical to the subject'* (Biology syllabus, page 32). Notice that the skills are marked out of 10. This simple decimal scale will make easier the conversion of scores (achieved by each individual student in each practical) into a final overall SBA score.

An example of an SBA practical testing ORR, MM and AI

Title: *The effect of temperature on catalase activity*

Aim: *To find out if temperature has an effect on the activity of catalase enzyme*

Apparatus/Materials:
4 test-tubes, large beaker to act as a water bath, stopwatch, 200 cm^3 hydrogen peroxide (H_2O_2) of 1.5 mol dm^3, forceps, thermometer, measuring cylinder, 10 cm^3 of potato extract in a Petri dish, squares of filter paper of size 1 cm^2, warm water, tap water, ice cubes

Method: 1. Prepare a water bath at a temperature of 60°C.

2. Measure out 20 cm^3 hydrogen peroxide and pour it into a test-tube.

3. Place the test-tube in the water bath for 3 minutes.

4. Dip a piece of filter paper in the potato extract using a pair of forceps.

5. Shake off any excess liquid.

6. Hold the filter paper above the hydrogen peroxide in the test-tube.

7. Drop the filter paper into the hydrogen peroxide. It should sink.

8. Record the time taken in seconds for the paper to rise to the surface.

9. Repeat using a fresh piece of filter paper.

10. Repeat steps 1–10 with three other temperatures between 10°C and 45°C.

Skills tested:

Manipulation/measurement

Use of basic laboratory equipment

1 Accurately measuring 20 cm³ of hydrogen peroxide (resting on flat surface; avoiding spillage; reading the meniscus at eye level).   3

2 Accurate timing of the activity in each test-tube.   3

3 Preparation of the water bath.   2

4 Accurately measuring the temperature of the water bath (immersion of bulb completely in water; lack of contact of the bulb with the beaker; immersion time adequate for equilibration; reading at eye level).   2

TOTAL   **10**

Observation/Reporting/Recording

There are three aspects to this skill. Hence, when devising a mark scheme, all three aspects should be examined.

Reporting

Format: clearly stated aim; apparatus and materials listed; method is stated in logical steps; table of results present; graph of results present.   2

Discussion/Conclusion present:

Complete description and correct sequence of procedure.   2

Language – past tense, Standard English, passive voice.   1

Observations

Bubbles form on paper, paper rises.   1

Recording

Table:

- Appropriate title (capital letters, underlined, placed on top, descriptive)   1
- Appropriate headings for columns (physical quantity in heading, units stated)   1
- Appropriate content   1
- Completeness/neatness   1

Graph:

- Title (below/above graph, in capitals, descriptive)   1
- Axes (*x*-axis –temperature, *y*-axis – time)   2
- Scale (more than half of graph paper used)   1

- Accuracy (points plotted accurately)   4

- Curve (joined appropriately to show cause and effect)   2

TOTAL   **20**

In the above, the marks would be scaled down to 10, i.e. divided by 2.

Analysis and Interpretation

Analysis and Interpretation has four aspects: background information, explanations of results, conclusions and limitations. A common heading – 'Conclusion' - might normally be used to cover all four aspects. However, some teachers may prefer to divide it into 'Discussion' and 'Conclusion' as this may help the student to organise his/her thoughts.

Background information   2

Catalase breaks down hydrogen peroxide to water and oxygen.

Enzymes are proteins.

Increasing temperature causes molecules to move faster and speeds up reactions.

High temperatures (over the optimum) cause enzymes to be denatured.

Explanation of Results   3

Oxygen produced by the breakdown of hydrogen peroxide by catalase causes paper to rise.

The lower the temperature, the less enzyme activity there is, the less oxygen is produced, so it takes longer for the paper to rise.

At higher temperatures, e.g. 60°C, enzymes are denatured, little or no oxygen is produced, so this takes the longest to rise.

At optimum temperature, the most oxygen is produced over the shortest time, so the enzymes work best.

Conclusion   3

The enzyme catalase was affected by temperature changes.

As temperature increases, the rate of breakdown of hydrogen peroxide increased until it passed the optimum temperature. It then decreased.

The optimum temperature for this activity was __.

Limitations (of method)   2

Not enough time allowed for the test-tube to stay in the water bath, in which case the temperature of the test-tube may actually be lower than the water bath.

Same hydrogen peroxide used to repeat reaction at same temperature (lower concentration was now used).

TOTAL   **10**

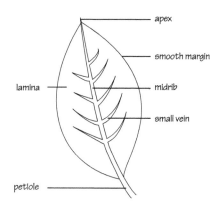

Drawing of the upper surface of a whole Ixora leaf x 1

An example of an SBA practical testing Dr

Title: *Drawing an* Ixora *leaf*

Aim: *To make an accurate representation of a leaf, drawn to scale*

Apparatus/Materials:

hand lens, whole *Ixora* leaf,

Method: **1.** Observe the leaf with a hand lens.

2. Make a large labelled drawing of the leaf.

Skills tested:

Drawing

Clarity   2

Clean continuous lines; neat lines showing midrib; no unnecessary details (only a few small veins shown); no shading/cross hatching.

Accuracy   3

Faithfulness of reproduction (apex is pointed, leaf blade (lamina) is oval; broad flat leaf; thick stalk; margin smooth); proportions correct; positioning of parts correct.

Labelling   1

Label lines parallel to top line; labels in pencil; no crossing over of label lines; lowercase lettering throughout; label lines touching the structure are labelled; no arrowheads on label lines.

Labels accurate   2

Apex, smooth margin, mid-rib, small vein, petiole, lamina – and correctly spelt.

Title   1

Placed underneath drawing; capital letters, underlined; should state 'drawing showing ...'; should state name of structure

May include view, e.g. top, front, dorsal, ventral, etc.

Magnification   1

Can be written as x2, etc. at end of title. A scale bar may be drawn.

TOTAL   10

An example of an SBA practical testing AI

Title: *Urinalysis*
(Note: This activity uses food tests but relates them to the kidney.)

Aim: *To analyse three samples of urine and evaluate the state of health of the patient.*

Apparatus/Materials:

artificial urine, control (dip a tea bag into 1 l of water; adjust to get colour of urine), sample A (100 cm³ control, 1 g sugar, 1 g chewable vitamin), sample B (100 cm³ control, 1 g albumin powder or 1 egg white (stirred), 3 drops of dilute acetic acid), Benedict's solution, 10% copper sulfate, KOH, 6 test-tubes in a rack, 3 x 1-cm³ syringes, Bunsen burner, tripod, gauze, large beaker, universal indicator paper

Method:
1. Test the control sample and samples A and B for glucose using 1-cm³ samples.

2. Test the control and samples A and B for protein using 1-cm³ samples.

3. Test the pH of the control and samples A and B using universal indicator paper.

4. Note the odour of the control and samples A and B.

5. Record your results in a table.

Skills tested:

Analysis and Interpretation

Background information   2

Glucose is not found normally in urine. Insulin is produced in the body, which helps to remove glucose from blood. Protein not usually found in blood. Protein molecules too large to pass through glomeruli. The foul odour of urine depends on bacteria being present. Fruit odours may be due to diabetes. pH is affected by many factors in diet.

Explanation of results   5

The control is used as a comparison for the other two samples. It contains no protein, no glucose, and represents a healthy person.

Sample 1: Glucose is found in the 'urine', which relates to diabetes, as insulin is produced. The person may be under stress, or may just have eaten a meal very rich in carbohydrates.

Sample 2: Protein is found in the 'urine', which relates to damage to the glomeruli, Bowman's capsule or tubules, or to disease. Acidic pH is due to diet, medication, etc.

Conclusion   2

The control contains no protein or glucose, so is healthy.

Sample 1 – Glucose is present, so the person is probably diabetic.

Sample 2 – Protein is present, which suggests damage to the kidney.

Limitations   1

You cannot tell the amount of protein or glucose present accurately – just that they are present or absent.

TOTAL   **10**

An example of an SBA practical testing PD

The scientific method

Experiments are performed in order to find out an answer to a particular scientific question. A question may have arisen from observations made repeatedly. In order to answer the question, a hypothesis is developed – a possible answer to the question. This hypothesis then needs to be tested. If the results of the test explain the original observations, the hypothesis may be true. You begin to draw conclusions. But, the hypothesis must be tested again – ideally many times – before a conclusion can be drawn. As a scientist, only when you perform an experiment several times can you be fairly sure that your conclusions are reliable. If you perform the experiment just once, you might introduce errors through clumsiness.

The above procedure – from observing, to questioning, to hypothesising, to testing, to retesting, to concluding – is commonly called 'the scientific method'. It is the basis upon which all scientists have made discoveries and drawn conclusion that have stood the test of time.

Why use a 'control'?

When testing a *hypothesis* it is important to have a control. The control acts as a neutral situation to compare other results against. In the experiment below, the control contains distilled water, while the other tubes contain different types of solution. If the results of the tube containing distilled water (a neutral substance) and the tube containing trypsin are different, then it can be concluded the difference is due to the trypsin.

When performing an experiment to compare a control to another situation, you must make sure that only one factor is being varied. This is necessary to make a valid conclusion based on the results. You may want to state your precautions clearly – how have you made sure that only one aspect of the experiment is being varied? This can help you plan and design a better experiment.

Limitations

Limitations are variables that you may not be able to control. These can therefore affect the experiment in some way. For example, if you are testing to see if a plant photosynthesises faster in red light than in white light, a bulb must be placed a certain distance from the plant. Inevitably, heat would be given off by the bulb and the temperature around the plant may change. This can affect the experiment, and is a limitation. You might solve this issue if a heat filter can be placed in front of the light bulb. If you cannot deal with the limitations of an experiment, you just need to recognise that they are there and state this fact.

Assumptions

When planning an experiment, certain assumptions are made. These are statements that are accepted as true. For example, if you are doing a transpiration experiment using a potometer, it is assumed that all the water that a plant takes in is lost through transpiration. But even though most of the water is lost by transpiration, the plant does actually use a bit of the water. Another assumption that is commonly made is that we test a leaf for starch to indicate that photosynthesis has occurred. But some leaves, e.g. those of chive and onion, contain glucose. A majority of leaves do store starch. Be sure that your assumptions are reasonable before you start your experiment, otherwise the experiment is unlikely to be wholly valid. You should therefore state your assumptions if possible.

Observation:

Vonnie observes that her mother places slices of fresh green paw-paw/papaya on meat for about half an hour before cooking it.

Vonnie's scientific question is:

Why does my mother place the green paw-paw on the meat?

Developing the Hypothesis:

This involves listing the variables which may affect what goes on with the paw-paw and the meat. Vonnie reckons that one variable in particular may be important.

Variables: **1.** Acidic substances and **2.** enzymes are present in the paw-paw that may break down proteins in the meat.

Hypothesis:

Pawpaw contains an enzyme that breaks down proteins in meat.

(Remember that a hypothesis must relate to the observation, and must make biological sense. Also, the hypothesis should consist of a single variable that can be tested.)

The experiment is therefore set out as follows:

Aim: *To test green paw paw slices to see if they contain enzymes which break down protein.*

(Remember – the aim must relate to the hypothesis.)

Apparatus/Materials:

a source of protein, e.g. egg white cubes (which may be coloured red with dye) or cubes of hot dog, fresh green paw-paw juice, boiled green paw-paw juice, commercial trypsin solution, distilled water, measuring cylinder, 4 test-tubes in a rack, ruler, scalpel, mortar and pestle, filter paper, hand lens, 4 Petri dishes

Method: (to be written in active voice, present tense)

1. Cut 16 cubes of hot dog or egg white, with sides of 1 cm.

2. Crush slices of paw-paw in a mortar using a pestle, then filter.

3. Label four test-tubes A, B, C and D.

4. Measure 5 cm^3 of distilled water in the test-tube labelled A.

5. Measure 5 cm^3 of fresh paw-paw juice in the test-tube labelled B.

6. Measure 5 cm^3 of boiled paw-paw juice in the test-tube labelled C.

7. Measure 5 cm^3 of trypsin solution in the test-tube labelled D.

8. Place four cubes of hot dog or egg white into each test-tube.

9. Leave for 30 minutes.

10. Pour each of the test tubes' contents into separate Petri dishes.

11. Using a hand lens, observe the consistency of the hot dog or egg white cubes.

12. Tabulate the results.

Expected Results:

Test tube	Expected observations
A – distilled water	Should remain firm and in one piece.
B – fresh paw-paw juice	Should be soft with small pieces broken off the cube.
C – boiled paw-paw juice	Should be firm and in one piece.
D - trypsin	Should be soft with small pieces broken off the cube.

Interpretation of Expected Results:

1 Fresh green paw-paw slices contain enzymes that digest protein.

2 This causes the meat to become soft.

3 The meat in the test-tubes with fresh green paw-paw juice and the enzyme trypsin would therefore become soft.

4 If the paw-paw is cooked, the enzymes would be denatured and would not work.

Variables: Manipulated variable = type of solution; responding variable = texture of meat

Control: test-tube with water

Constant variables:

Listed under precautions

Precautions:

To ensure that the experiment only contains one variable, the control test-tube (containing only distilled water) and the three other test-tubes, must all be experimented on under the exact same conditions:

• the same volume of solution is used in each test-tube

• meat cubes are left in the solutions for the same length of time

• the size of the cubes should all be the same

• the same number of cubes are placed in each test-tube

• the size, shape and material of the test-tubes should all be the same.

Limitations:

In this activity, the pH in some tubes (B and D) may not remain constant. The temperature may not have remained constant throughout the experiment.

Skills tested:

Planning and Design

Hypothesis: is it acceptable?	1
Is the aim related to the hypothesis?	1
Materials and apparatus listed/appropriate	2
The method is suitable – it is scientific	2
A control is included	1
Expected results and interpretation are stated	1
Limitations are noted	1
The experiment is presented in a suitable format	1
TOTAL	**10**

Other points to note when conducting the SBA

On ORR

- When drawing apparatus/equipment, only sections should be drawn.

- 'Diagrams' also include other types of illustrations such as maps, and can sometimes have shading, cross-hatching, etc, depending on the circumstances.

In addition to tables and line graphs, information may also be presented in the following ways:

- Information can be recorded in the form of pie charts. A key is necessary and the student may use colour. The title of the chart should be placed underneath it.

- Histograms can be used to show frequency of distribution. The bars are continuous. The title of the histogram is placed underneath it.

- Bar charts can also be used. These contain one non-numerical set of data. Bars must be of even width, and they may have a key. The title of the bar chart is placed underneath it.

On MM

- Preparation of biological materials is also assessed in some SBA practicals, e.g. potato strips for osmosis, temporary plant slides, etc.

- Careful handling of organisms is also assessed.

Activity	Chapter	Page	ORR	MM	AI	PD	Dr
1: Using a dichotomous key to identify small animals	1: The Diversity of Life	15	✓		✓		
2: Constructing a dichotomous key to identify leaves	1: The Diversity of Life	16	✓		✓		✓
1: Investigating abiotic and biotic components of a habitat	2: Living Organisms and the Environment	27	✓		✓		
2: Investigating the main components of soils	2: Living Organisms and the Environment	28	✓	✓	✓		
3: Investigating the water-holding capacity of different soils	2: Living Organisms and the Environment	29	✓	✓	✓		
4: Comparing the humus (organic matter) content of two soils	2: Living Organisms and the Environment	30	✓	✓	✓		
5: Estimating the numbers of static organisms in an area	2: Living Organisms and the Environment	34	✓				
6: Estimating the numbers of a mobile species	2: Living Organisms and the Environment	35	✓				
7: Estimating the numbers of a population of aquatic organisms	2: Living Organisms and the Environment	35	✓				
8: Investigating changes in abundance across an area	2: Living Organisms and the Environment	36	✓				
1: Constructing a food web	3: Food Chains and Food Webs	41	✓		✓		
2: Examining the role of decomposers	3: Food Chains and Food Webs	42	✓		✓		
1: Investigating carbon dioxide exchange in an ecosystem	4: Cycles in the Environment	53	✓		✓		
1: Using the microscope to look at a prepared slide	7: Cell Structure and Function	87	✓				
2: Making a slide of onion cells	7: Cell Structure and Function	88	✓				✓
3: Making a slide of leaf cells	7: Cell Structure and Function	89	✓				✓
4: Making a scale drawing of a leaf	7: Cell Structure and Function	93	✓	✓			✓
1: Demonstrating diffusion in a jelly	8: Movement of Substances in Cells	99	✓	✓	✓		
2: Comparing diffusion with mass flow in a gas	8: Movement of Substances in Cells	99	✓	✓	✓		
3: Investigating the effects of osmosis in onion epidermis cells	8: Movement of Substances in Cells	104	✓	✓	✓		✓
4: Investigating the effects of osmosis on potato tuber tissue	8: Movement of Substances in Cells	106	✓	✓	✓	✓	
1: Testing leaves for starch	9: How Plants Obtain Nutrition	112	✓	✓	✓		
2: Testing leaves for reducing sugars	9: How Plants Obtain Nutrition	114	✓	✓	✓		
3: Drawing a leaf section	9: How Plants Obtain Nutrition	118	✓				✓
4: Measuring the rate of photosynthesis using pondweed	9: How Plants Obtain Nutrition	121				✓	
5: Investigating the effects of lack of minerals on the growth of seedlings	9: How Plants Obtain Nutrition	124				✓	
1: Testing for some organic compounds	10: Chemicals of Life	133	✓	✓			

Activity	Chapter	Page	ORR	MM	AI	PD	Dr
1: The effect of temperature on the activity of amylase	11: Enzymes	141	✓	✓	✓		
2: The effect of pH on the activity of amylase	11: Enzymes	143				✓	
3: The effect of substrate concentration on the activity of the enzyme catalase	11: Enzymes	143	✓	✓	✓		
4: Investigations involving catalase activity	11: Enzymes	144				✓	
5: Finding out whether an enzyme is involved in the browning of yam	11: Enzymes	144				✓	
1: Experiment to find the energy content of food	12: How Humans Obtain Nutrition	154	✓	✓	✓		
2: Comparing the energy content of different foods	12: How Humans Obtain Nutrition	155				✓	
3: Making a Visking tubing model of the gut	12: How Humans Obtain Nutrition	162	✓	✓	✓		
1: Measuring the rate of oxygen uptake during respiration	13: Breathing and Respiration	169	✓	✓	✓		
2: Demonstrating that heat is produced by respiration	13: Breathing and Respiration	170	✓	✓	✓		
3: Showing the products of anaerobic respiration in yeast	13: Breathing and Respiration	173	✓	✓	✓		
4: Demonstrating the effects of anaerobic respiration in muscle	13: Breathing and Respiration	174	✓		✓		
5: Examining the lungs of a mammal	13: Breathing and Respiration	178	✓	✓	✓		✓
6: Comparing the carbon dioxide content of inhaled and exhaled air	13: Breathing and Respiration	181	✓	✓	✓		
7: Do other organisms produce carbon dioxide?	13: Breathing and Respiration	181				✓	
8: Making a bell jar of the thorax	13: Breathing and Respiration	182	✓		✓		
9: Dissection of a fish to show the gills	13: Breathing and Respiration	183	✓	✓	✓	✓	
10: Interpreting smoking data	13: Breathing and Respiration	187	✓	✓	✓	✓	
1: Does diffusion take longer in structures with a smaller surface area to volume ratio?	14: Transport in Mammals	194	✓	✓	✓		
2: Dissecting a sheep's heart	14: Transport in Mammals	196	✓	✓			
3: The effect of exercise on heart rate	14: Transport in Mammals	198	✓	✓	✓		
4: Examining a slide of blood vessels	14: Transport in Mammals	200	✓				✓
5: Examining a slide of blood cells	14: Transport in Mammals	204					✓
1: Observing water movement through a stem	15: Transport in Plants	210	✓				✓
2: How does the wind affect the rate of transpiration?	15: Transport in Plants	213	✓	✓	✓		
3: Investigating water loss from leaves	15: Transport in Plants	213	✓	✓	✓	✓	
4: Investigating light intensity and transpiration	15: Transport in Plants	214	✓	✓	✓	✓	
5: Looking at leaf surfaces	15: Transport in Plants	214	✓	✓	✓		✓
1: Testing plant organs for different foods	16: Food Storage	226		✓	✓		
1: Dissecting a kidney	17: Homeostasis and Excretion	233					✓
2: Testing 'urine' samples	17: Homeostasis and Excretion	237				✓	

Activity	Chapter	Page	ORR	MM	AI	PD	Dr
1: Testing the skin's sensitivity to touch	18: Sensitivity and Coordination in Animals and Plants	252	✓	✓	✓		
2: Does insulation reduce the loss of heat from a hot 'body'?	18: Sensitivity and Coordination in Animals and Plants	254	✓	✓	✓	✓	
3: Does a wet insulating surface increase heat loss from a 'body'?	18: Sensitivity and Coordination in Animals and Plants	255	✓	✓	✓	✓	
4: Testing reaction times	18: Sensitivity and Coordination in Animals and Plants	258	✓	✓	✓		
5: Testing the effects of caffeine on the heart rate	18: Sensitivity and Coordination in Animals and Plants	261				✓	
6: Finding out if woodlice prefer light or dark conditions	18: Sensitivity and Coordination in Animals and Plants	263	✓	✓	✓		
7: Investigating which part of a shoot is sensitive to light	18: Sensitivity and Coordination in Animals and Plants	272	✓	✓	✓		
8: Use of a clinostat to show geotropism in roots	18: Sensitivity and Coordination in Animals and Plants	273	✓	✓	✓		
9: Using a clinostat to show phototropism in roots	18: Sensitivity and Coordination in Animals and Plants	274				✓	
1: Dissecting a chicken wing	19: Support and Movement	287	✓	✓	✓		
1: Researching data on HIV in your country	20: Reproduction in Animals	303	✓	✓	✓		
1: Observing insect-pollinated flowers and wind-pollinated flowers	21: Reproduction in Plants	309	✓				✓
1: Measuring growth of a plant	22: Growth	315	✓	✓	✓	✓	✓
2: Examining a seed of a dicotyledonous plant	22: Growth	318	✓	✓	✓		✓
3: Determining whether temperature affects germination	22: Growth	319	✓	✓	✓		
4: Examining a germinating seedling	22: Growth	320	✓	✓			
5: Performing food tests on cotyledons	22: Growth	320	✓	✓	✓		
1: Researching data on mosquito-borne diseases	23: Disease in Humans	333	✓	✓	✓	✓	
2: Observing the stages in the life cycle of a mosquito	23: Disease in Humans	334	✓	✓	✓	✓	
1: Investigating variation in humans	26: Genetic Variation	368	✓	✓			

Appendix 3: Exam Tips

Very few people enjoy revising for examinations. Being very well organised with your revision will mean you revise for as little time as possible. Revising for hours on end without a break is not productive – your concentration will not last and you will not retain the facts.

Where to revise
Somewhere quiet away from distractions. You need good lighting and you should be comfortable.

When to revise
As soon as possible after school finishes each day, before you get too tired to work effectively.

How to revise
Have a plan and stick to it. Think about how many topics you need to cover. Make a list of all the topics and mark those that you find difficult. Revise a mixture of easy and more challenging topics each day. This will give you the satisfaction of making progress every day. When you have revised something, tick it on your list. This will show clearly how much still needs to be done.

Divide your time
Working for a whole evening without a break will achieve less than dividing your time into segments. Split your evening into revision time and leisure breaks. Take at least 10 minutes off for each hour worked. When you start again, take 5 minutes to review the previous topic, then move on.

Group revision
Some students find that revision is less boring and more effective in a group. Ask each other questions, chose topics in turn, share good ideas.

Organising the information
Think of ways to organise the information you need to revise. You can use summaries, checklists, file cards, sticky-notes, flow-schemes and key words.

Past paper questions
Read examination papers from previous years. This will show you the styles of questions and indicate which topics are more likely to be repeated. Look carefully at the marks printed on the paper and the spaces left for answers. Fill the space provided and make points according to marks available. For example, make two points if there are two marks available.

Instructions on the paper
Read the instructions carefully. If the paper tells you to choose one question from a section, don't answer them all. Someone always makes this mistake, don't let it be you!

Mathematical requirements
You will be awarded marks for every part of the answer that is correct, even if the final answer is wrong. Always have a go and show your working.

Examination day
Write any vital formulae or definitions on scrap paper and look at them until you are called into the exam room. Throw the paper away before you enter. As soon as you are told to begin, write the information down and put a cross through it. You can refer to it when you need to but it will not be mixed up with your answers.

Index